MEDIEVAL CHRISTIANITY IN PRACTICE

MEDIEVAL
CHRISTIANITY
IN PRACTICE

Miri Rubin, Editor

PRINCETON READINGS IN RELIGIONS

PRINCETON UNIVERSITY PRESS

PRINCETON AND OXFORD

Published by Princeton University Press, 41 William Street,
Princeton, New Jersey 08540

In the United Kingdom: Princeton University Press, 6 Oxford Street,
Woodstock, Oxfordshire OX20 1TW

Library of Congress Cataloging-in-Publication Data
Medieval Christianity in practice / edited by Miri Rubin.
        p.   cm.   —(Princeton readings in religions)
Includes bibliographical references and index.

ISBN 978-0-691-09058-0 (hardcover : alk. paper) —ISBN 978-0-691-09059-7 (pbk. : alk. paper)
  1. Christian life—History—Middle Ages, 600-1500—Sources.   2. Europe—Religious life and
customs—Sources.   I. Rubin, Miri, 1956–
BR253.M36 2009
270.3–dc22
2008053041
British Library Cataloging-in-Publication Data is available

This book has been composed in Berkeley

Printed on acid-free paper. ∞
press.princeton.edu

Printed in the United States of America

1 3 5 7 9 10 8 6 4 2

PRINCETON READINGS

IN RELIGIONS

———

Princeton Readings in Religions is a series of anthologies on the religions of the world, representing the significant advances that have been made in the study of religions in the last thirty years. The sourcebooks used by previous generations of students, whether for Judaism and Christianity or for the religions of Asia and the Middle East, placed a heavy emphasis on "canonical works." Princeton Readings in Religions provides a different configuration of texts in an attempt better to represent the range of religious practices, placing particular emphasis on the ways in which texts have been used in diverse contexts. The volumes in the series therefore include ritual manuals, hagiographical and autobiographical works, popular commentaries, and folktales, as well as some ethnographic material. Many works are drawn from vernacular sources. The readings in the series are new in two senses. First, very few of the works contained in the volumes have ever been made available in an anthology before; in the case of the volumes on Asia, few have even been translated into a Western language. Second, the readings are new in the sense that each volume provides new ways to read and understand the religions of the world, breaking down the sometimes misleading stereotypes inherited from the past in an effort to provide both more expansive and more focused perspectives on the richness and diversity of religious expressions. The series is designed for use by a wide range of readers, with key terms translated and technical notes omitted. Each volume also contains a substantial introduction by a distinguished scholar in which the histories of the traditions are outlined and the significance of each of the works is explored.

Medieval Christianity in Practice is the fourteenth volume in the series. It has been designed, organized, and edited by the eminent scholar of Medieval Europe, Miri Rubin. The forty-three contributors include many of the leading scholars of Medieval Christianity from North America, Europe, and the Middle East. Each scholar has provided a translation of a key work, a number of which are translated here for the first time. These works were produced in a variety of languages of western Europe between about 600 and 1500 CE, in regions ranging from Scandinavia in the north to Iberia in the south, from Ireland in the west to Poland in the east. They include life-cycle rituals, pastoral sermons, instructions for Crusaders, prayers, miracle stories, charms, traveler's tales, and spiritual exercises. Each chapter begins with the text itself, followed by a commentary in which the translator provides historical context, identifies points of particular importance, and offers suggestions for further reading. Miri Rubin opens the book with a general introduction to the world of Medieval Christianity.

The volumes *Zen in Practice* and *Yoga in Practice* are forthcoming in the series.

Donald S. Lopez, Jr.
Series Editor

CONTENTS

———

Work and Travel

Churches, Parishes, and Daily Life: Consecration

Churches, Parishes, and Daily Life: Pastoral Care

Churches, Parishes, and Daily Life: Preaching

Churches, Parishes, and Daily Life: Confession and Penance

Churches, Parishes, and Daily Life: Prayer

In Pursuit of Perfection: At the Edge of the World

Rituals of Power

CONTRIBUTORS

Debby Banham is an Anglo-Saxon historian with research interests in medicine, diet, and agriculture. She teaches for the University of Cambridge and Birkbeck College, University of London. She is the author of several articles and is currently working on a book entitled *Anglo-Saxon Farms and Farming*.

Caroline Barron is a Professorial Research Fellow at Royal Holloway, University of London. Her research interests include medieval London and other English towns, the reign of Richard II, women in medieval England, and all forms of medieval piety. Her most recent book was *London in the Later Middle Ages* and she is currently working on the early life of Sir Thomas More and the English cult of the Italian servant saint, Zita of Lucca.

Alexandra Barratt is Professor of English at the University of Waikato, New Zealand. Her research focuses on women in medieval England as readers and writers. Her books include the anthology *Women's Writing in Middle English* and she is currently working on her next book, *Anne Bulkeley and Her Book: Fashioning Female Piety in Early Tudor England*.

Peter Biller is Professor of Medieval History at the University of York. His books include *The Measure of Multitude: Population in Medieval Thought* and *The Waldenses, 1170-1530*.

Renate Blumenfeld-Kosinski is Professor at the University of Pittsburgh and an expert on late medieval women's vernacular writing. She has written and edited several volumes, among which are *Note to a Woman Born* (1990), *Reading Myth* (1997), and *The Writings of Margaret of Oingt* (1990).

Daniel Bornstein is Professor of History and Religious Studies at Washington University in St. Louis, where he holds the Stella K. Darrow Chair in Catholic Studies. A specialist in the religious culture of medieval and Renaissance Italy, he is the author of *The Bianchi of 1399: Popular Devotion in Late Medieval Italy* and a score of articles on female sanctity, parish priests, lay confraternities, and relics. Most recently, he edited the volume on *Medieval Christianity* in Fortress Press's seven-volume *A People's History of Christianity*.

Alain Boureau is Directeur d'études at the Ecole des Hautes Études en Sciences Sociales, Paris, and he heads the Scholastic Anthropology Groups there. He is the author of many books and articles, including *L'Empire du livre*, *La Religion de l'Etat*, and *Satan hérétique. La naissance de la démonologie dans l'Europe médiévale (1260-1350)*, which was translated into English as *Satan the Heretic: The Birth of Demonology in the Medieval West*.

Olivia Remie Constable is a Professor of Medieval History at the University of Notre Dame. She specializes in the history of the medieval Mediterranean world, with a particular interest in social, economic, and cultural contacts between Christians, Muslims, and Jews. She has published two monographs, *Trade and Traders in Muslim Spain* and *Housing the Stranger in the Medieval Mediterranean World*, and a collection of primary sources in translation, *Medieval Iberia: Readings from Christian, Muslim, and Jewish Sources.*

Rita Copeland is Professor of Classical Studies, English, and Comparative Literature at the University of Pennsylvania. She has written extensively on the histories of rhetoric, literary theory, translation, allegory, education, and intellectuals in the Middle Ages. Her books include *Pedagogy, Intellectuals, and Dissent in the Later Middle Ages*, *The Cambridge Companion to Allegory* (with Peter Struck, forthcoming), and *Medieval Grammar and Rhetoric: Language Arts and Literary Theory, AD 300-1475* (with Ineke Sluiter, forthcoming).

Peter Cramer is the author of *Baptism and Change in the Early Middle Ages, c. 200-c. 1150*. He teaches History at Winchester College, UK.

David d'Avray is Professor of History at University College London. His books have dealt with medieval preaching, death and monarchy, and the influence of symbolism on medieval marriage structures: *The Preaching of the Friars: Sermons Diffused from Paris before 1300*, *Death and the Prince: Memorial Preaching before 1350*, and *Medieval Marriage: Symbolism and Society*. His current research themes are rationalities, royal annulments and dispensations, and *longue durée* themes in papal history.

Eamon Duffy is Professor of the History of Christianity at the University of Cambridge. Among his books are *The Stripping of the Altars* (1992) and *The Voices of Morebath* (2001).

Sharon Farmer is Professor of Medieval History at the University of California-Santa Barbara. She is the author of *Communities of Saint Martin* (1991) and *Surviving Poverty in Medieval Paris* (2002).

Harold S. Fox (died 2007) was a historical geographer and Professor at the Centre for English Local History of the University of Leicester. He wrote on a wide range of themes in agrarian and social history of medieval England. He co-edited *Change in the Countryside: Essays on Rural England, 1500-1900* with R. A. Butlin and *Cornish Lands of the Arundells of Lanherne, Fourteenth to Sixteenth Centuries* with O. J. Padel. His last book was *The Evolution of the Fishing Village: Landscape and Society along the South Devon Coast 1086-1550*.

Penelope Galloway received her D.Phil. from Oxford University and taught at the University of Bristol. She is an expert on the Beguines in the French Low Countries.

György Geréby is Associate Professor and Head of the Medieval Studies Department at the Central European University, Budapest. He has published on medieval and late antique philosophy, theology, and spirituality, and is preparing a commentary on the *Protoevangelium of James*.

Joseph Goering is Professor of History at the University of Toronto. His research interests center on medieval law and theology, both scholastic and popular, and on ecclesiastical institutions. He has recently published *The Virgin and the Grail: Origins of a Legend* (New Haven, 2005).

Sarah Hamilton is Senior Lecturer in Medieval History at the University of Exeter. She is the author of *The Practice of Penance, 900-1050* (Royal Historical Society Studies in History New Series) and of various articles on early medieval penance. She is currently researching the history of early medieval excommunication.

Thomas Head is Professor of History at Hunter College and the Graduate Center of the City University of New York. His books include *Hagiography and the Cult of the Saints: The Diocese of Orléans, 800-1200* and, as editor, *Medieval Hagiography: An Anthology*. He is currently working on books concerning the development of the cult of the saints in western Christendom and on the Peace of God in eleventh-century France.

Yitzhak Hen is Professor of Medieval History at Ben-Gurion University of the Negev. His research focuses on cultural and religious issues of the early medieval West. His publications include *Culture and Religion in Merovingian Gaul, The Sacramentary of Echternach, The Royal Patronage of Liturgy in Frankish Gaul,* and *Roman Barbarians: The Royal Court and Culture in the Early Medieval West.* He is currently writing a book on Western Arianism.

Dominique Iogna-Prat is a Director of Research of the CNRS. He is the author of many studies, among them *Order and Exclusion* (2002) and *La maison Dieu* (2006), and he co-edited the collection *Marie: le culte de la Vierge dans la société médiévale* (1996).

Katherine L. Jansen is Associate Professor of History at the Catholic University of America and the author of *The Making of the Magdalen: Preaching and Popular Devotion in the Late Middle Ages.* A researcher in the fields of medieval history, Italian history, women and gender, and religious culture, she is at work on a book entitled *The Practice of Peace in Late Medieval Italy* and is co-editor of two volumes: *Charisma and Religious Authority: Jewish, Christian and Muslim Preaching, 1200-1500* with Miri Rubin and *Medieval Italy: Texts in Translation* with Joanna Drell and Frances Andrews.

Peter Jones is Fellow and Librarian at King's College, Cambridge, UK. His research interests include knowledge and practice in late medieval medicine, the use of images in healing, and amuletic objects and texts. He is the author of *Medieval Medicine in Illuminated Manuscripts.*

William Chester Jordan is Dayton-Stockton Professor of History and Chairman of the History Department of Princeton University. He is the author of, among other books, *The Great Famine: Northern Europe in the Early Fourteenth Century.* His present research focuses on Benedictine and Cistercian abbeys in the thirteenth century.

Gabor Klaniczay is Professor of Medieval Studies at the Central European University, Budapest, and Permanent Fellow at the Collegium Budapest Institute for Advanced Study. His research interests are in the historical anthropology of late medieval sainthood, belief in miracles, and witchcraft. His books include

The Uses of Supernatural Power and *Holy Rulers and Blessed Princesses*. He is currently working on visions.

Christiane Klapisch-Zuber is Directrice d'études (retired) at the École des Hautes Études en Sciences Sociales in Paris. Her research interests span social and family history in medieval and Renaissance Italy (mainly Tuscany), and social history of art. Her books include *Women, Family, and Ritual in Renaissance Italy*, *L'ombre des ancêtres*, and *Retour à la cité: les magnats de Florence, 1340-1440*. Her current work involves the transmission between artists in Florence and the history of images of pain in European art.

Aviad Kleinberg is a Professor of Medieval History at Tel Aviv University and the director of Tel Aviv University Press. His books include *Prophets in Their Own Country: Living Saints and the Making of Sainthood in the Later Middle Ages*, *Flesh Made Word: Saints' Stories and the Western Imagination*, and *Seven Deadly Sins*.

Robert E. Lerner is Professor of History, emeritus, and Peter B. Ritzma Professor in the Humanities, emeritus, at Northwestern University. His books include *The Heresy of the Free Spirit in the Later Middle Ages*, *The Powers of Prophecy*, and *The Feast of Saint Abraham*. His current work pursues themes and texts in the religious and intellectual history of western Europe in the thirteenth and fourteenth centuries.

Sara Lipton is an Associate Professor of Medieval History at the State University of New York, Stony Brook. Her research focuses on religious identity and practice, Jewish-Christian relations, and text and image in the high Middle Ages. Her first book, *Image of Intolerance: The Representation of Jews and Judaism in the Bible Moralisée* was awarded the John Nicholas Brown Prize by the Medieval Academy of America. She is currently completing a book entitled *Jews, Vision, and Witness in Medieval Christian Art*.

Rob Meens teaches medieval history at the University of Utrecht. His research focuses on the early Middle Ages, particularly in the fields of culture and religion. He has published on topics such as penance, sanctuary, dietary rules, and popular religion. He also has an interest in manuscripts and textual traditions.

Piroska Nagy is Professor of Medieval History at the University of Quebec in Montreal. Her research interests concern cultural and intellectual history of the medieval West. She is the author of *Le Don des Larmes au Moyen Age (Ve-XIIIe s)*. Her current work concerns the history of affective phenomena in relation to medieval conceptions and practices of body and mind.

Janet L. Nelson is Emeritus Professor of Medieval History at King's College London. Her main research interests lie in the fields of early medieval political, social, and gender history. Her books include *Charles the Bald* and *The Annals of St-Bertin*. She is currently working on a biography of Charlemagne.

Frederick S. Paxton is Brigida Pacchiani Ardenghi Professor of History at Connecticut College. He is the author of *Christianizing Death: The Creation of a Ritual Process in Early Medieval Europe*; *Anchoress and Abbess in Ninth-Century Saxony: The Lives of Liutbirga of Wendhausen and Hathumoda of Gandersheim*; and numerous articles on sickness and healing, death and dying, hagiography, codicology, and canon law. His reconstruction of the Death Ritual at Cluny in the central Middle Ages is forthcoming in 2010.

Virginia Reinburg is Associate Professor of History at Boston College. She is the author of articles on prayer, pilgrimage, and the liturgy in late medieval and early modern France, and the co-editor of two exhibition catalogues on medieval art.

Brigitte Resl is Professor of Medieval History at the University of Liverpool. She is author of *Rechnen mit der Ewigkeit* (1996), a study of charity in late medieval Vienna. Her most recent book is *Understanding Animals, 1150–1350* (2008).

Hans-Jochen Schiewer is President of the Albert-Ludwigs-Universität in Freiburg i. Br. and Professor of Medieval German Literature and Language. His research interests focus on religious literature, courtly literature, and the transmission of Middle High German texts.

Shulamith Shahar is Professor emerita of Tel Aviv University. Her books include *The Fourth Estate: A History of Women in the Middle Ages, Childhood in the Middle Ages, Growing Old in the Middle Ages: "Winter clothes us in shadow and pain"*, and *Women in a Medieval Heretical Sect: Agnes and Huguette the Waldensians*. She has translated the works of medieval authors into Hebrew, including Peter Abelard and Guibert de Nogent, and is currently working on a translation of Christine de Pisan's *Livre de la Cité des Dames*.

Walter Simons is Associate Professor of History at Dartmouth College. His research involves religious movements and their social context in the high and late Middle Ages. His most recent books include *Cities of Ladies: Beguine Communities in the Medieval Low Countries, 1200-1565* and the *Cambridge History of Christianity*, vol. IV: *Western Christianity, 1100-1500*, edited with Miri Rubin (Cambridge, 2009).

Julia M. H. Smith is Edwards Professor of Medieval History at the University of Glasgow, and publishes on politics, gender, hagiography, and the cult of saints in late antiquity and the early Middle Ages. Her publications include the *Cambridge History of Christianity III: Early Medieval Christianities, 600-1100* with Thomas F. X. Noble, *Europe after Rome: A New Cultural History 500-1000*, *Gender in the Early Medieval World: East and West 300-900* with Leslie Brubaker, *Early Medieval Rome and the Christian West*, and *Province and Empire: Brittany and the Carolingians*.

Susanna Throop is Assistant Professor of History at Ursinus College. She is interested in interdisciplinary perspectives on religion, emotion, and violence in medieval cultures. She is working on two forthcoming volumes: *Crusading as an Act of Vengeance, 1095–1216* and *Vengeance in the Middle Ages: Emotion, Religion and Feud* (co-edited with Paul Hyams).

John Van Engen is Professor of Medieval History at the University of Notre Dame. His research interests extend to the whole of medieval religious and intellectual history. He has recently published *Sisters and Brothers of the Common Life*, and is at work on the first translation from Middle Dutch of the writings of Alijt Bake (1415-55), as well as on a synthetic study tentatively entitled *The Spirit of Twelfth-century Europe*.

Nancy Bradley Warren is Professor of English and Courtesy Professor of Religion at Florida State University. She is the author of *Spiritual Economies: Female Monasticism in Later Medieval England* and *Women of God and Arms: Female Spirituality and Political Conflict, 1380-1600*. She is currently finishing a book on confessional, national, and historical boundaries, which examines continuity and change in medieval and early modern manifestations of female spirituality.

Joseph Ziegler teaches medieval history at the University of Haifa. He studies various aspects of the nexus of medicine and religion in the later Middle Ages. He is the author of *Medicine and Religion, c. 1300: The Case of Arnau de Vilanova* and has co-edited *Religion and Medicine in the Middle Ages* with Peter Biller. He is engaged in research into the rise of learned physiognomy from 1200 to 1500.

MEDIEVAL CHRISTIANITY IN PRACTICE

INTRODUCTION

Miri Rubin

No single volume can do justice to the variety of experiences of Christian life in medieval Europe. The contours of variety are themselves diverse. They arise from the fact that then, as now, western Europe was a vast continent of regions. In each, Christianity arrived, was absorbed, and became institutionalized under different circumstances. Think, for example, of the difference between the history of Irish Christianity and that of Italian cities; of the relatively late spread of many ecclesiastical forms to Scandinavia, and Hungary and Poland, or the unique concerns of Iberian Christians, who lived cheek by jowl with Jews and Muslims. While the liturgy was Latin, so many religious activities around it were conducted and experienced in the vernacular, above all, religious instruction; no one doubted that it had to be conducted in the *lingua materna*.

This volume follows *Religions of Late Antiquity in Practice*.[1] It picks up some of the trails laid down by the earlier volume, and follows them into western Europe, which over the following centuries developed a Latin theology and religious practices expressed in a variety of languages and regions that make up the Europe of today. This volume does not deal with the practices of Christians of the Greek-speaking world, within which dwelled communities that expressed their religious experiences in Coptic, Syriac, Armenian, and Persian.

Over the long period this volume covers—roughly 600 to 1500—a predominant feature of ecclesiastical organization was that which saw the spread from around 1050 of bureaucratic structures for the supervision and guidance of religious life in parishes, grouped in dioceses all over Europe. Thus, from the papal court, canon law—the law of the Church—was promulgated, but this was then applied and enforced through bishops in their diocese, who in turn passed on instructions and mandates to parish churches. Not only canon law, but modes of devotion, cults of saints, excommunications, and taxation also moved along the administrative channels. Although most business of instruction, scrutiny, and correction of religious life was conducted locally, for most people in the parish, deanery, or diocese, there were occasions that linked very clearly the pinnacle of the ecclesiastical structure with its many local parts: the excommunication of a ruler affected all religious activities in his domain, as it did in the case of Emperor

Henry IV in 1076, and when King John of England was excommunicated in 1215. The ideas expressed by an important theologian, and in which sources of possible error were identified by scholars close to the pope, could lead to summoning for scrutiny and correction, as was the case with the views of Berengar of Tours (ca. 1000–1088) on the Eucharist, condemned in 1055, and again in 1077 and 1078. Conversely, people appealed to the papacy against rulings of their local courts in cases of marriage litigation, from as far as Wales, and a saint who became famous in her lifetime in fourteenth-century Sweden, St. Birgitta of Sweden (1303–1373), soon animated a Europe-wide following and created a new order, the Briggittine order. Innovations, modes of worship that developed in particular regions out of local concerns and styles of devotion, could in turn be promoted into European observances, though we tend to know more about the success stories—like the new monastic order of Cluny in the tenth century, or the new feast of Corpus Christi in the fourteenth—rather than about the failed attempts at new cults and celebrations. European means of communication enabled a wide range of relations with and responses to religious life—reactions that were full of praise or disapproval, that aimed to correct or to emulate. Ideas were turned into recommended modes of practice, but these only really took meaningful form once they passed through the local mill of testing: in a local language adapted to local weather and landscape, conditions of life, memory, and traditions. A cult of a new saint, a doctrinal statement, the attempt to extirpate a common practice—like the cult of the Holy Greyhound in thirteenth-century France—such initiatives were tested by men and women, young and old, by those whose lives Christian practice defined and whose souls it promised it could save.

Practice is the inspiring term that defines the aim of this series. The history of religion has for centuries concentrated on ideas, doctrines, and philosophies with which scholars, often priests and men in religious orders, felt at ease. *Practice* is the favored term of the social historian, always something of a historical anthropologist, whose inquiries into the ways in which concepts and ideas about the world—in a word, culture—is lived, by embodied persons, within the material cultures that provide the frame for their lives. To focus on practice is not to ignore ideas, it is rather to trace them, to observe their realization, the interpretative process by which people discussed, say, the meaning of Virgin birth, or of the Mass, and the practices they designed to follow from such understandings. Studying practices allows us who are interested—students of religion, students of the Middle Ages—to encounter the contributions and reactions that ideas and recommended procedures elicited from those who had to live by them. And so this volume will observe priests at work, friars preaching, parishioners at prayer, people of all ranks at penance. Inasmuch as practice is akin to processes of domestication and of familiarization, it seems right to observe it through the life cycle, which is often close to family life. The first section thus follows the life cycle from birth to death, through important rituals that were marked by a multitude of celebrations, with varying degrees of clerical involvement. We will then move to observe religious practice attached to work; religious practice will emerge in the calls for protection by merchants on journeys, or in the needs of fishermen for special arrangements of prayer when away from their villages. The people studied in these two sections will be revisited as parishioners. The parish provided many recommended procedures; the parish was, after all, the unit within which sacraments were received and most worship experienced. Yet as rich and as accessible as parish life was, people still chose further activities, often in smaller groups—such as fraternities—or attached to a particular devotional theme: the crucifixion, the Eucharist, the rosary. Some detached themselves from the parish, inasmuch as they joined a religious order whose structures removed the member from most local ties; some even cut away in groups seeking perfection, in rigorous, often itinerant and extremely poor and demanding lifestyles. These are the subjects of the section on the Pursuit of Perfection.

This volume brings together the contributions of forty-three scholars, experts on the life of medieval Europe. They have been encouraged to choose texts, and some images, encountered in the course of their research, and to turn them into illustrations of religious practice. The resulting volume follows a number of trails: events of the life cycle, the world of work and production, the rhythms of participation in parish life, the practices of people who sought to go beyond the basic requirements of parish religion, and some occasions on which religious ritual enhanced representation of power and authority. The texts collected here are culled from a large variety of sources: prayer books, chronicles, diaries, liturgical books, sermons, accounts, hagiography, handbooks for the laity and clergy, and romance. They thus also demonstrate the vibrant state of medieval studies, which these days thrive on multidisciplinary approaches to religious cultures and on an expansive idea of the sources that reveal the practices and ideas of medieval life.

It is hoped that this volume will offer an occasion for discussion, instruction, and exchange. It is meant not solely, or even mainly, for study of the Middle Ages. It may give rise to new encounters with the lives of individuals and communities, at their best and at their worst, through the rich languages and gestures of religion, and the aspirations for well-being and consolation they so powerfully express.

Note

1. Richard Valanstasis, ed., *Religions of Late Antiquity in Practice* (Princeton, 2000).

The Life Cycle: Baptism

1

Baptismal Practice in Germany

Peter Cramer

AN ASPECT OF BAPTISM:
Exorcism of Those to Be Baptized, and the Blessing of the Easter Candle on Holy Saturday, from the Eleventh-Century Metz Pontifical

First: how the children are catechized [instructed]. At break of day on the Saturday, holy and most solemn day, the appropriate ornaments and instruments are brought into the church and put in place. After the third hour [about 9 a.m.] on Saturday, those who are to be baptized proceed to the church, with their godfathers and godmothers, and they are put in order by an acolyte, according to the written form. Males on the right, females on the left. Then, those who can do so recite the Lord's Prayer and the Creed, or in the case of those who cannot, their godfathers and godmothers make the recitation on behalf of those they are going to receive [from the font that night during the vigil]. The priest, making a cross with his thumb on the forehead of each, says: "In the name of the Father, the Son, and the Holy Spirit." Afterward, placing his hand on the head of each one, he says this prayer to the children to be catechized:

"It is no secret to you, Satan, that punishment is on its way, torments are looming up and judgment day is imminent, the day of perpetual punishment, the day at hand and like a fiery furnace when you with all your angels will go down to burial without end. And so, damned and rightly damned, pay honor to the true and living God, pay honor to Jesus Christ his Son and to the Holy Spirit in whose name and by whose strength I command you, whoever you are, to step back and quit this servant of God, N.,whom today our Lord Jesus Christ has stooped down to call to his grace and benediction and his font with the gift of baptism, and who will in this way be made a temple through the water of regeneration, in the remission of all sins, in the name of Our Lord Jesus Christ, who will come to judge the living and the dead and the things of this world in fire. Amen."

This done, the priest touches each one on nostrils and ears with spittle from his mouth, and speaks the following words into the right ear of each:

"Effeta," which means: "Open,"

Then into the nostrils:

"Into the smell of sweetness,"

And into the left ear:

"And you, devil, begone. The judgment of God is on its way."

After this he touches each one on the chest and between the shoulderblades, making a cross with holy oil with his thumb, and calling each of them by name, he says:

"Do you renounce Satan?"

Response: "I renounce him."

"And all his works?"

Response: "I renounce them."

"And all his pomps?"

Response: "I renounce them."

"And I daub you with the oil of salvation in Christ Jesus our Lord into eternal life. Amen."

This done, he walks in a circle, and placing his hand on each of their heads, he chants with raised voice:

"I believe in one God."

And he turns towards the females and does the same.

Now it is the archdeacon who speaks:

"Pray, you who are chosen, bend your knees."

Then, after a pause:

"Rise, end your prayer as one voice, saying: Amen."

And all should respond: "Amen."

Again, they are instructed by the archdeacon in these words:

"Let the catechumens leave. If any among you is a catechumen, let him leave. All catechumens are to go outside."

A deacon then says: "Dearest children, go back now to your homes, and wait there for the hour to come when it is possible for the grace of God to work about you through baptism."

On the same day, at the seventh hour [about 1 p.m.], bishop, priests, and deacons make their way into the sacristy and put on the most solemn vestments, with which they will celebrate the holy vigil, and, the candle lit in a purified place, the bishop, or a priest, blesses it, making the cross over it with low voice, but loud enough to be heard by the by-standers. While this is going on, the schola cantorum sings the seven penitential Psalms, beginning:

"The Lord be with you."

Response: "And with your spirit."

[Blessing of the candle]

"God, world's founder, light's author, and the stars' architect, God who threw a curtain of sharp-sighted light over a world in shadow, God, through whom it was and by whose unspeakable power it was, the very clarity of all that is had its beginning, calling on you in your works, in this holiest night-wake, we offer humbly to you in your majesty out of the stock of gifts you gave to us, a thing not polluted by the fat life of flesh, a thing unspoiled by profane oil, untouched by sacrilegious fire, but a thing made out of wax and tow, this is our offering, made with the humility of a religious devotion, burnt in honor of your name. It is necessary that we defer open-handed to the great mystery and awe-inspiring sacrament of this night when the inveterate shadows have felt day brought against them by the miracle of the resurrection of the Lord, and death once damned by unending night is dragged dumbstruck, snared by the light of God's victory into truth's radiant font, so that what was once damned, chained by the fatal presumption of the first man, who veered off the path, is now bright with liberty brought on by this miracle-working night. And so, coming down gleeful-spirited into this reverence-inspiring feast, as human devotion requires, we exhibit to you the leaping light of these flames, and with this unfolding of an encompassing faith, the praise of your creature is also sung. The light from the flame we know to be the same as the light in which the power of God's majesty deigned to appear to Moses, the same as the light that showed a luminous saving path to the people escaping the land of slavery, the light which saved the lives from flattery and vanity of the three boys the tyrant had put in the furnace. For, as the horror of the shadow-world is shut out by the preceding grace of light, so, Lord, by the light shed with the majesty of your command the burden of sins is put away. And now that we are standing agape at the beginning of this stuff, it is the moment to praise how the bees begin. The bees are frugal in taking their harvest, chaste in childbirth. They fashion cells by pouring waxy fluid, with an art the mastery of human skill could never equal. They pick flowers using their feet and on the flowers there is not so much as a bruise. They do not bring forth a fetus, but by stirring with their mouths they produce a swarm from an already-conceived fetus, just as with wonderful example Christ proceeded from the Father's mouth. It proved fertile this virginity without childbearing of the bees. The Lord saw fit to imitate it when he decided to take for himself a mother of flesh. This was out of love for the same virginity. Such offerings as these, fitting to your sacred altars, are the offerings with which the Christian religion, unhesitating, rejoices in you. Who lives, etc.

"Lord holy Father, all-powerful eternal God, in your name and the name of your Son Our Lord Jesus Christ and that of the Holy Spirit we bless this fire and sanctify it together with its wax and all the food of offering it has in it and we signify by the sign of the cross of Christ your most high Son, that we, kindled within or without, burn nothing harmful but that it [the fire] warm or light all that is needed for human use, and that what is kindled or warmed from this fire be blessed and useful to all human salvation; that all that burns here burns for you not the strange fire of errant Nadab and Abihu, but that we have the strength to immolate before you with Aaron and his sons Eleazar and Ithamar the roasted peace-bringing victims. And always with the fire of the

Holy Spirit scorch off our vices and throw light on our hearts with the brightness of your understanding and clarify our souls with the warmth of faith. Through the same, etc."

Then the candle held in the reed-stick is lit from the Easter candle, and carried by the bishop or the abbot or prior, and all proceed from the place of benediction at the same time toward the church with the same candle, in silence, singing nothing, and all the people follow, as above.

> Source: J. B. Pelt, *Études sur la cathédrale de Metz.*
> *La liturgie (v–viii siècles)* (Metz, 1937), 174–75.

The liturgical book from which this passage is taken was used in the cathedral of St. Stephen of Metz, certainly in the thirteenth century, very likely earlier. The manuscript dates from the eleventh century, and is a version of the Romano-Germanic Pontifical, a handbook for the bishop's reference based on the Roman rite but adapted for local German use in the tenth century. MS 353 from the Bibliothèque municipale in Metz has nothing unusual in it. I have chosen it because it is a witness to common liturgical practice, but known to have been in use in a given place.[1] It is in this sense characteristic, and it is an aspect of the practice of liturgy that it must be seen and felt to have behind it the authority of the normative. The blessing of the Easter candle, which appears in the so-called Gelasian Sacramentary, stands in the Pontifical as a prelude to baptism. It is steeped in the senses that apply in baptism itself. The blessing is made on the afternoon of Holy Saturday, after the public reconciliation of penitents on Holy Thursday and the liturgy of the cross on Good Friday, and before the baptism of the vigil of the Saturday night. It is important to see the passage in its context. A ritual of the baptism of infants is—to observe just one element of the sequence—articulated, by shared conceptual matter and linguistic echoes, with the reconciliation of adult penitent into the social body. With this, we are immediately face-to-face with the puzzle of how a ritual of enlightenment (baptism) can be effective with children as its subjects. And in this eleventh-century version of the Romano-Germanic Pontifical, we are faced with the phenomenon of a social nexus through which belief works, but at a time when the sense of the isolation of the soul before its Judge is awakening: one thinks of the gathering habit of private prayer represented by Jean de Fécamp and St. Anselm of Canterbury.

We are told that the people (*populus*) are present at these rites. They process for example with the newly lit taper from the place of blessing (perhaps the church of Notre-Dame-la-ronde in the case of Metz). The social resonance effected by the separation of the catechumens on Saturday morning after their exorcism must have made a striking piece of theater, involving not only the commotion of the children's departure from the cathedral but also the period of some hours' waiting for their return. This is social theater.

For all the effect of such self-evident, though suggestive, miming—the general sense of an expulsion from Eden associated with original sin would have been plain—the question is bound to arise: how much of a liturgy like this one was understood, and by how many? It is, after all, in Latin. An Italian with little or no Latin education might have understood much of it, a Spaniard less (but still quite a bit), a Saxon or Franconian or Swabian less still.[2] The precise answer to the question is likely to remain problematic. A part of the answer lies, no doubt, in the relation of images—in paint, wood, stone, cloth, or glass—and gesture to liturgical wording. It also lies in the propagation by vernacular versions of liturgical motifs, in vernacular versions of saints' lives for example. Image and vernacular version, perhaps gesture too, are at play in this benediction of the wax, which is at its heart a reenactment of the Harrowing of Hell and so implies a calling on the Last Judgment. The Harrowing of Hell is taken up by the vernaculars, and in its old, fourth-century or older, Latin version contains echoes of the text translated here. The liturgy is, at this point and in many places, a contraction of the peripateia of story into the one- or two-liner of a benediction. What appears a humorous, hilarious even, exchange between an appalled Satan and a panicky toadying Hell—whose responsibility it was to consign all humankind to death once and for all, and who has to explain how it is that a light is showing at the door and that a figure who apparently has, of all things, risen from the dead (Lazarus was bad enough, practically winging it back to life, now another one) is threatening to break in—this exchange is caught with marvelous brevity, if one knows it's being caught, in "the hard-bitten old shadows" that feel day coming at them, and "death, dragged dumbstruck, snared by the light of God's triumph into truth's radiant font."[3] And it is hard to imagine this *populus*, this liturgical crowd, would not have known some telling of this story. With this, one guesses at the presence of farce, a gallery of knockabout characters lurking just beneath the solemn liturgical surface, a shard of jest within earnest, related perhaps to the *risus paschalis*, the "Easter laughter."[4] And then how struck one is by the distance of this episode in Hell to the delicate operation of the bees whose ancestry is not in an Apocryphal Gospel (the Harrowing of Hell occurs in the Gospel of Nicodemus), but in Virgil's *Georgics*. Salvation, it might be hazarded, is the routing of story—the reversal of the decline from the stasis of Eden by divine interruptions of brevity; routings of the *antiqua nox* ("the age-old night") that Gregory the Great (ca. 540–604) keeps coming back to in his *Moralia in Job*, with the stress this phrase places on the long-term obduracy of the other side, a routing by suddenness. Much the same suddenness as that of the *subito*, "suddenly," after long lament, which closes the first of the seven penitential psalms, sung by the choir before the first prayer over the wax:

Confundantur et conturbentur vehementer omnes inimici mei; revertantur et confundantur subito—Let all my enemies be confounded and brutally scattered; turned back and confounded *straightaway*. (Ps. 6:10)

I have understood practice to be a kind of perceptive alertness to forms, to shifts in style (from exorcism to tactful banishment of the child-catechumens to Old Testament fire-epic to fragile bees), even to shifts in mood. (Think of the

great sequence of moods involved in the passing from the Friday of death to the
Sunday of resurrection, of which baptism is the part that speaks for the whole.) It
might be said: ritual is a heightened alertness to the obvious, so heightened that the
obvious becomes surprising once more. The sublime in repetition, its intimation of
the inevitable, the determinate, of a claustrophobia related to achieved forms, is the
backcloth against which the concrete particular is exposed as a one-off, as it is
"done again in memory." Walking in a cloister, an architecture of finished forms,
nothing can fittingly be done other than submit to a chance encounter and strike
up conversation, or behold a scene of Annunciation—or some other exact unbid-
den vision. In the Quattrocento the hand of Gabriel breaks in on the line that marks
the outer edge of the Virgin's cell: the hand breaks the Word open into episodic
time.[5] In the liturgy, the crescendo of God's actions, the *opera dei* as the prayers
put it, drops abruptly into the gentle garden scene of the bees, confirming the suspicion we already had that the diverse fires are problematic echoes, not routine it-
erations of the same thing. Each belongs to a context that isolates it, so that ab-
straction from context tends to stress singularity further. The fire of Daniel's
furnace is not the fire of the Burning Bush. And so St. Paul's instinct is borne out,
that the reenactment in ritual of events sets the rite aside from the world, without
losing a jot of its sensuality.

Notes

1. The wording of this Metz version of the blessing is very slightly different from that of the parent
manuscript of the Romano-Germanic Pontifical, now in the library of Montecassino, but probably
made in Mainz: C. Vogel and R. Elze, eds., *Le Pontifical Romano-germanique du dixième siècle*, 3 vols.
(Rome, 1963–72). For the same prayer in the Gelasian Sacramentary (of the eighth century), see L. C.
Mohlberg, ed., *Liber Sacramentorum Romanae Ecclesiae Ordinis anni circuli* (Rome, 1960), 68–70.

2. See R. McKitterick, *The Carolingians and the Written Word* (Cambridge, 1989), esp. 1–22.

3. C. Tischendorf, *Evangelia apocrypha*, 2nd ed. (Leipzig, 1876). On the wide subject of the vernacular versions, and the development of a theatrical treatment, see references in *Lexikon des Mittelalters*,
under "Descensus ad inferos."

4. Abelard, *Hymnarius paraclitensis*, ed. J. Szövérffy, 2nd ed. (Berlin and Leiden), vol. 2, 106,
no. 42 (for the nocturn of Good Friday), verse 4:

> Tu tibi compati sic fac nos, Domine,
> Tuae participes ut simus gloriae,
> Sic praesens triduum in luctu ducere,
> Ut risum tribuas paschalis gratiae.

Mikhaïl Bakhtine, *L'oeuvre de François Rabelais et la culture populaire au moyen âge et sous la renaissance* (Paris, 1970), gives material that makes us ask whether there was (is?) an element of carnival
within liturgy. And the strain of jest in earnest picked out by E. R. Curtius, *European Literature and the
Latin Middle Ages*, trans. W. R. Trask (London, 1953), 417–35.

5. Henry Brown pointed out this breathtaking precision of gesture to me.

Further Reading

P. Cramer, *Baptism and Change in the Early Middle Ages, c. 200–c. 1150* (Cambridge, 1993).

S. A. Keefe, *Water and the Word: Baptism and the Education of the Clergy in Carolingian Europe*, 2 vols. (Notre Dame, IN, 2002).

E. C. Whitaker (ed.), *Documents of the Baptismal Liturgy,* 3rd edn. revised and expanded by M. E. Johnson, Alcuin Club Collections 79 (London, 2003).

— 2 —

Cathars and Baptism

Shulamith Shahar

The same heretic said that our baptism [that of the Roman Church] has no value whatsoever because it is performed in material water, and because many lies are said at this baptism. The boy is asked: "Do you want to be baptized?" And it is answered for him: "I want." However he does not want; yet he cries when he is being baptized. He is also asked if he believes in this or that and it is answered for him: "I believe." Yet he does not believe because he does not have the use of reason. He is asked if he repudiates the Devil and his false shows, and it is answered for him that he does. However he does not repudiate the Devil because when he begins to grow up he lies and performs the works of the Devil. Our baptism [he said] is good because it is in the Holy Spirit and not in water. And also because they [the baptized] are adults and have the use of reason when they are being baptized. [He said] that by their baptism they became the sons of God and received his sonship, and that no believer of theirs became the son of God and received God's sonship unless he was baptized in their baptism.

Source: J. Duvernoy, ed., *Registre de l'Inquisition de Jacques Fournier évêque de Pamiers (1318–1325)*, vol. 2 (Toulouse, 1965), 410.

She said that when her mother-in-law was consoled, she, the speaker, had a son two or three months old who was sick. Guillaume Buscailh[1] said to her: "Do you want us to have one of these Good Men [speaking of the heretics] receive your son into their sect if he is about to die? Because if he is accepted by them and dies he'll be an angel of God." When she asked him what she would have to do with the infant after he had been accepted by the heretics, Guillaume told her that after that she would not give the boy either milk or anything else, but would let him die thus. Hearing this she said that in no way would she refrain from giving the breast to the boy as long as he lived, since he was a Christian and had committed no sin unless through her. She [said] that she believed that if she lost the boy, God would have him. Guillaume answered her that God would better have the soul of the boy if he were accepted by the Good Men than otherwise as he would then be an angel of God. The boy was not consoled because this being said Guillaume left her.

Source: *Registre de l'Inquisition de Jacques Fournier*, vol. 1, 499.

She said that about sixteen years earlier [she did not remember the time otherwise], she lost her daughter Marquèse. She mourned her and was very sad, and so was her husband. Gaillarde, widow of Raymond Escaunier, came to their house in Arques and comforted them on the death of Marquèse.

Source: *Registre de l'Inquisition de Jacques Fournier*, vol. 2, 403.

While this was going on, Jacqueline, the daughter of the speaker, who was not yet one year old, fell gravely ill. Her husband absolutely wanted her to be consoled. The speaker agreed with her husband, and they made the said heretic André come from the house of Raymond Maulen. He was staying there awaiting the death of the said Gaillarde so that in case she committed some sin according to the heretics he would reconsole her. The said heretic came in the night and consoled the girl who was in agony. He made many inclinations and elevations putting the book over the girl's head. After the consolation he said that from then on she should not be given to eat or drink milk or anything else that was born of flesh, and that if she survived she should be fed lenten food only. Present at the consolation were the speaker, her husband, and Bernard Vital of Arques. She did not remember who else was present. Her husband was very glad about his daughter's consolation and said that if she died in that state she would be an angel of God, and that he and the speaker could not give the girl as much as the said heretic gave her by consoling her.

This done and said the heretic together with her husband and Bernard Vital left the house. After they had left she suckled her daughter because, as she said, she could not see her die like that. When her husband came back he told her to take care not to give the girl milk because she had been received and it was said that she would be lost if she ate or drank milk after the consolation. She answered him that she had suckled her after the consolation, and her husband was very grieved and shaken by this. Pierre Maury comforted him saying it was not his fault, and that those who had to be by then Good Christians could not be so. To the girl Pierre Maury said: "You have a bad mother!" He also said to the speaker that she was a bad mother and that women were demons. Her husband wept, threatened and vituperated her. Since then for a long time he had not loved either the girl or the speaker until he recognized his error. The girl survived for about a year and then died. She was not reconsoled.

Source: *Registre de l'Inquisition de Jacques Fournier*, vol. 2, 414–15.

These four texts are extracts from the register of Bishop Jacques Fournier, Inquisitor in Pamiers in the county of Foix from 1318 to 1325. The first extract is from the testimony of Sybille Gousy-Peyre about the sermons of the Perfect Pierre Authié that she attended. Her detailed testimony, which constitutes a highly informative source of Cathar doctrinal teachings of the period, caused irreparable damage to the Cathar Church. In her interrogation, which started in November 1322, she testified about what she had heard and about events that had taken place about sixteen to eighteen years earlier.

The next three extracts refer to events in the personal lives of a woman called Mengarde Buscailh and the same Sybille. What they told the Inquisitor about the sickness and death of their infants illustrates the problem they had to face as Cathar Believers in such cases.

The Cathar doctrine was one of dualism: a cosmic dualism as well as an anthropological one. The absolute dualists among the Cathars, who constituted the dominant group, believed in the existence of two separate and eternal Principles not equal in their power: the Principle of Good and the Principle of Evil. The present wicked world in its entirety was created by the Devil, whose corruption and rebellion were caused by the Evil Principle. It lies within the power of man to assist God.

The Cathars were divided into Believers (*Credentes*), and a small minority of Perfects (*Perfecti*). The Believers lived fully in the world and listened to the doctrinal and moral teachings of the Perfects, supporting them and protecting them. They were allowed to assist in the Cathar supreme ritual, the *consolamentum*, but they were not members of the Cathar Church.

A person could receive the *consolamentum* after a period of probation and instruction. By this rite, defined as a spiritual baptism, he was admitted to the Cathar Church and became a Perfect. His sins were remitted and he was considered liberated from the chains of the flesh already in this world. His soul was reunited with his spirit, and it was believed that after the death of his body his soul would not have to transmigrate into a new body, but would return to its original source and find salvation. The *consolamentum* was both a sign of salvation by divine grace and a precondition for it. The Perfects renounced the world and undertook a strict moral code. They had to live a life of absolute chastity and poverty, keep long fasts and totally abstain from eating any products of coition: meat, milk, cheese, and eggs.

The *consolamentum* received by a person after undergoing a period of testing and instruction was not the sole occasion on which *consolamentum* was received. There was also a *consolamentum* on deathbed, and most Believers chose to postpone the *consolamentum* to their final hour. After the reception of this *consolamentum*, the recipient had to keep a fast (*endura*): absolutely refrain from eating and drink only water. This practice of *endura* by the moribund probably appeared first in the last quarter of the thirteenth century. Its purpose was to safeguard the integrity of a *consolamentum* administered to a person who had not demonstrated during his lifetime the ability to endure the Perfect's abstinence and might have been unable to maintain his purity. This *consolamentum in extremis* was also believed to ensure salvation.

The practice of administering the *consolamentum* to dying infants developed in the second decade of the fourteenth century. In the years 1308 to 1310 most of the Perfects of Languedoc were burned at the stake by order of the Inquisition. The sect lost its leaders and was in its final decline. Jean Duvernoy rightly argued that the practice of consoling dying infants was exercised only by the last Perfects of the region who were the least initiated ones.[2] What was a blatant digression from the original Cathar teachings was not the administration of the *consolamentum* to the dying in itself (with the tacit or declared promise of salvation), but the fact that it was conferred on infants. One of Pierre Authié's main arguments against Catholic baptism was that the baptized infant lacked "the use of reason" (extract 1). Other Perfects, however, encouraged parents to let their dying infants receive the *consolamentum*, thus forcing them to make a heartbreaking choice.

The rate of infant and child mortality was very high. Many parents lost more than one child. The question of the consolation of little Jacqueline (extract 4) was raised after Sybille and her husband Raymond had already lost one daughter (extract 3). A possible explanation of the development of the practice is that the Perfects who advocated it were aware of a deep-seated need of parents to ensure that the soul of the dead child be at peace and find salvation. The Catholic Church offered baptism for infants. The Cathar Church up to the fourteenth century did not offer a parallel rite. It is well known that Catholic parents were both grieved and scared when their infant died without baptism. They were grieved because of the loss of the child, and because a child dying without having been baptized was believed to go to Limbo where he would be forever denied the light of the vision of God. They were scared because according to popular belief (shared also by the Cathar adherents of Languedoc), the infant who died before entering the community of the living, which in the context of Christian culture meant before being baptized, would return to haunt the living.

The deposition of the two women clearly shows, in blatant contradiction to the theory of Philip Ariès and his followers, that parents did love their young children, cared for them, and mourned them if they died. Little Marquèse's father was sad at her death and mourned her no less than did her mother (extract 3). Yet there was a difference in what men and women were ready to do in order to ensure the salvation of the soul of an infant. Men urged the mothers to have their infants consoled. The mothers were not, however, capable of doing what this *consolamentum* entailed: refraining from giving the breast to the infant in agony and letting him die of hunger. It was they who cared daily for the infant, fed him, and were closer to him and knew him better than his father. It was *they* who were required to do it, that is, let the infant die of hunger. It appears that motherly feeling was stronger than religious norms and beliefs. They refused to obey.

Notes

1. Her brother-in-law.
2. J. Duvernoy, *Le Catharisme: La religion de Cathares* (Toulouse, 1976), 170.

Further Reading

Malcolm Barber, *The Cathars: Dualist Heretics in the High Middle Ages* (Harlow, Essex, 2000).
R. I. Moore, *The Origins of European Dissent* (London, 1977).
W. L. Wakefield and A. P. Evans, eds. and trans. *Heresies of the High Middle Ages: Selected Sources Translated and Annotated* (New York and London, 1969).

The Life Cycle:
Confirmation and
Coming of Age

——3——

The Early Medieval *Barbatoria*

Yitzhak Hen

Prayers for Him, Whose Beard Is Shaven for the First Time

O God, through whose providence every adult creature rejoices in growth, be gracious to this servant of yours, flourishing in youthful blush of age and shaving his bloom for the first time. May he be strengthened in every respect through the support of your protection; may he live to old age and may he rejoice in your protection in this life and in eternity. Through our Lord.

O omnipotent and eternal God, creator and begetter of all things, and who promotes the growth of everything that was born, bless this servant of yours, who offers you the first fruits of his youth, and grant him long life and strong faith. Lead him, O Lord, in the way of your laws, so that passing through it he may please you and reach old age. [. . .]

O God, who ordered that in the time of adolescence a beard of first bloom will also adorn the face, we implore the holy spirit of the prophets with singing mouth, and beg that the beard will be blessed; humbly we pray over the beard of this servant of yours, that was never cut by a knife before, but in front of your holy altars, sitting humble in head and soul, the hand of the priest will bless the beard on his face, like holy Aaron who was anointed with holy oil. [. . .]

Source: *Liber sacramentorum Gellonensis*, 374, ed. A. Dumas and J. Deshusses, CCSL 159 (Turnhout, 1981), entries 2499–2501, 380–81. The Latin of the text, and especially of the third part, is extremely corrupt.

Ever since the publication of Arnold Van Gennep's pioneering study of rites of passage, historians have become increasingly aware of the fact that many societies had a mandatory public rite for men that marked the transition between childhood and adult life. Such a rite symbolized the separation from the previous world of childhood, and it was the turning point that distinguished a man from a boy. As far as the Middle Ages are concerned, dubbing, that is, the ceremonial arming of the young knight, has commonly been accepted by historians as the rite of passage par excellence for free young men. Such a rite was already described by Tacitus (d. ca. 116/120) in his survey of the Germanic people:

> . . . one of the chiefs or his father or his relatives equip the young man with a shield and a spear. This corresponds with them to the toga, and is youth's first public distinction.[1]

Tacitus's description, however, must not be taken to imply that from this time onward a similar arming ceremony was celebrated among the inhabitants of western Europe. In fact, there is hardly any evidence from the early Middle Ages that the ceremonial arming of the young warrior was performed as a rite of passage, and it seems that in several Barbarian kingdoms of the post-Roman world a different rite marked the young man's passage from childhood to adulthood. This rite, commonly known as *barbatoria*, was celebrated when the beard (*barba* in Latin) of the young man was shaved for the first time.

The *barbatoria*, one should stress, was not a Germanic novelty. The Greeks and Romans celebrated similar ceremonies long before the Barbarians took over the western provinces of the Roman Empire. The Hellenistic poet Callimachus (d. ca. 240 BC), for example, mentions a similar ceremony that was performed in Delos, during which the shaved hair was offered to the gods,[2] and several Roman authors refer to a similar rite that was performed in Rome. Given the fragmentary and biased nature of our sources, it is impossible to gauge whether the *barbatoria* was ever celebrated by Barbarians before they first encountered Roman civilization. Whether it was an indigenous Germanic rite or a Roman import, by the time the Barbarians settled in western Europe (i.e. the late fifth and early sixth centuries), the *barbatoria* was already part and parcel of their cultural heritage, and was celebrated as a solemn rite of passage.

It is very difficult to discern how exactly the *barbatoria* was celebrated, since no detailed description of it appears in any of our sources. From the little evidence that survives, it seems that the ceremony was performed on the young man at the age of twelve, in accordance with what several Barbarian law codes define as the borderline between childhood and adulthood. The ceremony was usually performed by the boy's father or, in the case of a fatherless child, by the closest male relative, and it was probably attended by other members of the family and close friends. It is not at all clear whether an actual shaving of the beard took place during this ritual. First, as rightly pointed out by P. Guilhiermoz, in order to shave a beard one needs to have one,[3] and it is doubtful whether many young men would grow a beard by the age of twelve. Moreover, shaving in the Middle Ages was a slow and painful process, evidently unsuitable for a solemn rite. Thus, it seems that during the ceremony itself only a symbolic gesture of patting or touching the cheek was performed,[4] and the actual shaving of the young man (if indeed he had a beard) took place before or after the solemn ceremony.

Apart from symbolizing the passage from childhood to adulthood, the *barbatoria* possessed some deeper social meanings. When performed by someone other than the father, cutting a young man's beard created a special bond between the two—a bond as strong and as significant as adoption. Thus, loaded with social and legal implications, the *barbatoria* was a symbolic gesture that created a new social and legal order, by forming spiritual kinship and sponsorship. No wonder that unauthorized *barbatoria* was unequivocally banned by law.

With the conversion of western Europe and the gradual Christianization of everyday life, the ceremony of *barbatoria* was also Christianized, as the prayers cited above clearly bear witness. This process of Christianization was long and slow. First allusion to it appears in a poem by Paulinus of Nola (d. 431), which refers to a young man who shaved off his first hair and offered it at a tomb of a saint.[5] The best evidence for the Christianization of the *barbatoria* comes from a series of eighth-century liturgical sacramentaries from the Frankish kingdom. The prayers that appear in these texts clearly suggest that *barbatoria* in Frankish Gaul was given a Christian context and, subsequently, was performed in church and in the presence of a bishop. Furthermore, it is highly probable that a mass was sung in honor of the occasion, and it might have been accompanied by the confirmation of the young man. Similar efforts to Christianize the *barbatoria* are also attested in Visigothic Spain and in early Byzantium.[6]

From the prayers cited above, all of which beseech God to bless the young man with a long and successful life, it is obvious that efforts were made to accommodate the *barbatoria* within a Christian milieu. Whether the impetus for this process was given by the people, who were Christian and therefore wanted to Christianize a major social ceremony, or by ecclesiastical officials, who strove to get control over the crucial moments of the individual's life cycle, is difficult to ascertain. Nevertheless, the efforts to Christianize a secular ceremony illustrate quite clearly the widespread Christianization of everyday life in early medieval Europe. It gives us a rare glimpse of how in the early Middle Ages the Barbarian tradition, the Roman past, and the Christian present were allowed to interact freely in order to create a new social and cultural system that relied on both Roman and Barbarian traditions and was deeply imbued with Christian meaning. The *barbatoria* as a case study reflects a larger process whose end result was what is commonly known as "medieval culture."

The *barbatoria* was known and practiced as a rite of passage throughout the early Middle Ages, up to the Carolingian period. In 738 Pippin III (d. 768) was sent by his father, Charles Martel (d. 741), to the Lombard King Liutprand (d. 744) to have his *barbatoria*. Adrevald of Fleury (d. 878), more than a century later, relates that:

> He [i.e., Charles Martel] sent his son Pippin, so that according to the custom of the Christians he [i.e., Liutprand] will be the first to cut his hair, and thus become his spiritual father.[7]

Adrevald, it seems, had no doubt that *barbatoria* was indeed a Christian rite and, moreover, he was well aware of the various social and legal implications that such a ceremony entailed.

The *barbatoria* of Pippin III is the last *barbatoria* documented in our sources. It is not at all clear why this ceremony died out, but it appears that from the time of Charlemagne (d. 814) onward the Franks preferred the arming ceremony over the first shaving of the beard. There is no evidence from the Carolingian period that *barbatoria* was still practiced in Francia, and gradually it disappeared even from the liturgical books. Charlemagne, it appears, preferred, yet again, a German practice to a supposedly Roman one, and his successors duly followed suit, at least as far as the rite of passage for the young Frankish man was concerned. Consequently, some of the prayers we have just mentioned were recycled by later generations and were used, for example, as part of the inauguration ceremony in the so-called Leofric Missal.

Notes

1. Tacitus, *De origine et situ Germanorum [Germania]*, chap. 13.1, ed. M. Winterbottom, in *Cornelii Taciti Opera minora*, ed. M. Winterbottom and R. M. Ogilvie (Oxford, 1975), 44.

2. Callimachus, *Hymns and Epigrams*, IV, 298–99, ed. and trans. A. W. Mait, The Loeb Classical Library (London and Cambridge, MA, 1921), 108–9.

3. P. Guilhiermoz, *Essai sur l'origine de la noblesse en France au Moyen Age* (Paris, 1902), 408.

4. The Mozarabic *Liber ordinum* uses the verb "tangere" (to touch); see *Liber ordinum*, ed. M. Férotin, Monumenta ecclesiastic liturgica 5 (Paris, 1905), cols. 43–46.

5. Paulinus of Nola, *Carmen XXI*, 377–78, ed. G. de Hartel, CSEL 30 (Vienna, 1894), 170.

6. See *Liber ordinum*, ed. Férotin, cols. 43–46; *Euchologion sive rituale Graecorum*, ed. J. Goar (Paris, 1647), 375–79; *Liber pontificalis*, c. 83, ed. L. Duchesne, 2 vols. (Paris, 1886–92) [reprinted with vol. 3 by C. Vogel (Paris, 1955–57)], vol. 1, 112.

7. Adrevald of Fleury, *Miracula sancti Benedicti*, I, 14, ed. O. Holder-Egger, MGH SS XV, 1 (Stuttgart, 1887), 483.

Further Reading

Robert Bartlett, "Symbolic Meanings of Hair in the Middle Ages," *Transactions of the Royal Historical Society* n.s. 4 (1994): 43–60.

Giles Constable, Introduction, in *Burchardi abbatis Bellevallis Apologia de barbis*, ed. R.B.C. Huygens, CCCM 62 (Turnhout, 1985), 47–130.

Paul Dutton, *Charlemagne's Mustache and Other Cultural Clusters of a Dark Age* (New York, 2004).

Yitzhak Hen, *Culture and Religion in Merovingian Gaul, A.D. 481–751*, Cultures, Beliefs and Traditions: Medieval and Early Modern Peoples 1 (Leiden, New York, and Cologne, 1995), especially pp. 137–43.

Bernhard Jussen, *Spiritual Kinship and Social Practice: Godparenthood and Adoption in the Early Middle Ages*, trans. Pamela Selwyn (Newark and London, 2000).

Régine Le Jan, "Frankish Giving of Arms and Rituals of Power: Continuity and Change in the Carolingian Period," in *Rituals of Power from Late Antiquity to the Early Middle Ages*, ed. Frans Theuws and Janet L. Nelson, The Transformation of the Roman World 8 (Leiden, Boston, and Cologne, 2000), 281–309.

Joseph Lynch, *Godparents and Kinship in Early Medieval Europe* (Princeton, 1986).

Arnold Van Gennep, *Les rites des passages* (Paris, 1909); English translation: *The Rites of Passage*, trans. Monika B. Vizedom and Gabrielle L. Caffe, with and introduction by Solon T. Kimball (London, 1965).

The Life Cycle: Instruction

— 4 —

Lollard Instruction

Rita Copeland

1.

For this cause a synful caytif, hauynge compassioun on lewed [*unlearned*] men, declarith the gospel of Mathew to lewid men in Englische, with exposicioun of seyntis and holy writ, and alleggith onely holy writ and olde doctours in his exposicioun, as Seynt Austyn, Seynt Ierom, Seynt Gregor, Seynt Ambrose, Seynt Crisostom, Seynt Bernard, Grosted [*Grosseteste*] and olde lawes of seyntis and of holy chirche wel groundid in holy writ and resoun; and alleggith also the Maister of Sentence [*Peter Lombard*] rehersynge olde seyntis and doctours, and also Rabanes on Mathew an olde monk and doctour [i.e., *Rabanus Maurus*], rehersynge copiously olde holy doctouris for hym, as Austyn, Gregor, Ambrose, Bede, Illarie, Crisostom, Ierom and many mo. This coward synful caitif alleggith Ierom on Mathew on the same text whiche he declarith; and therfore he alleggith thus "Ierom here," that is, on the same text. First in glos a word of text is vnderdrawen [*underlined*]; thanne cometh glos and the doctour seyinge [*following*] that is alleggid in the ende of the glos. And aftir that doctour, al the glose suyinge is of the next doctour alleggid, so that the glos is set before, and the doctour alleggid after, who it is and where [i.e., *the name of each authority is given after the gloss derived from him*]; and the same weye of Crisostom in his werk vncomplete on Mathew, euene on the text expowned: and the same maner of Rabanes, for he goth thourout on Mathew. Also in the sarmoun of the lord on the hil, that is on v, vj, and vij chapitris of Mathew, he alleggith Austyn on this maner "Austyn here," that is, on the same text expowned. Whanne he alleggith Gregor, Bernard or Austyn or Ierom in other bokis or other doctour or lawis, he tellith in what bok and what chapitre, for men schulden not be in doute of treuthe. And this synful caitif, seeld vnder synne as the postle seith, takith pleinly and schortly the sentence of this doctours with groundis of holy scripturis withouten any settyng to of other men. For the sekenesse of oure peple is so gret that the nylen suffir pore men lyvynge now to reprove her synnes and open here vycis, though thei tellen never so pleynly holy writ ensaumple of cristis lif and his postlis with pleyn resoun. Therfore men [laye] to oure seke peple the plastre of holy writ with thes doctours biforseid withoute more addynge, if god wol in any maner of his grete mercy make hem to knowe and amende her yvel lyvynge and acorde with holy writ byfore that thei of this lyf gon.

Source: From the prologue to the "long" exposition of Matthew, MS Oxford Bodleian Laud Misc. 235, fol. 2r; printed in Copeland, *Pedagogy, Intellectuals, and Dissent in the Later Middle Ages: Lollardy and Ideas of Learning* (Cambridge, 2001), 133–34.

2.

Oure lord Iesu Crist very God and very man, cam to serve those meke men, and to teche hem the gospel. And for this cause Seynt Poul seith, that he and other apostlis of Crist ben servantis of cristen men bi oure lord Iesu Crist. And eft he seith: y am dettour to wise men and unwise. And eft bere ye the chargis [*responsibilities*] an other of an other. And so ye schulen fille the lawe of Crist, that is of charite as Seynt Austyn expowneth. Herfor a symple creature of God, willinge to bere in party [*in part*] the chargis of symple pore men, wel willinge to Goddis cause, writith a schort glos in Englisch on the gospel of Ioon, and settith onely the text of holy writ and the opyn and schorte sentencis of holy doctours, bothe Grekis and Latyns, and allegith hem in general for to ese the symple wit and cost of pore symple men; remyttinge [*deferring/referring*] to the grettir gloos writun on Ioon where and in what bokis thes doctours seyen thes sentences. And sumtyme he takith the cleer sentence of lawis of the churche maad of seyntis, wel groundid in holy writ and pleyn resoun, to dispise synnes and comende vertues. First the text is set and thanne the sentence of a doctour is set aftir, and the doctour is aleggid in the ende of the same sentence.

Source: From the prologue to the "short" exposition of John, MS Oxford Bodleian Bodley 243, fol. 115v, printed in Copeland, *Pedagogy, Intellectuals*, 132.

3.

As it is demed a greet werk of mersi and of charite to teche a unkunnyng man the right and sykur weie, whanne many perels ben in wrong weies and nameli if the man mut go that unknowen weie, and ellis perische in greet meschef, so it is a fer gretter werk of mersi and charite to telle opynli the treuthe of the holi gospel to lewid men and symple lettrid prestis, sithen the gospel is the right weie to heuene, without which noon may come to heuene. Herfor a symple creature expowneth schortli the gospel of matheu to lewid men in englische tunge, that thei mow the beter knowe it and kepe it, and to be savyd with outen ende in the blisse of heuene. In this schort exposicioun is set oneli the text of holy writ with opyn sentensis of elde holi doctours and approuyd of holi chirche, and summe lawis of the churche groundid in goddis lawe and reason. The text of the gospel is set first bi itself, an hool sentence togidere, and thane sueth the exposicioun in this maner. First a sentence of a doctour declarynge the text is set aftir the text, and in the ende of that sentence the name of the doctour seynge it is set, that men wit certeynli hou fer that doctour goith. And so of all doctours and lawes aleggid in this exposicioun. Whanne y seie ierom [*Jerome*] here, ierom seith that sentence on matheu . . . [*continues with explanation of text and gloss layout, similar to explanations in the texts quoted above*]. At the begynnyng I purposide with goddis grace to sette pleynli the text of holi writ in the trewer sentensis of thes doctours in maner bifor seid. And blessid be al myghti god of his graciouse help. Y am not gilti in my silf that y erride fro this purpos. If ony lernd man in holi writ and holi doctours be displesid in ony sentence set here, for goddis loue loke he wel trewe originals, and y hope he schal not fynde greet defaute. Netheles if he fynde ony defaute, for goddids loue sette he in the trewer sentence and no thing ellis, that cristen puple may haue sikirle [*securely*] the trewer undurstondyng of holi writ. For y desire noon other thing in this werk, no but that cristen puple knowe and kepe treuli holi writ, and come bi goddis mersi ther bi to the endles blis of heuene. Ihu kyng of mersi of pees and charite that scheddith thi preciouse blood for the loue of mennis soules, graunte this ende. Amen.

Source: Prologue to the "short" exposition of Matthew,
British Library Additional MS 41175, fol. 1v

The religious practice represented in this selection is that of heterodox religious instruction, exemplified by Lollard glosses to the gospels. The English heretical movement known as Lollardy originated in the late fourteenth century, with the teachings of the Oxford theology master John Wyclif (d. 1384). From an academic circle at Oxford, the heretical movement spread into lay communities, where it gained considerable following among artisanal groups in towns, especially in the midlands, East Anglia, and Kent. The heterodox movement seems to have been attractive to laypeople for a variety of reasons. Some of the Wycliffite teachings carried anticlerical and especially anti-mendicant themes that resonated with existing anticlerical sentiments in England. Wycliffite writings also carried political themes that emphasized secular over ecclesiastical power. But perhaps most important, Wycliffite teaching stressed the capacity and entitlement of lay readers to interpret Scripture themselves, without clerical mediation and mystification. To this last end, Wycliffite scholars during the 1390s had undertaken a program of vernacular bible translation and bible commentary of unprecedented ambition. Two versions of an English translation of the entire Bible were produced, along with various kinds of commentary in English, based on standard patristic and later medieval commentaries. All of this material aimed at providing "open and clear" access to Scripture, offering what was seen as the "letter of the text," a "literal sense" unclouded by scholastic subtleties, and thereby appropriate to the needs and interpretive capacities of nonclerical and often nonliterate audiences.

The literal sense of Scripture had recently (since the thirteenth century) achieved a fashionable status as the object of academic exegesis, with theology masters like the Franciscan Nicholas of Lyra (1270–1349) declaring a strong preference for producing literal commentaries that eschewed allegorized readings. Wycliffite hermeneutics, in turn, preferred literal reading of Scripture on the grounds that it gave more direct access to the intended meaning of the Scriptural text, to a theological truth unsullied by self-interested exegetical meddling. But a literal approach to texts also had roots in a long secular tradition of elementary pedagogy, in which literal explanation of the classical texts that formed the school curricula was valued as the best way to introduce young students to Latin language and literature.

It was the genius of the early Wycliffite scholars, who produced the English translations of Scripture and English commentaries for the use of lay communities, to combine the recent theological respect for the literal sense of Scripture with the ancient pedagogical utility of literal explanation, and to present "literal" expositions of Scripture to lay audiences whose own literacy may have been at an elementary level. In other words, they converted a theological preference for the "plain sense" of Scripture to the pedagogical purpose of introducing the biblical text to vernacular audiences.

Some of the Wycliffite commentaries seem to have been designed for targeted pedagogical purposes. This is the case with the selections given above. These come from a group of texts known as the *Glossed Gospels*, which were composed sometime between the 1390s and 1407 (the year in which English Bible translation was outlawed by Archbishop Thomas Arundel's anti-Lollard statutes, the Oxford *Constitutions*), which aimed to make available in English authoritative commentaries on the "literal sense" of Scripture. The *Glossed Gospels* are commentaries on individual Gospels, and originally represented a comprehensive program of introduction to the Gospel texts, although not all of the original commentaries survive (presumably a long and a short commentary on each Gospel was planned, but only six commentaries survive among nine manuscripts). All the surviving *Glossed Gospels* use the same format: following the earlier translation of the Wycliffite Bible, they break the text into small sections of a few lines, each of which is followed by an English commentary derived from orthodox sources, typically the Fathers Bede, Pseudo-Chrysostom, St. Bernard, and Thomas Aquinas. Many of the sources are taken directly from Aquinas's own synthesis of biblical commentary, the *Catena aurea* ("Golden Chain").

The physical layouts of the *Glossed Gospels* manuscripts suggest the ways in which laypeople might have used the texts for group instruction and even, possibly, basic literacy acquisition for nonliterate participants in a local "conventicle" or group of Lollard sect members. The differences between text and gloss are clearly marked (usually by underlining of text), and the sources of the glosses are clearly indicated with the names of the authorities given at the end of the gloss and sometimes also in the margins to ensure that readers will distinguish between the Scriptural text and the glosses. But even more important, the *Glossed Gospels* offer certain precise indicators of their own pedagogical import. Six manuscripts of the *Glossed Gospels* preserve prologues addressed to communities of readers. The prologues explain the layout of the text in the form of a "user's guide," and also address the hermeneutical concerns that the lay readers may have about the doctrinal truth of the exposition, the reliability of sources, and the authenticity of the texts. For example, in the prologue to the long exposition of Matthew, the reader is told how to identify *lemmata* (the citation words that link the gloss to the section of text being commented on) by their underlining, how to identify the name of an authority from whom a particular commentary is taken (his name will be given after the gloss), how to distinguish between a gloss taken from one authority and a gloss taken from another authority (the name of each authority will appear after each gloss), and how to know where the text ends and the gloss begins. At the same time, these pedagogical signposts will also reassure readers of the truthfulness of what they are reading, "for men schulden not be in doute of treuthe." The prologue promises to give the "sentence" of the authorities "pleinly and schortly," and the foundations (the "groundis" of Scripture "withouten any settyng to of other men"), in other words, the plain meaning of Scripture and straightforward exposition without any excess of interpretation. This promise of an uncluttered "plain sense" is echoed in the prologue to the short exposition of John, which offers "cleer sentence of lawis . . . wel groundid in holy writ and pleyn resoun"; "plain reason" is often a Lollard term for the common sense derived from the literal level of Scripture, to which all reasonable people have access.

The kind of guide to reading offered in these prologues has hardly any parallel in clerical, academic Scriptural commentary. The authors of the *Glossed Gospels* seem to have taken some cues from Aquinas's explanation of his synthetic method in the *Catena aurea*, where Aquinas notes how he has indicated the names of the authorities whose testimonies he is quoting; but in Aquinas's *Catena* as in other university-level commentaries, such technical explanations serve to advertise innovations in the layout of commentary texts, not to instruct readers in the basics of how to read a commentary. By contrast, the *Glossed Gospels* prologues offer detailed instruction in codicological literacy, so that any lay Lollard group of readers can find its way around in the text, even in the absence of a clerical teacher.

We do not have a great deal of external evidence of how Lollard conventicles practiced group readings, although various trial records indicate that literate members of a small community read aloud to other members or even provided some literacy instruction. But the internal evidence of texts like the *Glossed Gospels* suggests that the authors of the vernacular texts directed to lay Lollard communities were sensitive to the pedagogical needs of lay readers unfamiliar with the conventions of academic theological texts. These commentary texts answer both the doctrinal interests and the basic instructional needs of the lay dissenters reading Scripture in English.

Further Reading

Rita Copeland, *Pedagogy, Intellectuals, and Dissent in the Later Middle Ages: Lollardy and Ideas of Learning* (Cambridge, 2001).

Henry Hargreaves, "Popularizing Biblical Scholarship: The Role of the Wycliffite *Glossed Gospels*," in *The Bible in Medieval Culture*, ed. W. Lourdaux and D. Verhelst (Leuven, 1979), 171–89.

Anne Hudson, *The Premature Reformation: Wycliffite Texts and Lollard History* (Oxford, 1988).

Shannon McSheffrey, *Gender and Heresy: Women and Men in Lollard Communities, 1420–1530* (Philadelphia, 1995).

Jo Ann Hoeppner Moran, *The Growth of English Schooling, 1340–1548* (Princeton, 1985).

Nicholas Watson, "Censorship and Cultural Change in Late-Medieval England: Vernacular Theology, the Oxford Translation Debate, and Arundel's Constitutions of 1409," *Speculum* 70 (1995): 821–64.

The Life Cycle:
Marriage and Its
Unmaking

5

Florentine Marriage in the Fifteenth Century

Christiane Klapisch-Zuber

RECORDS OF A FLORENTINE ARISTOCRATIC MARRIAGE

It is to be recorded [*Ricordanza*] that this day . . . of August 1432, we have taken the oath and as is the custom have gathered the men at St. Thomas in the Old Market [San Tommaso in Merchato Vecchio], a great gathering of men. We took the oath at Santa Maria sopra Porta, near the Guelph part, and chose that place as the most convenient for both sides: it is there that the oath was taken with great decorum, as demonstrated by the size of the crowd and its great happiness. The notary was. . . .

We organized a beautiful procession of women and a very fine feast at our house on that day, as is the custom. We sent the procession to the house of my father-in-law where a great and honorable feast was offered, with the music of pipes and trumpets, as is the custom. That evening, I went to the feast with my closest relatives, that is Piero di Cosimo di Giovanni di Bicci dei Medici, Filippo di Filippo di Sir Simone Tornabuoni, brother of my mother Sandra, and Tedice d'Antonio di Tedice first cousin of my father Giuliano, that is the son of Lady Cosa who is the sister of Lady Gostanza and mother of Giuliano my father, as well as Giovanni di Giannozzo Gianfigliazzi, son of Lady Tita sister of my father and daughter of Averardo my grandfather. And so all four accompanied me to the banquet that evening, and it was the most beautiful feast imaginable.

On the . . . of August, I sent the casket to my wife Gostanza: a large silver basin, a kerchief of silk full of pearls, a pearl necklace, a skein of thread, and two belts decorated with silver. I shall not mention here the contents of the casket, since these will be described below in detail.

It is to be recorded that on 14 June 1433 I brought my wife Gostanza to my home. I visited my father-in-law on that day to give him the ring, accompanied by much of the youth of Florence, all guests at the marriage feast. We gave a magnificent feast: we flattened the whole courtyard and the little garden, and erected a decoration, the like of which had not been seen for years: the sculptor Michelozzo

was in charge of it. At nightfall my wife arrived on horseback with knights, Sir
Lorenzo Ridolfi and Sir Palla di Nofri Strozzi. [. . .] My father-in-law, Averardo, offered a great meal for the citizens of the city.

On the morrow, my father-in-law sent to me a beautiful procession of
twenty-five women and we had a great, beautiful and joyous feast. On the following day, after dinner, the young folk of the wedding accompanied my wife
on horse to the Guicciardini family, with great joy and a beautiful feast and
jousts in the city streets.

Here follows the list of items given to Gostanza as she left to join her husband:

Sir Battista da Comapofregoso gave her a cloth of very fine and beautiful
linen, a silver mirror decorated in silver within a small case, and a testicle of
musk. He sent these presents on behalf of his wife Ilaria, who was then living
in Sarzana, while Battista was in Florence with his son, a young boy called Sir
Pierino. [. . .] Averardo thanked Battista for all the gifts.

Lady Caterina, wife of Cosimo dei Medici, gave Gostanza a pearl of little
value, mounted in gold.

Lady Ginevra, wife of Lorenzo dei Medici, gave her a little balas ruby set in
gold, this also of little value.

Lady Caterina, wife of Alamanno Salviati, daughter of Averardo my grandfather, gave Gostanza a sapphire set in gold, of little value.

Lady Tancia, wife of Antonio di Salvestro Serristori, daughter of my grandfather Averardo de Medici, gave her a sapphire of little value, set in gold.

Lady Vaggia, widow of Bernardo di Filippo Magalottie and daughter of Averardo my grandfather, who lived in the monastery of San Gaggio, gave her a
small sapphire set in gold; Giuliano my father had given this to her to offer to
Gostanza, as is the custom.

Giovanni di Giannozzo Gianfigliazzi, son of Lady Tita, daughter of Averardo my grandfather and sister of my father Giuliano, gave Gostanza a balas
ruby set in gold.

Francesco di Michele degli Arrighi, son of Lady Alessandra, aunt of Giuliano
my father, gave Gostanza a small sapphire of little value.

Lady Lena, widow of Jacopo Guidetti and sister of my mother Lady Sandra,
gave Gostanza a large but poor sapphire set in gold.

Andrea di Lippaccio dei Bardi, Associate of Averardo's bank, gave Gostanza a
very small balas ruby set in gold, of little value.

Talento di Talento dei Medici, who lives in the family home, gave her a very
small pearl, set in gold.

Mariotto d'Averardo gave Gostanza a small emerald of little value, set in gold,
which we gave him to bestow on her.

Ulvieri di Bruscolino da Gagliano, Steward of Cafaggiolo, gave her a pretty
stone, set in gold.

Lady Maddalena, wife of Averardo, gave Gostanza a balas ruby, set in gold,
which we gave her to offer as a gift to Gostanza.

Averardo my grandfather gave Gostanza a fine balas ruby set in gold, which he borrowed from Andrea di Lippaccio Nardi, his associate at the bank.

Giuliano my father gave Gostanza a piece of crimson velvet, which he had bought from Antonio Canacci, but for which he had not as yet paid.

Lady Sandra my mother gave her one hundred marten furs bought from Domenico the furrier, on credit.

And I, Francesco, when I gave her the ring, I also gave her two rings: a diamond set between two brilliants, and an emerald between two small rubies, both set in gold.

The Commune of Santa Maria a Campiano gave her a heifer for the wedding feast.

The Commune of San Piero a Sieve gave her a heifer for the wedding feast.

The Commune of Bruscoli gave her a heifer for the wedding feast, and sent a group of youngsters, led by Donato di Giusto di Bertino to the wedding.

The Commune of Castro gave her a heifer for the wedding, but it escaped when it was being delivered, yet it was found, though we did not kill it for the wedding feast.

The Commune of Ronta and our friends there gave one hundred new tailloirs for the wedding day.

The Abbot of Pacciano, Sir Francesco di Ser Lodovico della Casa, gave a heifer for the wedding feast, and it was a very nice one.

Giovanni di Giovanni da Castel San Giovanni gave one hundred and twenty flasks of vintage Trebbiano wine, for the wedding.

It is to be recorded that before I take Gostanza, Francesco di Sir Simone Tornabuoni negotiated the union between Piero Guicciardini and my father Giuliano and has brought both sides to agreement on the dowry, so that Piero will give me 200 florins in the form of trousseau. I shall describe this trousseau, as follows [. . .]:

A pair of painted chests, worth 62 florins
A piece of damask, 70 feet long, worth 55 florins
A robe of white cloth decorated with pearls, worth 25 florins
An undergarment in green cloth, worth 15 florins
A velvet headdress adorned with silver, worth 11 florins
A piece of cloth for handkerchieves, worth 10 florins
A light dress with silk sleeves, worth 8 florins
A coral wreath, worth 8 florins
A small book with the office of Our Lady, worth 6 florins
                                      200 florins in total

Items that were not included:

A green outer garment
A garment of purple cloth decorated in silver

A garment of cotton
17 blouses
30 large bands of cloth
40 head scarves
25 hoods
A headdress of white velvet with three ounces of silver
A blue velvet cap embellished with pearls
A green velvet cap with small buttons
A white cap made of several linked pieces
A green and black cap made of velvet
2 large hats decorated with silver
A brass basin and its pitcher
A small knife with silver haft
2 small knives in a case
A green belt decorated with silver
3 needle cases
3 purses
A pair of hose and pumps
A pair of mules
A pair of scissors
2 silver thimbles
2 ivory combs
2 mirrors
Ribbons and threads of various colors
A pair of large head cloths
A silk veil
A case for a small Book of Hours
A handkerchief
Some caskets

These small items were not counted toward the trousseau.

As to the rest of the dowry, which amounts to a thousand florins, Piero Guc-ciardini promised to give it to me in cash. We agreed in the presence of the said Francesco Tornabuoni, and since he did not have the sum in cash, he gave me credit notes to be drawn from the Commune's Bank [*Monte*], which I am to keep for six months, after which he agreed to pay me the 1,000 florins. Were he not in a position to do so, I could draw the interest from the *Monte*. Piero was satisfied that after six months he would incur that cost and we were in perfect accord. The sums drawn on the Monte and ceded to me are: 2,568 florins 12s 7d drawn on the Pisa fund and transferred to his daughter [Tessa], my wife[1]; 810 florins on the Florentine fund *Monte commune* transferred to her, and 600 florins from that fund falling due in 1435, also transferred to the same daughter.

Source: Florence, Archivio di Stato, Mediceo avanti il Principato, 148, no. 31, fols. 32r–35r. *Translated from the Italian by Miri Rubin.*

Francesco di Giuliano di Averardo de' Medici (1415–1441), the author of this extensive note (*ricordo*) about his marriage, was the second nephew of Cosimo di Giovanni di Bicci, that is, Cosimo dei Medici the Elder. Like Cosimo, Francesco belonged to the branch of the Medici known as Cafaggiolo, one of seven branches then in existence. This was the most rich and influential Medici branch. Its fortune was relatively recent; the great-great-grandfather of Francesco who was also grandfather of Cosimo, Averardo "Bicci" (d. 1363), bequeathed to his five sons no more than a few farms north of Florence. But two of the sons, Giovanni (Cosimo's father) and Francesco (great-grandfather of our Francesco), developed financial careers in another relative's bank and founded their own establishments, after that bank's dissolution, in the late fourteenth century. We know of the success of Cosimo's bank, but his cousin, Averardo di Francesco, appears as one of the most highly taxed Florentines in 1427, with a personal fortune of 15,000 florins. The Cafaggiolo Medici branch enjoyed immense political prestige, due to Cosimo's position as leader of a party that was establishing itself with vigor in city life in these years. Although a short while after the ceremony described here, Cosimo, and then Averardo and Giuliano, were sent into exile by their adversaries, they were called back a year later, in October 1434, by their "friends" in Florence, where the family's preeminence became established. Yet our Francesco was unable to maintain the prosperity of the bank, which he inherited after the death of his grandfather and father soon after their return from exile.

Francesco thus belongs to a rich and influential Florentine household. It is characteristic of the ruling class of Florence, and in it in 1427 three generations cohabited: his grandfather's and his father's, both with children, and in 1433 an old uncle, Talento. In 1433 three slaves were recorded, too. Francesco, who was born in 1415, was young at his marriage: only seventeen when the marriage was contracted, and eighteen during the celebrations that followed a few months later. Such youth at marriage was rare among men of his class. His wife, Gostanza Guicciardini—for whom he also uses the nickname Tessa—is somewhat older than he. Yet even this precocious union did not assure the continuity of this particular family line which died out with Francesco in 1441–42.

Like the *ricordi* written by many of his contemporaries, Francesco gives a detailed account of the stages and participants in the contracting of marriage, as well as of the gifts showered upon the couple, or more precisely, on the bride. The entries are made with little attention to style; they were to be read solely by close family and were aimed at establishing the legality of the union and to record the financial and moral commitments created by the occasion. Francesco deploys a relatively limited vocabulary—very fine, very large, and so on—in describing the series of feasts by which the couple was honored. This is a typical example of marriage customs of the patriciate, without the excess that characterized equivalent later ceremonies. The narrative follows the customary progression: on five occasions Francesco emphasizes that it was done "as is the custom." Francesco describes the two great public stages, a year apart, of the process that linked the families of well-off citizens through marriage. First, the *giuramento del parentado*, a solemn public oath in the presence of a large gathering of men, in front of a church, to witness the engagement of the families. A notary prepares the document that records this betrothal, in Italian, the *sposalizio*. The groom promises (through his father's words if he is a minor), to uphold the terms of the alliance. The betrothed woman is never present, and is engaged through her father. The warranties that attach to the financial clauses of the union are proclaimed—the dowry and the stages of its payment—as are the fines that would be imposed on the party that defaulted on its obligation. Such conditions rendered the rupture of the union highly improbable, for such a failure led to a drastic rupture between the families, a rupture almost impossible to mend.

In order to express their commitment to the union the parties were expected to show their happiness. The women of the groom's family, excluded from the solemn betrothal ritual, did participate in the festivities that followed. They made their way as a group from the groom's house to that of the bride-to-be, where a feast awaited them. During that evening the betrothed man and some of his close relatives—Francesco was accompanied by three uncles and a cousin—went to the father-in-law's house and participated in the party or ball offered in their honor.

Was this really an engagement? Certainly, some clerics were tempted to assimilate to the ceremony the importance attached to the reading of banns, or to the making of a promise to marry (*verba de futuro*). But this is not a personal consent, expressed by the parties face to face, in front of a priest. It is probable that before the giving of the oath Francesco had not seen his future wife, except by chance in the street or at mass. The exchange of keepsakes, of gifts, the sharing of a glass of wine or of an apple—all the many ways in which entry into marriage could be implied in other, less exalted, social milieus, were wholly anathema to the great families of Florence. There the girls were kept hidden in their chambers, away from public places, behind windows and balconies. The first real meeting with the betrothed was at the feast given by the father of the betrothed woman, after the engagement of the families was publicly proclaimed. Francesco then remembered that during that month of August 1432 he sent Gostanza the traditional chest containing the precious jewelry he bestowed upon her. He does not mention accompanying this gift.

Francesco also describes the second stage of the marriage, which followed a year later. On 14 June 1433, a Saturday, he went to the Guicciardini family, to give the ring to his wife. Like the promise of marriage, this too was solemnized without recourse to a church. Upon doing so he and his future wife enacted the crucial component of a Christian marriage—the words in the present tense—expressing their consent to the union. The feast then began, this time in the husband's wife home, to which the wife arrives that night, mounted and accompanied by two gallants. Giuliano's grandfather offers the meal and the evening party that follows. At this stage the Florentines' gifting of rings to the new wife takes place, gifts from relatives and friends. Francesco lists some fifteen participants in this ceremony, each giving the bride a ring, a precious stone set in gold, although Francesco dismisses most of the offerings as "of little value." The festivities continued until Monday, when the young wife returned to her father surrounded by a noisy entourage of her husband's friends. A few days later Francesco went to bring her to his home. He says nothing of the moment of marriage consummation, nor is there any mention of a blessing in a church. The families have enacted their union in the streets of their city; the public space was theirs to use, filled with entourages of young men, and assemblies of women.

Francesco's account describes a very medieval type of marriage. At this stage of development of marriage rituals, the alliance between the families, and the passage of the woman from one kin household to another, are clearly separate stages. The church plays no role in the process, except in having made it necessary for the betrothed to express their consent in words of the "present tense."

Translated from French by Miri Rubin

Further Reading

Christopher Brooke, *The Medieval Idea of Marriage* (Oxford, 1989).

Dale Kent, *The Rise of the Medici: Faction in Florence, 1426–1434* (Oxford, 1978).

Christiane Klapisch-Zuber, *Women, Family, and Ritual in Renaissance Italy* (Chicago and London, 1985), chaps. 9–11.

Raymond de Roover, *The Rise and Decline of the Medici Bank, 1397-1494* (Cambridge, MA, 1963).

— 6 —

Annulment of Henry III's "Marriage" to Joan of Ponthieu Confirmed by Innocent IV on 20 May 1254

David d'Avray

Innocent, bishop, servant of the servants of God, to his most dear son in Christ H., the illustrious king of England, health and apostolic benediction. It is fitting for us to grant consent easily to the just desires of petitioners, and to ensure that the desired effect follows wishes that do not diverge from the path of reason. Therefore, since the world has grown old in corruption, and many people not only presume the worst in matters of which they are ignorant, but do not even hesitate to misrepresent as evil things they know without doubt to be good, you, prudently bearing this in mind, and desiring to make sure that calumnies of any jealous men should not in the future become linked to your children, humbly implored us some time ago to take care to provide for your honor and that of your children by our paternal solicitude, concerning the fact that you had sworn to marry our most dear daughter in Christ Joan the illustrious queen of Castile—the daughter of the Count of Ponthieu at that time— who was then single, and that you thought fit to contract a marriage with her insofar as it was in your power, and finally, after finding that you were related to her in the fourth degree of consanguinity, when this marriage had by no means been consummated, you joined to yourself in matrimony in the sight of the Church our most dear daughter in Christ Eleanor, the illustrious queen of England, daughter of the count of Provence of famous memory. But when we had entrusted by our letters in mandate form to our venerable brothers N̲. the Archbishop of York and N̲. the bishop of Hereford that they should inquire diligently into the truth of the matter, and make sure to declare by apostolic authority, if this is what they decided was the right verdict, that the first marriage and the oath you swore to contract this first one do not hold good, and that the second one is valid, and curbing by ecclesiastical censure anyone who contradicts them, without the possibility of appeal: in the end, after the Archbishop had excused himself legitimately by his letter because he was unable to take part in the cognizance of this matter, the same bishop proceeded with this matter alone, as our letter allowed him to do, and, when the parties had been summoned, and the aforesaid Queen of Castile replied that she was related to you in the aforesaid degree of consanguinity, and that she would not send anyone to defend this case, and would not come, nor involve herself in it, and finally neither appeared in person nor through a representative, he [i.e., the bishop of Hereford], in the presence of your proctor, after taking cognizance of the merits of the same case, and observing the order of law, with the counsel of prudent men, pronounced as his sentence that the marriage between you and the aforesaid Queen of Castile had been null, because of the impediment of your relationship of consanguinity in the fourth degree with her, and that the marriage contracted between you and the aforesaid E. the Queen of England was legitimate, notwithstanding the oath you swore to contract the first one, [all this] just as is contained more fully in the letter written at the hearing.

Therefore we, after inspecting diligently the proceedings of the aforesaid bishop and the sentence, approving the same sentence, and making good from our plenitude of power any defect, if there should be one, and regarding the same sentence to be ratified and pleasing: acceding to your requests, we confirm by apostolic authority and strengthen by the protection of the present document the sentence together with the aforesaid proceedings. We have had the content of these proceedings and this sentence set out word for word, and they are as follows.

A whole archive of correspondence generated by the investigation then follows. We rejoin the text when the witnesses are giving evidence at the trial in Sens.

Lord Jean de Molins, a knight said on oath that Louis the Fat the King of France had two sons, King Louis and Pierre de Courtenay. Asked how he knows this, he replied that he heard it from his elders and trustworthy people, and that it is generally thought to be so[1] and this opinion was the common one. He also said that the issue of Louis was King Philip and Alaysia the Countess of Ponthieu, who was married to Guillaume Count of Ponthieu, whom he saw many times, as he did King Philip. Again, how he knows. He says that he heard it said by trustworthy elders and that it was the common opinion on the matter and still is. And he said that the whole court of King Philip called the daughter of the aforesaid Alaysia the grandchild of the aforesaid king and the daughter of the said Alaysia the sister of the aforesaid King Philip, and this was the general belief about the matter and common opinion. He said that Marie the Countess of Ponthieu was the issue of aforesaid Alaysia. Asked how he knew, he said that it was and still is the common opinion, and that he saw her and heard it said by her that she was the daughter of Alaysia and of Guillaume the Count of Ponthieu, and she was the heiress of the Count and Countess of Ponthieu, and succeeded to the County, and he said that he was frequently in the County with her. Again he said that Joan the Queen of Castile as she now is was the issue of Marie. Asked how he knows, he said that this is the common opinion on the matter, and he saw the same Joan with Marie her mother, and Marie treated Joan as her daughter, and Joan Marie as her mother, and he[2] was frequently with her, and Marie married her as her daughter to the King of Castile. He said too that Alaysia the Countess of Angouleme was the direct issue of Pierre de Courtenay and that Isabella Queen of England was the issue of the Countess of Angouleme, and that Henry King of England as he now is was the issue of Isabella. Asked how he knows, he said that he heard it said by his elders and trustworthy people, and it both was and is common opinion. Asked if he saw any of the aforesaid persons, he said yes. He saw the Queen of England, who acted as the mother of that King of England, and was treated as such by the King, and by everybody else. And he saw that King of England, who treated her as a mother. Asked in what degree the King of England and Joan the Queen of Castile were related, he said that it was in the fourth degree. Asked how he knows, he said that King Louis and Pierre de Courtenay the sons of Louis the Fat were brothers, and thus in the first degree. Alaysia Countess of Ponthieu the daughter of Louis, and the Countess of Angouleme the daughter of Pierre de Courtenay were first cousins, and thus in the second degree. Marie Countess of Ponthieu daughter of the aforesaid Alaysia, and Isabella the Queen of England, daughter of the aforesaid Countess of Angouleme, were blood relations in the third degree, and thus King Henry of England as he now is, the son of the aforesaid Isabella Queen of England, and Joan the Queen of Castile, the daughter of Marie the Countess of Ponthieu, are related in the fourth degree. Asked whether, through the things he had heard and understood or saw, he firmly believes it to be so, he said on oath that it is. Asked if he was born in the Kingdom of France, he said yes. Again he said that he was brought up with and lived a long time among those people of the aforesaid families (*generibus*). Asked how old he is, he said: fifty.[3]

Source: London, British Library, Cotton Cleopatra E, fols. 194v–195r; printed in D. L. d'Avray, "Authentication of Marital Status: A Thirteenth-Century English Royal Annulment Process and Late Medieval Cases from the Papal Penitentiary," *English Historical Review* 120 (2005): 987–1013; at 998–99.

These two extracts come from an enormous papal bull, annulling the first marriage of Henry III of England to Joan of Ponthieu. It was as far removed from a "quickie" annulment as can be imagined. The grounds for declaring the marriage null were obvious, one might have thought. Yet Henry III went through an elaborate and time-consuming formal procedure to get documents no one could challenge proving his first marriage invalid and his second marriage a true one. The document raises many questions for the historian, but first some background facts.

Marriage and the Continental Plans of
Henry III of England

In the mid-1230s Henry was interested in marrying a wife who would help him with his continental ambitions. His father King John had ruled much more of France than the French king, but lost most of it, most notably Normandy. Henry was not reconciled to an insular role. In his early years he hoped to recover the lost lands (later he had hopes of the Sicilian crown). Ponthieu bordered with Normandy. So he arranged a marriage to the heiress. They actually got married, by proxy, which was valid in itself; consent was the key thing. However, there was an impediment of which Henry was well aware: they were related within the forbidden degrees of kinship. Without a dispensation, the marriage could not stand. By "marrying" Joan, Henry could however stop her from marrying anyone else while he obtained a dispensation. Before he did so he changed his mind. It was clear that the French monarchy would work against the dispensation and another attractive possibility had opened: marriage to the daughter of the Count of Provence, a match that would put him at the center of European politics. So he told his agents to stop trying to get a dispensation and instead he married Eleanor of Provence. Joan married the King of Castile so everyone was happy.

Years later, however, Henry went to immense trouble to get a formal annulment of the first marriage. Its invalidity may seem obvious to the modern scholar who is familiar with medieval Church law; even so, some of Henry's contemporaries might have used it to suggest that his children by Eleanor of Provence were bastards.

A Mobile Archive

This bull of 1254 is like a Russian doll, a document containing documents containing documents. It is the archive of the whole process—a mobile archive. The whole archive is contained on one side of parchment, but the parchment is as big as the top of a desk, and its whole contents fill a couple of dozen pages of typescript. Here there has been space to print a fraction of it only: the deposition of the first witness to kinship and most of the initial papal bull that started the process off formally.

Rescript Government

Informally the process would have been started by a request from Henry III to the papacy. That was normal with papal government. Popes seldom started cases. People came to them. They developed mechanisms for dealing with such demand (and this archive bull shows those mechanisms at work). In the entire history of the medieval papacy, response was more common than initiative, though there were exceptions—Gregory VII in the eleventh century, John XXII in the fourteenth, perhaps Innocent III at the beginning of the thirteenth. Mostly, however, it was "rescript government," writing back to a letter received, like the government of the Roman emperors from whom the papacy borrowed legal and administrative method (improving it, one could argue).

Diplomatics

To read this initial bull right one needs to know what is common form, stereotypical language and structure not custom built for this particular occasion. A discipline called "Diplomatics" studies such questions, together with the genesis of documents and their setting in life—it is a transferable skill that everyone should borrow from professional medievalists to use in everyday life. Thus, for instance, the lines about granting requests of petitioners, the corruption of the world, and misrepresentations of evil men are the *arenga*: generalized moralizing platitudes placed near the beginning of solemn papal, royal, and imperial documents in the Middle Ages, not representative of the pope's personal state of mind. The next part is the *narratio*, which fills in the background. (Historians have sometimes gone far astray by mistaking a narration for a decision.) Much of the wording probably comes straight from a letter of Henry to the papacy; the pope then goes on to say what he did about the request. A new section, the actual decision, called the *dispositio* in Diplomatics, begins with the words: "Therefore we, after inspecting diligently . . ."

Formal Legal Rationality

The mobile archive shows us proceedings conducted with strict respect for ecclesiastical law. For instance, Joan (now Queen of Castile) had to be formally summoned to the hearing in Sens. It was a formality. She had no intention of going. But if she had not been summoned, if she had not refused, been "contumacious," the proceedings would not have met the formal requirements of the law.

Patterns and Peculiarities in World History

If we stand back from the details, this document tells us a lot: it presupposes a number of facts about the Middle Ages, facts we think about too little because they are too big to see. Some of the big facts about the Middle Ages are unique in world history. Historians tend to think that all the things they study are unique in world history, but that is because they do not look at other times and places where they would find some uncannily similar patterns, as great men like Max Weber and Marc Bloch did because they took the trouble to look. The structures are not quite like natural scientific laws, which do not admit of exceptions or allow a shortlist of alternatives. In the social world we find patterns like: "If men have the power (and they usually do), those who can afford it either have the power to divorce wives easily, or to have several wives, or to have a primary wife and several secondary wives (official concubines), or some combination of the foregoing." The laws of history are "hypothetico-disjunctive laws"; they involve "*ifs*" and "*either . . . ors*." And they allow for exceptions. The foregoing law works for all literate societies except the medieval west and the societies it has influenced—in proportion to that influence.

Control of Royal Marriage by Independent Religious Authority

It is an extraordinary thing that late medieval kings accepted the jurisdiction of a foreign government over their private, personal lives. The papacy was not a state (outside its central Italian domains), but a government it certainly was. Monarchs put their sex lives in its hands. Patterns normal for great men in other societies, semiofficial concubinage or easy divorce or both, were delegitimized in the last three medieval centuries to a point that Indian kings or Chinese emperors would hardly have accepted. The rules of legitimate sexual union were made by a religious power, not a local holy man in the monarch's own domains but by an institution with rules and laws and an administration to enforce them.

The distinction of "State and Church" or "Politics and Religion" is itself rather peculiar in world history. In most societies—classical Greek, Islamic, Hindu, Confucian, Bantu—the distinction is alien. There may be a division of labor, but it is like the division in modern societies between judiciary, legislative, and executive rather than like "the separation of church and state." That distinction goes back to ancient Roman Law and the growth of Christian structures within the pagan empire.

The modern idea of a wall between politics and religion takes the antithesis much further but can still be regarded as an exotic extension of a pagan Roman and Christian tradition. It goes beyond distinction to postulate absolute separation. When it comes to questions like marriage there is no neat division of spheres; once religious and political authority are distinct, one or the other makes the binding law, and in most modern societies it is the state.

In the later Middle Ages religious and political authorities were distinct but both had to deal with the law of marriage, and it was the religious power that made the rules. It had not always been so. Until the eleventh century the Church did not have exclusive jurisdiction in questions involving the validity of marriage. In the course of that century a Church monopoly was achieved. Even then, kings and nobles found it easy to get out of a marriage because the forbidden degrees were so wide as to be almost impossible to check. If they wanted out of their marriage, they might well discover that their wife was related to them within the forbidden degrees and get an annulment. If she wasn't, they could invent a genealogy that would be extremely hard to check.

Only in the early thirteenth century was the loophole closed, and it became hard to get out of a marriage even if you were a king. The contrast with the twelfth century is striking. Eleanor of Aquitaine easily got her marriage annulled by a French synod in circumstances of dubious legality, and promptly married Henry II of England though he was also related to her within the forbidden degrees. Henry III's "marriage" to Joan was a non-starter to anyone who knew the circumstances and canon law, yet he still went to immense trouble to get a painstaking and authoritative papal document to prove it. After Philip Augustus of France's attempt to get his marriage to Ingeborg of Denmark annulled, kings seem to have given up the easy option of getting a pliable local synod of bishops to annul a marriage—bishops whose appointment they may have secured and whose life they could make difficult—without consulting the papacy. It is not quite clear why kings stopped doing this. However that may be, papal trials such as this one are quite unlike anything in the history of any other civilization.

Central Government at a Distance

This trial is also remarkable as an example of central government at a distance. Here again it is representative of a structure found in the last three centuries of medieval history and seldom if ever in the history of other civilizations. The usual pattern with authorities both secular and religious is for a local regional ruler, let us call him a satrap, to stand for the central ruler in that locality and to deal with all cases except the most important. The most important cases may be taken to the center on appeal. Or the central ruler may deal with local cases when he happens to be in the locality, as with the Roman Empire. Or, and again this was the case with the Roman Empire, anyone may in principle take a case to the center, but there is no system to deal with such petitions rationally.

In the second half of the twelfth century, in western Europe, two administrations worked out ways of combining central government with local knowledge. One was the English monarchy, with the system of stereotyped writs that could be purchased centrally and that started a protest ending with a jury of local people in the know deciding on the facts in front of an itinerant royal justice.

The other was the system of papal judges delegate. In this case the Bishop of Hereford acts with papal authority—just for this case. The system was a response to demand. It depended on a pool of ecclesiastics who would take time off to judge a case. Such men might be bishops like Peter of Aigueblanche who judged this case (his fellow judge the Archbishop of York excused himself) but also, for instance, university masters. There are reasons too technical to discuss here for thinking that the judges would normally be acceptable to both sides. The preconditions of the system only fell into place in the twelfth century; they included the exercise of administrative imagination, for the system worked on a bureaucratic shoestring and high papal prestige.

Kinship

Papal prestige was becoming crucial to the kinship system in northern Europe, in France and England at least. In Italy and Flanders clans were still important. In Anglo-Saxon England they may have been, too. In England and France vassal-fief bonds and the power of monarchy relativized kinship beyond the micro-structure of very close relatives. For instance, family had been important in the Anglo-Saxon system of fines for crimes; if the kin paid them, blood feud was avoided (blood feud was not anarchy but an ordered legal sanction). The system outlasted the Norman Conquest, but the new alien population doubtless made it too complicated, and anyway the monarchy took over crime and hanging replaced fines to the family. Of course there was still an extended kinship system if only for inheritance purposes, though feudal lordship interfered with it. However, there was another kinship system, based around the forbidden degrees as defined by the Church. The lines were redrawn in 1215, making the system more workable and preventing easy annulments. Previously even sixth cousins had been out of bounds. After 1215 one could marry a fourth cousin. If a couple were thinking of marriage, each counted up the generations, starting with a parent, to their common ancestor. One was four degrees removed from a great-great-grandparent, five from a great-great-great-grandparent. The number of degrees would need to be worked out for each side, so that one's relationship would be understood in the form: four degrees on one side, five degrees on the other. The working of the system would have impressed the Church's conception of kinship on lay minds. In the second extract translated here we see the process at work: Molin has to structure kinship in his mind in such a way as to demonstrate the blood relationship to the judges. As he did so, his own kinship mentality was being developed or reinforced.

This document takes the reader in many interpretative directions. It shows us a world of law and marriage very different from ours, but rational and well organized in its own terms.

Notes

1. "Ita fama est."

2. This more probably refers to the witness, but could also mean "she," i.e., Joan.

3. D. L. d'Avray, "Authentication of Marital Status: A Thirteenth-Century English Royal Annulment Process and Late Medieval Cases from the Papal Penitentiary," *English Historical Review* 120 (2005): 1005.

The Life Cycle:
Death and Burial

— 7 —

Agius of Corvey's Account of
the Death of Hathumoda, First Abbess
of Gandersheim, in 874

Frederick S. Paxton

19. . . . I am about to relate a little thing that may seem extraordinary to certain people, but it is an indicator of her feeling toward me. If ever in her pain she could by no other means be induced to eat something, she could be persuaded by this ruse—if she believed that I had prepared or sent the food. On that day as well, when I came, she ordered me to be called when her food had been brought to her. In my presence she ate as much as she could, and, for the sake of consoling me, strove in every way to keep herself cheerful, so that I conceived the greatest hope that she would escape death. With the coming of the evening, however, things took a turn for the worse and transformed my joy into sorrow, hope into desperation. For her condition began to worsen and increase so steadily, minute by minute, that we gave up all hope of her recovery and waited only for her death. In the affliction of her feebleness, however, both her kindness toward you, holy sisters, and your affection and devotion toward her were made manifest.

20. First of all, her venerable paternal aunt, although already quite old and feeble from age, assisted her tirelessly, insofar as grief permitted her to be present. The prioress of the monastery, although pressed by all sorts of business, did not leave her bedside for so much as a moment. The dean and the sacristan, as much as their duties allowed, came to see her frequently and wanted to be there at all times. The rest of the sisters, who were not all able to be continually in the cell with her, either occupied themselves with psalms or prayers for her in the church or camped out outside the doors of her cell. Her own sisters' affection toward her, though already known to all previously, shone forth then especially. They stood by the bed, tirelessly assisted the one lying there,

arranged the bed[clothes], supported her head with pillows, kept her body elevated, rubbed her hands, warmed her feet and stomach, cooled her fever with
a fan, wiped the flowing sweat with linen cloths, regulated the hot water for
washing, prepared food and brought it to her, and anticipated everything that
needed to be done.

21. Above all her admirable mother, hiding her anguish inside with a quite
placid countenance, applied herself to consoling the sorrowing ones, restrain-
ing those crying, soothing each with her conversation. Desiring to console others, she was herself inconsolable. With how many sighs did she run back and
forth between the church and her daughter's bed? How many tears did she
shed in front of the tomb of the saints, not to be deprived of her child; to die
herself in her daughter's place so that the one who came into this light first
might be also the first to leave it? At different times she stopped on the way to
her daughter and turned back her step. So did this most strong of women
doubt what she ought to do. The one alternative called this way; the other that.
This one urged her to be with her dying daughter; that one warned her not to
wish to see what she would be unable to see without anguish. What could she
do? Maternal affection called her one way, sorrow the other. She would enter
the cell, but could not bear seeing her daughter burning up in death. She would
leave, but could not bear being away from her suffering child.

22. And now the body began to fail little by little, and yet Hathumoda's mind
remained firmly fixed on heaven. She often sang along with us the same psalms,
and often different ones, as well as certain verses from here and there in the
Psalter, so linked to one another in conjoined order that they could not be
doubted to have been inspired to her holy mind by the same spirit through
which they had been written. Between the psalmody and the prayers the Lord
was always in her mouth, Christ always in her heart, and unless she closed her
eyes for a bit as if sleeping, she always either sang psalms or spoke about the salvation of her soul. She continuously confessed her sins committed in thought,
word, or deed, and asked if she should hope for any forgiveness for them. She
often spoke, too, about the division or separation of the good and the bad on
the Day of Judgment, of the sheep on the right and the goats on the left, asking
if the saints could really succor others or good people help anyone. Just before
she died, commending herself into my hands and to my faith, she asked also to
be commended, through my modest agency, to the saints whose relics are held
at my monastery.

23. Meanwhile, with her eyes so fixed upon that which she seemed to see, as
though she were in the midst of those events about which she was speaking,
as though she were already standing before the judge's tribunal, she became
afraid and began to tremble. And we were able to gather from the changes in her
face and from other things that she foresaw and prophesied that at that time she
saw and heard future events. You will remember, I am sure, you who were there

then, that on the night before she died we were with you, together there; and when I sat down, and you were standing by her bed, she seemed to be murmuring something or other about me. You supposed that she was asking after me, as was her usual custom. Whence, when you said that I was there, she said that she had sinned along with you because of me, and that I incurred an offense on my own account, because I was staying with you longer than was my custom. To which I said that none of what she said was so; that it would only please my fellow monks if I were able to bring her solace and consolation, that being the purpose for which I had been sent. "It is not so," she responded. "It would be better for you to hasten to go and return." And what she said was absolutely true. For indeed, having returned to my monastery, I found out that it was so. For my brothers knowing not yet how grave her sickness was, had borne heavily my delay in returning and, as I learned afterward, at the same exact time of night when she had said this, the lord abbot had spoken with some of my brothers to that effect. That likewise urges us to believe that she also beheld something extraordinary on the occasion when she cried out repeatedly "Well done, well done!" and often repeated "Listen, listen!" as if to encourage us who were present to listen, since she herself no longer could.

24. Bishop Marcwardus was present at that time with his priests, and everything that seemed necessary for one about to die was carried out with great solemnity: the anointing with holy oil, the final reconciliation, the last communion. Psalms were sung continually, litanies and prayers recited, the Gospel read. Nothing that ought to be done before the departure of the soul was omitted. And now, all of her limbs already having slowly gone dead and her speech diminished, her breathing gradually began to fail, while she nevertheless, devoutly kissing the wood of the holy cross that I held there, focused her eyes on it and murmured from time to time, with what strength she was able, something of the psalms. We agreed then that we would, having begun the Psalter at the beginning, pay careful attention to the verse and psalm being chanted at the moment of her death. It turned out to agree right well with her holy merits. For when we were reciting the penultimate verse of Psalm 40, "Thou hast received me because of my innocence, and established me in thy presence for ever," she, as I faithfully believe, returning her blessed soul to heaven, at last breathed her last breath. We nevertheless waited, as is customary, to see if she would breathe again, until we completed the next two psalms, which were most appropriate to her death. When the manifest signs of her death appeared, all the bells of the church resounded testifying to this.

25. It can hardly be told how many people flowed into the church and how groaning and mournful was the crowd of sisters who gathered to commend her blessed soul to the Lord. Having finished with many tears, some remained in that place to wash the body in the customary way. The rest meanwhile occupied themselves in the church with psalms and prayers. I am about to relate a wondrous thing, but nevertheless true according to the testimony of the sisters who

were there. When the blessed body had been arranged on the stool for the washing, she, like a person still alive, raised her eyes and moved her lips as if speaking to the attendants. After the worthy body was washed and wrapped in a shroud, it was carried to the church on the shoulders of priests accompanied by a worthy choir of virgins singing psalms and led by the young girls carrying candles. How many sighs, what great mourning, what lamentation there was there, who can recall without tears, who can relate without weeping? All was tumult. . . .

26. Now truly when the time came to bury her, no one ought ever to hear greater mourning, greater lamentation. In no way could it be kept in check, although the priests, in accordance with the duties of their ministry, prayed for it for God's sake. But just when it seemed somewhat contained, then surging up greater from the crowds, who in troops never stopped flowing, it somehow began again. This occurred not only on that and the following days, but almost up until the thirtieth day, the more noble people, also as if in hordes, converging from here and there, not so much mourning her, since they knew that she was then more truly alive in God, as grieving that they had lost so great a good and so great a solace; that the holy congregation was bereft of such a mother; that so worthy a mother was deprived of such an offspring; and that her most glorious sister, our lady the queen, would be abandoned by such a sibling. . . .

28. . . . It ought not to disturb you, most holy sisters, that a woman of such sanctity should pass over to the Lord through a very difficult and protracted death. The Lord truly wished it, so that it would not be said to her in the future, "She was not among men, and with men she did not suffer." The Lord wished this so that if anything still needed to be cleansed from the taint of human fragility in her mind, it would be purged through a more severe death. As for the rest, as I believe and pray, she is now following Christ the bridegroom as his most chaste bride wherever he might go. She sings that new canticle that no one is able to sing except the one hundred and forty-four thousand, who were purchased among men, the first fruits given to God and to the lamb, in whose mouth no lie was found, and they are [present] without stain before the throne of God.

29. This holy and incomparable woman died on the second day of the week, the third of the kalends of December, in the year of the incarnation of our lord 874, in the seventh indiction. She lived as abbess for twenty-two years. All the years of her life were thirty-four; our Lord Jesus Christ reigning, who lives and reigns forever and ever. Amen.

Source: Agius, Vita et obitus Hathumodae, ed. G. H. Pertz, Monumenta Germaniae
Historica, Scriptores IV (Hannover, 1841), 165–89. Pertz edited the Life of
Hathumoda from a manuscript written by Andreas Lang, abbot of St. Michael's on
the Mountain in Bamberg (1483–1502), which contains the sole surviving copy
of the ninth-century original. For a translation of the complete text as well
as the poem of consolation that accompanies it, see Anchoress and Abbess in

Ninth-Century Saxony: The Lives of Liutbirga of Wendhausen and Hathumoda of Gandersheim, translated with an introduction and notes by Fredrick S. Paxton (Washington, DC: Catholic University of America Press, 2009).

Eight hundred seventy-four was a hard year in northwestern Europe. Famine and disease may have killed as much as a third of the population of Gaul and Germany. One of the dead was Hathumoda, the thirty-four-year-old daughter of Liudolf and Oda, count and countess of East Saxony. Since her twelfth birthday, Hathumoda had headed a community of cloistered women, founded on family land at a place called Gandersheim. The cloistered community of Gandersheim is best remembered as the home of the tenth-century Latin poet, dramatist, and historian Hrotsvit. Hrotsvit only wrote a few conventional lines on Hathumoda's death, but Agius, a monk and priest of the abbey of Corvey, had more to say. He knew Hathumoda well and was present at her death. Shortly thereafter, he wrote a prose *Life* of the abbess as well as a poetic *Dialogue* to console the women of Gandersheim for their loss.

Most cloistered women in ninth-century Saxony were daughters and widows of the aristocracy. They took no permanent vows, and could own and inherit property, live in private quarters, and maintain servants. But Hathumoda's community was different, at least according to Agius. The women of Gandersheim strictly adhered to the ideals of poverty and the common life, and Hathumoda's behavior as abbess was in perfect conformity with the Benedictine Rule. Hathumoda did not, however, establish Gandersheim as a Benedictine convent. Nor did she obtain a royal grant of immunity to protect the house from the interference of local bishops or the depredations of warriors, although she sought one for years. She did not even live to see the completion of the abbey buildings at Gandersheim or the dedication of its church in 881. While tending to the ill and dying members of her community during the epidemic of 874, she too succumbed to the fever and died.

The ritual accompaniment to Hathumoda's death (chap. 24), which was the product of several centuries of liturgical innovation and elaboration, had only just taken on a more or less definitive form. It began with an anointing with oil, cleansing the body of the effects of sin. It soothed the conscience with a final confession and absolution, and it fortified the soul with the bread and wine of communion. Finally, it surrounded the dying person with sung prayer, anointing her with sound, shielding her from malignant spirits, and creating an audible womb for the birthing soul so it could pass through death to eternal life. It must have been an impressive ceremony, especially when led by a bishop.

Rituals, however—in their uniformity, regularity, and generality—tell us more about what clerics did for the dying than what the dying, or their families and community, might do in the face of death. Fortunately, Agius did not limit his description of Hathumoda's passing to the death ritual as such. Hathumoda's final illness lasted many weeks, during which her fever waxed and waned, and she experienced frequent visions, both soothing and frightening, as her condition worsened. In her last days, the whole community was drawn into the tumult of her passing. Her own sisters, also cloistered at Gandersheim, attended her constantly, and her mother came in from outside, along with an aged aunt. The prioress and other officers of the house let their regular duties go in order to stay continually by her side, and the women of the house fell asleep in the hall outside her door. Agius reports this all in considerable detail.

Agius's writing is highly rhetorical and was meant to serve a number of purposes—convincing the royal family to take Gandersheim under its protection, for one (which they did in 877) and claiming Gandersheim as a dependent of Corvey, for another. Agius also clearly hoped to contribute to a house literary tradition. Fifty years earlier, the Corvey monk Paschasius Radbertus had appended a short consolatory poetic dialogue to his *Life* of the abbey's founder, Adalard. But just as Agius's dialogue between himself and the bereaved women of Gandersheim outdoes Paschasius by some 582 lines, his portrait of Hathumoda in the *Vita* outdoes his model in terms of the art by which he describes his heroine's last days.

Or is it artlessness? That is, did the fact that Hathumoda's life played out within the confines of a cloistered community of women provide too little material for a standard hagiography? Or was it that, since priests could enter Gandersheim only to care for the sick and dying, Agius concentrated almost exclusively on those aspects of her story? Whatever the case, Agius's account preserves rare glimpses of responses to death and dying from a writer unwilling to ignore them in the service of hagiographic convention. He not only describes the deathbed ritual and the general response of the community to Hathumoda's death already referred to, but also notes the importance of the preparation of food for the sick, a skill apparently shared by both Hathumoda and Agius. Agius's description of the agony suffered by Hathumoda's mother, who cannot decide if she should be at her daughter's bedside or in church beseeching the saints for aid, is extraordinary. So are his references to Hathumoda's fear of judgment and uncertainty over the power of the saints to aid the dying and the dead and the calm that descended over the attendants as they chanted the psalms in anticipation of her last breath. Agius's portrayal of Hathumoda's funeral and burial lacks some of that specificity, but it still paints a vivid picture of how a ninth-century Christian community might have responded to the death of an important figure. It also gave him an opportunity to introduce the consolatory themes that he would elaborate more fully in his poetic *Dialogue*.

Further Reading

Frederick S. Paxton, *Christianizing Death: The Creation of a Ritual Process in Early Medieval Europe* (Ithaca, NY, 1990).

———, "Forgetting Hathumoda: The Afterlife of the First Abbess of Gandersheim," in *History in the Comic Mode*, ed. Rachel Fulton and Bruce W. Holsinger (New York, 2007), 15–24.

— 8 —

A Royal Funeral of 1498

Alain Boureau

PIERRE D'URFÉ ON CHARLES VIII'S FUNERAL, 1498

And when the body will come to Notre-Dame-des-Champs,[1] above the coffin where the body lies, a platform will be erected, on which will be a bed of display (*lit de parement*) where will be laid the statue [= the effigy] of the aforesaid Lord in its royal clothing, as follows:

First, on this bed, there will be a blanket; upon the aforesaid blanket a shroud made from Dutch linen, falling largely apart; upon the aforesaid shroud, there will be a great sheet made with fifty *aunes*[2] of black velvet; and upon the aforesaid blanket, another one, that is to say twenty-five *aunes* of the best gold-woven material, of the best quality that is to be found . . . And upon this golden sheet, there will be two pillows made of gold-woven woolen cloth, placed one under the head, the second under the feet of the king's statue, which will be laid upon the bed and ornamented as follows:

First, the aforesaid lord's face, represented as living (*fait au vif*) will have the hat turned down and the crown on the head;[3] [the effigy] will bear sandals made of blue satin sprinkled with fleur-de-lis and a crimson taffeta dress embroidered and edged with gold ribbon, and, upon this, a blue satin gown sprinkled with fleur-de-lis, embroidered and edged with gold ribbon. And upon this, it will bear a blue velvet mantle sprinkled with fleur-de-lis, embroidered and fur-lined with ermine, slit on the right side, with a Florence gold clip above the slit. [The effigy] bears on the right hand, the royal scepter and on the left one the hand of justice; the sign of the order [of Saint Michael] will be suspended at its neck. And it will have its two hands gloved; the right one will bear a gold ring and will be placed above the left one on the chest. . . .

And when the vigil service will be sung in the church of Notre Dame of Paris, the lords who bear the banner, the ensign, the pennon and the guidon[4] will leave by their seat the aforesaid banners, ensigns, pennons, and guidons and there will be a hole where they will plant them. They will take them over the following day, when they will fill their place in the order, at their hour of duty. And when the body leaves, they will bear those signs in person through the whole city, just as they entered the city and from the very limit of the city, because the statue [the effigy] of the king will always be visibly borne all the way to Saint-Denis in France. . . .

When it comes to the *Libera*, at the end of the mass, the body will be carried from the choir to the grave to place it in the earth, with everyone keeping to his proper place. And then, when the feet of the body have been placed in the tomb, the stewards will be called by the herald-at-arms; said stewards will come one after the other to put their batons on the tomb; and this being done, the heralds and sergeants-at-arms will place their coats of arms and maces on the tomb. Then he who carries the guidon will rest his lance on the tomb with the greatest reverence possible; at this point, the body will be half within the grave. The same will be done by the Lord d'Allegre, who carries the ensign; and afterward, when the body is entirely in the grave, he who carries the panon will do the same. And the last will be the First Chamberlain, who similarly will lower the banner and place it on top of all the other things, over the head of the body. And then the Master of the Horse, who will have laid his sword on the body at the opening of the tomb, will thrust the point on high and cry *Vive le Roy!* After this cry the heralds will take back their coats of arms and put them on; and similarly the First Chamberlain must raise up again the banner, because it never dies; and a monk from the Church [Saint-Denis] shall come, and take it in his hands, to place it where it belongs . . .

The aforesaid organization was actually fulfilled and executed, as well at Amboise and then during the circuit, when the body left from there, going from one place to the other all the way to Notre-Dame-des-Champs, near Paris, where our lords from the Court, from the University, from the *Chambre des Comptes*, the Provost of Paris, the Provost of the Parisian Merchants and the aldermen of the city exposed their rules and privileges, each for a specific place, in order to escort the body from Notre-Dame-des-Champs to the Hanging Cross[5] and Saint-Denis.

> Source: *La vraie ordonnance faicte par Messire Pierre Durfé, chevalier, grand escuyer de France, ainsi que audit grant escuyer appartint de faire, pour l'enterrement du corps du bon roy Charles huitiesme, que Dieu absoille. Et ladite ordonnance leue et auctorizée par Monseigneur de la Trimoille, premier chambellan et lieutenant du roy à accompaignier ledit corps. Et aussi par le conseil de messeigneurs les chambellans et autres*, Paris, Pierre Le Caron, 1498, 12 folios, in-4°. This rare edition was reproduced by Théodore Godefroy, *Le Cérémonial de France* (Paris, 1619), and Le Caron, *Histoire de Charles VIII* (Paris, 1684); by Arthur Franklin in 1875; and most recently in Alain Boureau, *Le simple corps du roi* (Paris, 1988; repr. 2000), 94–114.

Death came to King Charles VIII of France unexpectedly in 1498, when he was twenty-seven years old; he bashed his forehead against a low-hanging lintel in his

castle at Amboise. His corpse was exposed a few days in the hall of the castle before being put into a coffin, for a long circuit with stages at the collegiate church of
Saint-Florentin at Amboise, then—on 22 April—at Notre-Dame of Cléry (where
the king's heart was left), and finally in Paris (first Notre-Dame-des-Champs, then
Notre-Dame-de-Paris and finally Saint-Denis, for the burial on 1 May 1498. The
text describes this last stage. It is the end of the prescriptive part of the text, which
is written in the future tense. The last paragraph shifts to the description in the
past tense.

The funerary services commenced immediately and lasted for more than three
weeks, including a long procession that ended with the burying of the corpse at
the abbey of Saint-Denis, near Paris. These proceedings were prescribed and described, with a surprising amount of unexpected details, by Pierre d'Urfé, Charles's
Master of the Horse who planned and executed the bulk of the arrangements.
Our text is taken from his description, which was printed in that same year, 1498.

From the early fifteenth century on, and especially in France, medieval royal funerals had gradually become occasions of great solemnity and publicity. Throughout human history, the solemnity of monarchical funerals has reflected personal
expressions of self-esteem, power, succession, or religious attitudes toward death.
Egyptians pyramids were essentially royal tombs. In medieval France, the abbey
of Saint-Denis became a royal necropolis; nearly all monarchs were buried there
and Saint Louis had the disposition of the tombs in the crypt reorganized. France's
late development of funerary ceremonies can be explained by the general attitude
of the Christian Church toward death. Augustine had insisted that funerals mean
nothing; the vanishing of the body was no more than a provisional absence, since
all corpses would be resurrected before the Last Judgment. Funeral offerings,
meals, and celebrations should therefore be abolished. But in spite of those warn-
ings, kings could not simply accept this lack of consideration for their bodies. During the High Middle Ages, masses for the dead and specific prayers for the souls
were increasingly stipulated in the last wills of aristocrats and kings. Corpses began to be fragmented so that they could benefit from prayers in different and eminent places of worship. For instance, when Philippe III of France died in October
1285 at Perpignan, his body was eviscerated and boiled. The flesh and viscera were
buried in the nearby cathedral of Narbonne. The abbey of Saint-Denis received his
bones, while his heart was given to the Dominican convent of Saint-Jacques in
Paris. This type of bodily division became common, and not only for kings,
throughout all of western Europe. More importantly, this bold manipulation of
corpses created the conditions of possibility for another, more political, process that
took place in France.

At the heart of this process was royal succession, the moment of transition between one reign and another. Such moments were occasions of great social fragility
in medieval society. During the twelfth and thirteenth centuries, kings could sup-
press those risks by associating their elder son to the monarchy, as a *rex designatus*,
a designated king. By chance, all French kings had produced male heirs who lived
long enough to become real successors. The principle of succession became so

solid that Philip Augustus could actually abstain from this anticipated coronation. But things changed when King Louis X died without any surviving male heir (1316) and when the Hundred Years' War endangered the accustomed principle of succession. From Charles VI's death (1422) on, royal funerals became instrumental, self-conscious exercises in power management within the monarchy.

This, at least, was Ralph Giesey's interpretation. His *Royal Funeral Ceremony in Renaissance France*, published in 1960, holds that French royal funerary ritual evolved out of a combination of adherence to established precedent as well as circumstantial innovation; it was not simply the reflection of a conscious, humanistic attempt to mimic Imperial Roman funerals. Giesey argued persuasively that French funerals retained many of the same basic traditions out of reverence for a metaphysical sense that the incorporeal dignity of the king of France never dies (*dignitas non moritur*). The meaning of a political doctrine, first elaborated in England, was thus translated into ceremonies in France. The king, by exception, had two bodies: one was natural, subject to illness and death; the second one was immortal, without decay. The uninterrupted series of mortal kings acceded to the perennial monarchical corporation made from one single member, the perpetual king.

Although notable innovations were introduced over time, a fundamental commitment to this idea lingered, perhaps exemplified most vividly in the use of the effigy, which first appeared in 1422. It was then a substitute for the real body, permitting the long process of the funerals. This effigy, with a mask and two hands out of a mantle, was posed upon the coffin. Afterward, the effigy replaced and ultimately excluded the corpse; during François I's funerals (1547), meals were served to the effigy, which presented the royal face with open eyes. The structural play between the corpse and the effigy was crucial: the first stage in the ceremony exhibited the dead corpse. Then, the corpse disappeared and was replaced by the effigy. Finally, the corpse was buried, in the absence of the effigy, the role of which was filled by royal insignia and by the vocal proclamation of an oxymoron "*Le roi est mort. Vive le roi*" (The king is dead. Long live the king). The absence of the successor during the ceremony, a shocking difference with "normal" funerals, had, too, political meaning; even if the king's son succeeded his father instantly at the very moment of his death, there could be only one king at the same time and the effigy was a transitional monarch.

Pierre d'Urfé's narrative thus offers a confirmation and, at the same time, sheds doubts on this functional interpretation of ceremonies by R. Giesey. On the one hand, the formal process was in its intermediate stage between the invention of the effigy and the perfection of the system in 1547. On the other hand, certain erratic details do not fit the functional grid: Louis d'Orléans, Charles VIII's successor, was absent partly because, as a former enemy of the king, he had much to fear at the court and much to gain at preparing his access to the throne. The circuit from Amboise to Saint-Denis went through Cléry, the most cherished sanctuary devoted to Mary, where Charles's father Louis XI chose to be buried, against the monarchical system, which implied, for both kings, that individual piety mattered as much as political propaganda. Many details in Pierre d'Urfé's description aimed at protecting the dead king's former friends and servants. In conclusion, we could say that a ceremony is not a simple tool of power, first because such a complex formalization could not be perceived by ordinary people, and, even when perceived by a limited elite, it could not automatically produce consent. Second, a funeral ceremony, because it implies many stakes, both individual and collective, about memory, salvation, honor, has necessarily many meanings. Agency remains active even in the deadly ceremonies of death.

Notes

1. Notre-Dame-des-Champs is not the present-day church in Montparnasse; it was located on Rue Saint-Jacques, outside the walls of Paris, on the southern edge of the city, just like Saint-Denis was outside Paris, on the northern edge.

2. Aune: about 1.3 meter.

3. Parisians could see only the effigy; the coffin had been completely covered by layers of luxurious material. But it was the coffin that was escorted to Saint-Denis.

4. *Banner, ensign, pennon, guidon*: types of military standards connected to the king (the guidon was triangular, the pennon long and thin, the banner square). The royal ensign with fleur-de-lis came to be more important than the traditional oriflamme, which was kept in Saint-Denis.

5. This cross marked the frontier between Paris and Saint-Denis.

Further Reading

Alain Boureau, *Le simple corps du roi: L'impossible sacralité des souverains français, XVe–XVIIIe siècles* (Paris, 1988; repr. 2000).

Elizabeth A. R. Brown, *The Monarchy of Capetian France and Royal Ceremonial* (Aldershot, 1991).

Alain Erlande-Brandenburg, *Le Roi est mort: Études sur les funérailles, les sépultures et les tombeaux des rois de France jusqu'à la fin du XIIIe siècle* (Geneva, 1975).

Ralph E. Giesey, *The Royal Funeral Ceremony in Renaissance France* (Geneva, 1960).

Ernst H. Kantorowicz, *The King's Two Bodies* (Princeton, 1957).

Work and Travel

— 9 —

Charms to Ward off Sheep and Pig Murrain

William C. Jordan

For murrain of sheep. First have a mass in honor of the Holy Spirit sung, and an offering made. Afterward the sheep should be gathered together in a cote (*domo*) or outside, and the priest should read these four gospel passages: "In the beginning" (*In principio*); "After Jesus had been born" (*Cum natus*); "Lastly (Jesus) showed himself to the Eleven themselves while they were at table" (*Recumbentibus*); and "And Jesus was saying this" (*Loquente*). Then he should sprinkle holy water over them, saying "In the name of the Father and the Son and the Holy Spirit, Amen." Afterward he should say this charm (*carm*): The holy little girl (*filiole*) was sitting on a sea isle, when the Lord God came upon her; then the Lord God said, "Little girl, why do you sit here?" "Sire," said the little girl, "I am keeping my sheep from the pox (*verole*) and the scab (*clousike*)." Then the Lord God said to her, "Protected are these sheep from the pox and from the scab." And straightaway they were protected, for God who formed and made the whole world told her this. Moreover, my lady, Saint Mary, implored him. And truly, indeed, these sheep are protected, as one sees, from the pox and the scab, just as you told her. Amen. Then, while kneeling, the Our Father is said three times along with the Ave Maria in honor of our Lord; and then this collect is said: "Almighty God, Holy Father, who has granted compassion for the toils of men and of the brute animals, we humble petitioners implore you, through our Lord Jesus Christ, that you not permit us to lose those things that we need (*usibus*) without which the human condition is not sustained."

Source: *Cartulary of Eynsham*, ed. H. E. Salter, 2 vols., Oxford Historical Society 49 and 51 (Oxford, 1907–8), I, 18, no. xx.

Blessing of barley for diseased pigs: "Ineffable God, inestimable and invisible God, you have created all things from nothing, we humbly implore your pity and mercy in the holy and formidable name of your Son, that you bless this barley, your creation, and infuse it with the power of your beneficence, that this blessing and holy act render the animals which, for human needs, you deigned to grant us, without impairment when they gather it up and eat it and keep them unharmed from any outbreak of disease." And then these four gospel passages should be read over the barley: "In the beginning" (*In principio*); "Mary surnamed the Magdalene" (*Maria Magdalene*); "Lastly (Jesus) showed himself to the Eleven themselves while they were at table" (*Recumbentibus undecim discipulis*); "If anyone loves me" (*Si quis diligit me*). Then let holy water be sprinkled over the barley and let it be given to the animals.

Source: Tony Hunt, Popular Medicine in Thirteenth-Century England:
Introduction and Texts (Cambridge, 1990), 99.

Countryfolk—elite and modest alike—relied on an accumulated treasury of expert knowledge to deal with the routine crises of rural husbandry. They or the men and women working for them knew how to spot the early signs of plant and animal diseases, and, if necessary, they authorized and took drastic actions in order to prevent the spread of these diseases. These actions, in the case of animals, included the substantial culling of herds and flocks and the disposal, by burning, of affected carcasses. To prevent both small and great disasters and to curtail them, if they did occur, they also appealed, in the Christian Middle Ages and in other times and cultures, to God and the saints or other instantiations of the holy, for help.[1]

Texts that throw some of the most revealing light on rustics' appeals for divine protection and help were penned by clerics but presume the effective role of parishioners. It is assumed, for example, that the ceremonies of the Rogation Days (springtime processions and blessings of fields) were a cooperative enterprise of clergy, lords, and peasants, even though the texts prescribing the ceremonies are in Latin and are formally liturgical. Sometimes clerics note occasions in which individual peasants brought their sick animals to shrines or churches for blessings or sprinkling with holy water.[2] These brief notices, typically in Latin until the end of the thirteenth century, provide us only with those details that the clerical scribe thought important. They may remind a modern reader of the often-terse notations in monastic miracle collections recalling the visits of sick pilgrims to local shrines. The monks needed to make records of these visits in case miracles subsequently took place. If miracles did occur, longer accounts might be penned or the original accounts supplemented. The information was thus purposively accumulated for fiscal purposes (sometimes there were offerings to be recorded) or to be available to recall to subsequent pilgrims the power of various shrines and relics, to stimulate additional visits to them, and even to help make a formal case to higher ecclesiastical authorities about the holiness of the various men and women venerated at the sites. Despite the slanted nature of the records, there seems little reason to doubt that the records also preserve hints of peasants' genuine interests and concerns.

Another class of texts offers much more extensive information on the array of practices in which countryfolk participated during periods of agricultural crisis or in anticipation of it. These are descriptions of entire ceremonies of supplication. They were recorded, perhaps originally designed, for reuse, if not for annual reuse as on Rogation Days, at least for when disaster hit or threatened in the future. The texts describing two of these ceremonies are offered in translation at the beginning of this section.

The ceremony titled, "For Murrain of Sheep," is found in a form that dates from the late thirteenth century and was written into the cartulary of the Benedictine monastery of Eynsham in central Oxfordshire. It is macaronic in form, that is to say, in a mixture of languages. Sometimes macaronic texts are parodies or burlesques, making fun of pretentiousness by inserting utterly silly foreign or foreign-sounding words and phrases into speeches, but just as often they are highly serious, and the shifts from one language to another have a complex significance.[3] That is the case in this supplication ceremony text. It begins with the instructions in Anglo-Norman French, the written vernacular language of the highborn monks at Eynsham, who were probably (to varying degrees) functionally trilingual in Latin, French, and English. Then the language shifts to Latin, the language of formal medieval Catholic worship, in order to describe in greater detail the ritual gestures and blessings to be performed over the animals. Then comes a "charm" that is to be said, shifting back, in the French vernacular. Within the charm itself there are words that are as much English as French, especially *clousike*, which describes one form of the diseases or symptoms that affected the animals. Following the French charm, the text of the supplication ceremony has additional instructions in French on the prayers to pray, the prayers themselves being indicated in Latin words. Finally—and solely in Latin—comes a closing collective prayer ending with a conventional liturgical Latin formula.

The ceremony described in the text is for the murrain or sickness/dying (from Latin *mori*, to die) of the sheep. The specific disease to be protected against is one that appears to have as its principal symptom ulcerous formations (scab and pox) on the sheep's skin. The fear and existence of similar diseases among human beings gave rise to similar responses, as in a contemporary Jewish exorcism in Old French but written in Hebrew letters.[4]

In the part of the world where the text of the supplication ceremony under discussion here was available for use, infection among diseased animals like sheep was likely to spread rapidly. The reason is hinted at by a phrase employed in the text. The sheep are to be gathered together in the cote (*in domo*). In the region of England from which this text comes sheepcotes were permanent stone buildings, usually long and narrow, with doors and thatched roofs and holes for ventilation along the interface of roof and stone wall.[5] They "provided winter shelter for animals," protected places for "ewes at lambing time" in the spring, shaded areas for milking in summer, and were storehouses for "hurdles and stakes, tar and grease for smearing the sheep as a precaution against scab, various medicines, reddle for marking, supplementary feed for lambs, such as draff (brewers' dregs) or grain, or containers for the cows' milk sometimes given to lambs." These cotes were "a considerable investment," but they served two other important tasks besides protecting the sheep from harsh winter weather. Over the winter, manure accumulated. About every two weeks (judging sometimes, to be sure, from later evidence) the floor would be strewn with straw and marl or pebbly clay, so that by the "end of winter a rich mixture would have been available to spread on the land." (A nineteenth-century writer says, "such deposits could be six feet deep.") The high ammonium content of the urine soaked accumulation probably inhibited the growth of many pathogens—a second, if indirect, benefit to the rustics and the sheep.

On an estate of an abbey of Eynsham's stature, it is estimated that one such sheepcote could hold three hundred to five hundred sheep with one square yard per sheep. Infectious diseases of epizootic extensiveness were rare, but when they hit it was inevitable that in close quarters such as these sheepcotes, the results were serious. In order to protect against this happening or to bring the diseases to an end a ceremony such as the one being described would be employed. (Dispersing the sheep in wintertime was not a viable alternative. The flock would have suffered grievous losses.)

The ceremony begins with mass in church. Congregants would have been called to mass by the ringing of a bell or bells, perhaps bells specifically inscribed with formulas meant to bring an end to bad weather, demonic forces, or any other causes of the diseases; inscriptions to this effect are well-attested on surviving bells from this period.[6] The mass at which the congregants gathered was a votive mass, one asking for the Holy Spirit to descend and consecrate the occasion. The mass of the Holy Spirit would traditionally have incorporated a special hymn, more technically a sequence, a hymn sung between the reading of the epistle (and its response or gradual) and the gospel. This was the *Veni Sancte Spiritus*, "Come Holy Spirit, Come," which was believed to be particularly efficacious when sung in votive masses in times of pestilence.[7]

Congregants then made an offering in church. The peasant laborers were to have the sheep in the cote or, perhaps if weather permitted, gathered in a field. The priest read four gospel portions (identified by their incipits): John 1, Matthew 2:1, Mark 16:14, John 8:30. Sequences like these, most often beginning with John 1, the prologue of the fourth gospel, occur everywhere in talismans, amulets, and similar texts.[8] With whatever passage they begin with, such sequences make an argument.[9] The first passage, "In the beginning was the Word: the Word was with God, and the Word was God. He was with God in the beginning. Through him all things came to be, not one thing had its being but through him. All that came to be had life in him," emphasizes the absolute creative power of God to give life.

The next passage takes the listener to the nativity and incarnation: "After Jesus had been born in Bethlehem in Judea." The sequence of the two passages suggests that the second is an example or an instantiation of the claims of the first. God has the power to give life. He gives his own life in Jesus in the incarnation from the Virgin. The argument continues in the next gospel citation, "Lastly (Jesus) showed himself to the Eleven themselves while they were at table. He reproached them for their incredulity and obstinacy, because they had refused to believe those who had seen him after he had risen." In other words, God's absolute power to give life (John 1) instantiated in the incarnation (Matthew 2:1) is lost on his disciples after his resurrection until he humiliates them (Mark 16:14).

The congregants and participants in this ceremony are, by implication, those who have doubted, who have lost hope in God's power to help them in protecting against or ending the animal pestilence. They need upbraiding, but if their faith is restored, the promise of reconciliation will be fulfilled. So, the fourth gospel passage articulates this sentiment: "As (Jesus) was saying this, many came to believe in him. To the Jews who believed in him Jesus said: If you make my word your home, you will indeed be my disciples, you will learn the truth and the truth will make you free."

In another text of a thirteenth-century supplication ceremony, like this one and also translated at the beginning of this section, "Blessing of barley for diseased pigs," a very similar series of gospel passages, making the same argument, is invoked. God is the giver of life; John 1 ("In the beginning was the Word . . ."), as in the first text, begins the sequence. The priest reads the passage. In this ceremony, the life-giving power of the Creator is instantiated not with reference to the incarnation but by allusion to the deliverance of Mary Magdalene, that most consummate sinner, from her sins and the deliverance of a host of other sinful women from theirs in Luke 8:2: "[C]ertain women who had been cured of evil spirits and ailments: Mary surnamed the Magdalene, from whom seven devils had gone out, and Joanna the wife of Herod's steward Chuza, and Susanna, and several others." Nevertheless, like the apostles after the crucifixion, there is a lack of confidence in the power of God; here the "Blessing of the barley" invokes the same gospel passage as the ceremony for the "Murrain of sheep," Mark 16:14, that is, the risen Christ's denunciation of the Eleven's "incredulity and obstinacy." To those who would come to believe (by extension, the congregants), a promise of reconciliation is proffered with as moving a gospel passage as that invoked in the "Murrain of sheep," John 14:23: "If anyone loves me he will keep my word, and my Father will love him, and we shall come to him and make our home with him."

In the texts of the supplication ceremonies for "Murrain of sheep" and for "Blessing of barley for sick pigs" now comes the priest's sprinkling of the beasts (or barley) with holy water, a gesture that reenacts ritually the bestowing of blessings on any number of occasions in Christian liturgical practice. But the ceremony does not stop here, at least not according to the text for "Murrain of sheep," as it seems to have stopped in the "Blessing of barley for sick pigs." There is a charm (carm).[10] Perhaps conservatives would have regarded the presence of a charm in such a ceremony as a superstitious addition. The shift into vernacular French from Latin suggests as much, and the presence of this text in a cartulary rather than a liturgical or quasi-liturgical book may point in the same direction. But there may be a positive reason as well for it to be preserved in a cartulary, that is to say, among copies of charters documenting the monastery's property holdings.

The charm was appropriate for a cartulary in that it was meant to cure or protect the sheep, livestock that were valuable property of the monastery.

Spoken, then, in vernacular French, the charm was accessible to the elite, but not presumably to peasants. However, the gist of its meaning, I suggest, would have been transmitted fairly easily to all the bystanders. Moreover, everyone in attendance from the lowliest to the most highborn would have known that the charm addressed the problem of disease—partly because the whole ceremony did so, but also because the charm itself makes this clear *in English*. The word *clousike*, as intimated earlier, is a macaronic compound of French *clou* (the head of a nail, scab, pustule)[11] and the native English *sike*, sickness, obsolete now in this meaning as an abstract noun, but still seen in adjectival form in the modern English word *sick*.[12] *Clousike*, scab disease, pustule disease: every listener, already primed for the allusion, would have caught the charm's repeated references to *sike, sike, sike*—disease, disease, disease.

The charming story (in both meanings of the word charming) has a little girl (*filiole*, the word can also mean godchild) at its center, a shepherdess, the very model of innocence.[13] It is possible that the reference is indirectly to the brief episode in Saint Margaret of Antioch's life as a tender of sheep.[14] The wide popularity of her cult in England is suggestive, even though there are few obvious commonalities between her *vita* and the narrative of the charm. But Saint Margaret of Antioch's purity and that of other exemplary female figures with whom there was some symbolic association with sheep or lambs, like Saint Agnes, seem to provide a reasonable background for interpretation of the charming *filiole*. The whole emphasis is on the divine grace rewarding her girlish innocence.

The shepherdess sits in an idyllic environment, a sea isle, a benevolent image that is employed in many other contexts. Sheep were often pastured on lake, river, estuary, and bay islands to protect them from predators and from straying.[15] A grassy sea isle or marsh island would have been imagined with this iconic picture of lush verdure and perfect safety from loss and predatory harm already in the countryfolks' heads. Men who lived on sea isles were believed to dwell in a favored environment, even living longer lives than ordinary men did, a belief that persisted throughout the Middle Ages; one need only think of the Isle of Sheep in the wildly popular medieval tale of the *Voyage of Saint Brendan*.[16]

The Lord God (*Damnedeu*) approaches the good little shepherdess and blesses her sheep. And he reminds the listener, as the prologue of the Gospel of Saint John earlier invoked also insisted, that he is the same God "who formed and made the whole world." His power has no limit, but to add to the persuasive rhetoric of the charm, the listeners are informed that the Blessed Virgin was interceding with God the Father on the *filiole*'s behalf as well.

This charm differs in many respects from what a modern reader might anticipate. It is a genuine narrative not a simple incantation. It has no immediately evident singsong pattern, despite the fact that the word *charm* in its Latin guise can also mean *song*. It does not sound like a traditional toddler's rhyming charm, like "Rain, rain, go away; come again another day," or a counting rhyme, "Eeny, meeny, mynee, mo," or a butter-churning (repetitive-task) rhyme, "Come, butter, come; come, butter, come . . ."[17] It is not even like contemporary macaronic incantations from the very same monastery.[18] Nonetheless, there may be vestiges of rhyming incantatory verse in the prose version of the charm recorded in the cartulary. These vestiges do not scan in so regular a fashion as much typical incantatory verse, but they are suggestive. For example, the passage, "Then the Lord God said to her, 'Healed are these sheep from the pox and the scab' " in the prose translation offered at the beginning of this section fails to capture the verselike quality of the charm.

> Damenedeu dunt a li dist,
> "Garri sent ces berbiz."

Or, again, the prose rendition, "God who formed and made the whole world, had said this to her," does not come close to the verselike incantatory quality of

> cun deu li auoyt ce dist,
> ke tut le mound format & fist

What the presence of verselike vestiges of incantation in the charm, as recorded by the monks, suggests is that it was adapted from an original free-standing rhyming charm (the rhyme is all on long *e*), but now, as we have it, flattened, though not quite completely, into prose and incorporated into a much more complex religious ceremony. The further significance of the translation of the mono-rhymed original into prose is uncertain. It might be argued that the singsong vernacular original was inappropriate once the content of the charm was introduced into a religious ceremony. It is no counterargument that a Latin mono-rhymed verse form, conventionally known as Goliardic, was routinely being used for hymns.[19] Few if any hymns, even mono-rhymed, mimic the singsong character of incantatory verse.

Whatever the case, the charm having concluded, the supplication ceremony over the animals continues with a series of prayers and prayer gestures, which, like the very participation of the priest, were intended to sanctify the charm: the Lord's Prayer, said three times while kneeling, and the Ave Maria, reemphasizing the intercessionary role of the Blessed Virgin, also said three times.[20] Finally, those literate in Latin—priest or priests, monks, local deacons, lectors, acolytes, perhaps some lay folk—prayed in explicit terms for a divine intervention that would preserve the livelihood of the whole community. And all would have joined together, over and over again, in saying, "Amen."

And, then, of course, the normal rhythms of country life would have resumed. Whether skeptical of the power of such ceremonies to prevent or bring an end to pestilences or confident of their efficacy, men and women returned to work. If the animals suffered in the aftermath, some of the witnesses to the ceremonies would have shaken their heads that anyone could have believed in their potency at all, and others no doubt would have found an explanation for the failure of the rites in the sinfulness and doubts of the community or of individuals within it who, like the eleven apostles whom the resurrected Christ upbraided, were simply too obstinate to believe. Medieval religion—in practice—was never simple.

Notes

1. Cf., for example, Lynn Remly, "Magic, Myth, and Medicine: The Veterinary Art in the Middle Ages (9th–15th Centuries)," *Fifteenth-Century Studies* 2 (1979): 203–9; Hervé Fillipette and Janine Trotereau, *Symboles et pratiques rituelles dans la maison paysanne traditionelle* (Paris, 1978), 269–80.

2. James Rattue, "An Inventory of Ancient, Holy, and Healing Wells in Leicestershire," *Transactions of the Leicestershire Archaeological and Historical Society* 67 (1993): 56–69; Nicholas Vincent, *Peter des Roches: An Alien in English Politics, 1205–1238* (Cambridge, 1996), 245; Kurt Müller-Veltlin, *Mittelrheinische Steinkreuze aus Basaltlava* (Neuss, 1980), 85.

3. Heather Barkley, "Liturgical Influences on the Anglo-Saxon Charms against Cattle Theft," *Notes and Queries* 242 (1997): 450–52; Donald Skemer, "Written Amulets and the Medieval Book," *Scrittura e civilta* 23 (1999): 274. For a magisterial study of macaronic sermon texts, see Siegfried Wenzel, *Macaronic Sermons: Bilingualism and Preaching in Late-Medieval England* (Ann Arbor, MI, 1994).

4. Menahem Banitt, "Une Formule d'exorcisme en Ancien Français," in *Studies in Honor of Mario A. Pei*, ed. John Fisher and Paul Gaeng (Chapel Hill, NC, 1972), 37–48.

5. This information and the various quotations in the next two paragraphs with respect to sheepcotes are from Christopher Dyer, "Sheepcotes: Evidence for Medieval Sheepfarming," *Medieval Archaeology* 39 (1995): 136–64.

6. Jacqueline Leclercq-Marx, "*Vox dei clamat in tempestate*: A propos de l'iconographie des vents et d'un groupe d'inscriptions campanaires (IXe–XIIIe siècles)," *Cahiers de civilization médiévale* 42 (1999): 185–87.

7. John Julian, ed., *A Dictionary of Hymnology* (New York, 1892), 1214.

8. Skemer, "Written Amulets and the Medieval Book," 270–73.

9. Barkley, "Liturgical Influences," 450–52. Cf. Thomas Hill, "The Æcerbot Charm and Its Christian User," *Anglo-Saxon England* 6 (1977): 213–21; Karen Jolly, *Popular Religion in Late Saxon England: Elf Charms in Context* (Chapel Hill, NC and London, 1996).

10. The editor expands *carm* to Latin *carmen*. The Latin word can mean song (just as a charm can be a chant, that is, an incantation). Vincent, *Peter des Roches*, 245n84, thinks that the *carmina* of this and similar texts are special collects, congregational prayers, but this usage, so far as I can tell, is not otherwise attested.

11. For a contemporary use of *clou* as scab in an exorcism formula, see Banitt, "Une Formule d'exorcisme," 39–40.

12. Afrikaans *lamsiekte*, lame-disease, adopted into South African English, preserves the otherwise obsolete meaning of Common Germanic "sick," illness.

13. William Paden, "The Figure of the Shepherdess in the Medieval Pastourelle," *Medievalia et Humanistica* 25 (1998): 1–14.

14. The editions and studies of Margaret and the tropes employed in her *vitae* are very large in number; readily accessible is the thirteenth-century hagiographical romance published in Brigitte Cazelles, *The Lady as Saint: A Collection of French Hagiographic Romances of the Thirteenth Century* (Philadelphia, 1991), 218–28.

15. Relics of this practice are evidenced in place names, as, for example, Magna Sheepy and Parva Sheepy (Great Sheep Isle and Little Sheep Isle) in Leicestershire and the Isle of Sheppey in Kent.

16. J. F. Webb, trans., *Lives of the Saints* (Harmondsworth, 1965), 40–41. See also, on the cliché "sea isle" in the Romance tradition, where this phrase describes King Arthur's sometimes idyllic, even magical, realm, Claude Luttrell, "Arthurian Geography: The Islands of the Sea," *Neophilologus* 83 (1999): 187–96.

17. That such forms were "available" for imitation, see the example reproduced and translated in David Howlett, "Israelite Learning in Insular Latin," *Peritia: Journal of the Medieval Academy of Ireland* 11 (1997): 138–39, and the repertory, conveniently collected and translated, in John Shinners, *Medieval Popular Religion, 1000–1500* (Peterborough, ON, 1997), 282–90.

18. For examples, see Robert Easting, "A Neglected Pair of Charms for Monastic Travellers in Middle English and Anglo-Norman," *Notes and Queries* 246 (2001): 103–5.

19. Cf. A. G. Rigg, "Metrics," in *Medieval Latin: An Introduction and Bibliographical Guide*, ed. F.A.C. Mantello and A. G. Rigg (Washington, DC, 1996), 108–9.

20. For similar invocations of the Ave Maria in amulets, see Skemer, "Written Amulets and the Medieval Book," 270.

Acknowledgments: Several people, besides those cited in the notes, have provided texts and references and have discussed aspects of the ceremonies discussed in this essay. They include Susan Einbinder, Gail Gibson, Richard Hoffmann, Emmanuel Kreike, Kenneth Levy, Emily Lyle, Jarbel Rodriguez, and Achim Timmermann.

Further Reading

Tony Hunt, *Popular Medicine in Thirteenth-Century England: Introduction and Texts* (Cambridge, 1990).

Karen Jolly, *Popular Religion in Late Saxon England: Elf Charms in Context* (Chapel Hill, NC, and London, 1996).

Derek Rivard, *Blessing the World: Ritual and Lay Piety in Medieval Religion* (Washington, DC, 2009).

—10—

Fishermen and Mariners

Harold S. Fox

MANDATE FROM POPE JOHN XXIII TO THE BISHOP OF EXETER
Dated at Fribourg, April 1415

A recent petition to the pope from the inhabitants of Exmouth has stated that the town is situated close up against the sea and depends upon the parochial chapel of St. John the Baptist at Withycombe Raleigh. Between town and chapel the high tides mean that on many Sundays and saints' feast days and holy days some of the parishioners are unable to go to worship without danger of death; in winter they are prevented from attending sometimes for a whole week; the road between the two is so rocky and muddy that the dead cannot be carried to the chapel for burial, children cannot be taken there for christening, nor women go there for churching; the town is situated so near the sea that pirates can take the whole of it and burn it if the inhabitants go to the chapel for divine offices. On account of all of this the inhabitants do not hear masses and other divine offices as often as they ought and have suffered and continue to suffer great dangers and losses when going to the chapel. Because of their special devotion to our Lord Jesus Christ and to the Most Blessed Virgin Mary His Mother, they desire to found a chapel dedicated to St. Anne the mother of Mary, with font, cemetery, and bell tower and to assign their obventions, tithes, and oblations for the income of a priest. Having received the petition of the inhabitants for license to found a chapel in the town, the pope hereby orders the bishop to permit it to be built at their expenses.

DEED OF THE BISHOP OF EXETER FOR THE DEDICATION OF REVELSTOKE CHAPEL AND ITS CHURCHYARD
Date Damaged, but Probably around 1430

To all the sons of the holy mother church to whom this writing may come, Edmund by divine mercy Bishop of Exeter sends greeting, grace, and blessing.

There has recently been presented to us a petition from our beloved sons Thomas Hereston, John Mederill, and Thomas Pey and the rest of the parishioners in the chapelry of St. Peter of Revelstoke, dependent on the mother church of St. Bartholomew of Yealmpton. . . . The petition states that the chapel of Revelstoke, together with a cemetery for burial of the dead (when it shall have been consecrated), is situated close to the shore of the high sea and has all the distinctive features of a parish, the right of burial excepted, and in it from time immemorial the parishioners have heard divine offices and have received the sacraments from a qualified chaplain at the cost of the vicar of Yealmpton, burial of the dead alone excepted, for which they have to go to the mother church of Yealmpton. . . . The parishioners are distant from that church by three English miles or more along which way there are many dangerous places on the roads, two tidal creeks, and a perilous river called the Yealm, so that sometimes in winter for days on end nobody is able to cross it without peril to his life, the bodies of the dead often remaining unburied to the great inconvenience and grievance of the parishioners. . . . Moreover, Revelstoke has a long seacoast so that, when the parishioners happen to be engaged in burying their dead, the enemies of the king and the realm can row or sail in and burn and spoil the whole district. The parishioners state that most of them are fishermen and laborers who work by the toil of their hands to obtain food and clothing for their families and are called away from their means of making a living because they have to attend funerals over such long and difficult roads. . . . We (the bishop), considering that the supplication of the parishioners is just, and having obtained the consent of all interested parties, now decree that the chapel of Revelstoke and the adjacent enclosure designated as a cemetery ought to be consecrated and dedicated. We also decree that the parochial chaplain of Revelstoke and his parishioners shall go yearly to the mother church of Yealmpton on the feast of St. Bartholomew the Apostle and on the day of the dedication of that church and shall pay, each year, as a token of subjection, 6 shillings and 8 pence.

Source: *The Register of Edmund Lacy, Bishop of Exeter, 1420–1455,*
ed. G. R. Dunston (Torquay, 1971), vol. 4, 315–18.

The two documents calendared above contain fifteenth-century petitions from people living near the coast of the southwestern English county of Devon. There is a basic similarity between them: both are requests for closer places of worship and burial, by people for whom the established parochial structure, already venerably old by the fifteenth century, had become inconvenient; and both are from people with maritime occupations, the 'fishermen and labourers' of Revelstoke (in the second instance), the inhabitants (in the first) of the 'vill' of Exmouth, by this time a compact shoreline settlement inhabited by people whose occupations included those of fisher, mariner, merchant and ferryman. Both represent the genuine voices of the people, not those of their lords or priests. The petition for Exmouth comes from "the inhabitants" and that for Revelstoke from three named men "and the rest of the parishioners." No doubt in both places there were individuals who were less devout than most. Yet the inhabitants of Exmouth were prepared to build a new chapel at their own expense and took the costly step of petitioning the pope, which they need not have done, for a request to their bishop would have sufficed; while the people of Revelstoke were willing to pay in perpetuity for their new graveyard. Voices of devout communities can be heard here.

All medieval people lived very close to nature. Dwellers inland were at the mercy of grain harvests, and death from starvation following a failure persisted in England until the seventeenth century. Those who lived close to the sea were especially affected by what we still today call "Acts of God." Their houses and their lives could be taken away by waves during a great storm and flooding or coastal erosion could severely damage settlements, as happened in the later fourteenth century at the ports of Lyme Regis (Dorset) when the town was severely battered and merchants killed or forced to withdraw; and at Ravenserodd on the Humber (Yorkshire). Drowning at sea was commonplace: medieval sources relating to the coasts are full of references to wrecks or parts of ships driven onto the shore by storms while, in 1433, a man who was asked during the course of an inquisition to remember the date of a certain event, said that he could recall it because his boat capsized on that day in Bigbury Bay, not far from Revelstoke. For mariners who ventured to foreign countries, there were pirates and hostile foreigners to contend with. What Devon mariners might have to suffer abroad is illustrated by their own treatment of a Genoese carrack, laden with a cargo of some value, which found itself in difficulties in 1418. According to a royal inquisition, local inhabitants living near the coast set upon the ship, some crew members were ill-treated, even killed, and the goods were stolen. The Hundred Years' War, caused initially by a dispute with France over the requirement of English kings to do fealty to French monarchs for land held across the channel, resulted in attacks on wealthy ports (e.g., Winchelsea, Sussex, in 1380; and Dartmouth, Devon, in 1404) and on many smaller coastal places. This is the context of the reference to "enemies of the realm" in the document relating to Revelstoke. Pirates also menaced the coast, as recounted in the document from Exmouth.

Closeness to death was one reason for piety and strong religious feeling among fishermen, mariners, and their families. No doubt they were also aware of the fishermen-apostles who were called by Jesus to become fishers of men; Revelstoke church was dedicated to St. Peter the Fisherman. They regarded the fish harvest, which could be variable from day to day and from year to year, as given by God and were, in the words of the people of St. Mawes (Cornwall), in a petition for a more convenient chapel, "glad to pray in public for good speed when they go out to sea and likewise to give God thanks for their prosperity when they return." In Devon in the early modern period, fishermen operating near the mouth of the River Lyn employed their own priest to say prayers in the open air, once in the morning and once in the evening. Piety was expressed, also, in the naming of ships: of the 350 or so ships assembled by Edward III for an expedition to France, 45 percent bore saints' names, with Mary the most popular and St. Nicholas also significant. Invocation of Mary, to whom the people of Exmouth were especially devoted, was clearly thought to be especially powerful when a ship was beset by storms, while medieval mariners would have known the legend of St. Nicholas (an obscure person whose written "life" was probably a conflation of several traditions relating to different individuals) in which he intervenes to save seamen during a storm.

The piety, and communality, of fishermen and mariners is also clear from their desire for their own places of worship, especially ones that seemed physically to associate with the waters by standing next to them. In many of Devon's coastal parishes, ancient parish churches tended not to be at the water's edge, for a variety of reasons connected with the history of settlement and need for security, and this was clearly inconvenient for those wishing to pray for safety before setting out to sea and to make votive offerings on their return. The first stage might be the setting up of a simple cross, like that on the rocks near Lanteglos (Cornwall) noticed by John Leland in the early sixteenth century and surviving until recently through the care of fishermen who renewed it if it was damaged by storms. The next stage was for a group of mariners and fishermen to build a small structure called a chapel of ease, perhaps not regularly served by a priest but convenient for individual or collective prayer; such are the well-known fishermen's chapels of Our Lady in the Dunes, at Dunkirk, at St. Brelade (Channel Islands), and also that dedicated to St. Nicholas on the shore of the River Teign (Devon). A final step might be for the maritime community to aspire to its own church served by its own priest, a difficult procedure because it meant a challenge to the old-established pattern of ecclesiastical provision. This is the background to the two petitions printed here. In both cases the topographical details are complex, and are not entirely clear at this remove. The basic points are that the people of Exmouth desired to have what amounted to a parish church of their own while Revelstoke already had quasi-parochial status, lacking only the right to burial, which was requested in the petition. Some commentators might add that new shoreline buildings such as these would have had multiple uses. Secular purposes might include use of the chapel as a landmark to guide shipping, as a place in which people sheltered while waiting for a ferry (such as the one that plied from Exmouth in the Middle Ages) or even for storage. Nevertheless, pious motives would have been at the fore, for piety, as we have seen, was strongly developed in coastal communities. The petitions printed here, and others like them, demonstrate those feelings and also show how groups of people were well capable of expressing religious choice and independent religious behavior during the fifteenth century.

Further Reading

P. F. Anson, *Fishermen and Fishing Ways* (London, 1932).

H.S.A. Fox, *The Evolution of the Fishing Village: Landscape and Society along the South Devon Coast* (Oxford, 1991).

Gervase Rosser, "Parochial Conformity and Voluntary Religion in Late-Medieval England," *Transactions of the Royal Historical Society* sixth series 1 (1991): 173–89.

— 11 —

Storms at Sea on a Voyage between Rhodes and Venice, November 1470

Olivia Remie Constable

On the evening of the fourteenth of November [1470], we [John Adorno and his father, Anselm Adorno] boarded a ship [in Rhodes] owned by a Basque or a Spaniard, on which there were five hundred containers filled with weapons and other armaments so that those on board would be armed not only to resist but also to attack any enemies. The Grand Master of Rhodes had provided the armaments, and had sent a number of his knights on board, including his nephew, the head prior of Apulia. The ship was to pursue war against the Turks, and also to load up with grain in Apulia, and therefore it was necessary to be well armed in order to be able to defend against enemies, mainly Turks and infidels, if we should encounter them. The ship's master took many days, sailing here and there across the sea by out-of-the-way routes, hoping to come upon enemies.

In this ship, we encountered many dangers at sea, of which two were so great as to make us fear for our lives. One day, during very clear weather, when we were near the island of Cythera, on the spot a fierce tempest came upon us, with a wild wind that caused our ship to heel over onto its side, as though led or moved by somebody's hand, making the ship pitch like a drunken man. We supposed that the ship would be overturned by these blows. Making a virtue of necessity, we commended ourselves to the All Powerful, then we fled from the savagery of the sea and sailed toward the port of Cythera.

Likewise, when we were in the Gulf of Venice, which was called the Gulf of the Adriatic in ancient times, in the middle of the night, our Spanish sailors— who were proud of their skill and overly confident—hoisted the studding sails over the main sail because a favorable wind came up. Then, unforeseen by our watchmen because of the darkness of the night, a fierce wind known as the *group* hit us with such force that our main sail was split in pieces, and rigging, masts, and anchor were torn away. It was such a strong wind, along with so much lightning, thunder, and rain that our ship was violently shaken and gave such a sound from the impact that we and the sailors equally believed that the ship was about to be broken into many pieces. We began to call upon God, in miserable voices, like those at the end of their lives, and to entreat him ardently, since we had no hope of assistance except through the grace of God. There was nobody among us then who did not think that we had come to our last hour. Oh, how our unhappy cries rose up! We all looked at each other, everyone weeping in sorrow. My father took me by the hand, as though wanting to say: "My son, I wish that we shall die together." Nevertheless, he continued praying to Saint Catherine, putting faith and hope in her, and even more in our Lord Jesus Christ, though the captain and sailors were entirely without hope. But our most merciful Lord Jesus delivered us from this terrible tempest, which lasted for three days and nights of thunder, wind, and rain. Indeed, we actually had a favorable wind—if it had been against us, we would have been lost or would afterward have found ourselves swept to another part of the world.

The Blessed Catherine, Bride of Christ and our most excellent advocate, helped us pilgrims to reach our port in safety. We appealed to her, and she pleaded with the Lord on our behalf. This was clear because we reached the port of Brindisi, safe and well, after surviving the huge storm, on her saint's day, the twenty-fifth of November. Oh, how we longed for this haven of safety in order to find refuge from the innumerable dangers of the sea, which were particularly great during the winter! The storms then are of the greatest violence, the winds are sharpest, there are the strongest rains and gales, the nights are always dark and very long, so that one must find passage over the high and wide sea, avoiding submerged reefs hidden from navigators. If these elements are dangerous to sailors in shallow water, then how much greater they are to those on the deep sea. There are no truer words than those of the [Roman author] Terence: "Those who never embark on a sea voyage have no idea of how many dangers they escape." Likewise, according to the [Greek] philosopher Pittacos, of all the elements, the sea is the most inconstant and the earth is the most stable.

Source: Jacques Heers, ed., *Itinéraire d'Anselme Adorno* (Paris, 1978), 367–70.

Medieval travelers faced many hardships on their journeys, and among these, they described storms at sea with particular frequency. Indeed, depictions of the travails of storm-tossed passengers are almost a trope of travel literature, not only in the Middle Ages but also in earlier periods. This passage, from the late fifteenth-century Latin narrative of a pilgrimage from Flanders to the Holy Land made by Anselm Adorno and his son, John, contains a characteristic description of the terrors of a tempest at sea.

Many medieval journals, letters, and other travel narratives mention an encounter with sudden and fierce squalls during maritime journeys, often with vivid descriptions of shrieking winds, a creaking wooden ship, cracking masts, helpless sailors, and panicking passengers. In such dire circumstances, the fear of shipwreck and imminent death almost always led travelers to appeal for divine assistance. There is no question that storms at sea were very real and terrifying events, but there is often a tension in their descriptions between graphic individual experience and consciousness of the universality of the peril. In this passage, the Adornos prayed fervently to Christ and St. Catherine in their time of need, but writing after the event, the author also displayed an awareness of genre in his somewhat pedantic insertion of quotations from the classical authors Terence and Pittacos.

Certain themes occur again and again in medieval stories of maritime disasters. Passengers not only appealed to God for aid, but they also found explanations for their experiences in Divine will. Medieval writers describing tempests at sea often recalled the biblical story of Jonah and cited the role of God in both calling and quelling storms. The English traveler Margery Kempe, en route to Santiago de Compostela across the Bay of Biscay in the early fifteenth century, probably had the paradigm of Jonah in mind when "before she entered the ship, she said her prayers that God would guard and preserve them from vengeance, storms and perils on the sea, so that they might go and return in safety. For she had been told that, if they had any storm, they would throw her into the sea, for they said it would be because of her."[1] Happily, both her journey to and from Spain was calm and uneventful. As in the tale of Jonah, storms could also be a sign of God's anger at some transgression. In 1393, a boat traveling from Cyprus was beset by storms every time it tried to leave the harbor, even in the clearest weather, until one of the passengers returned a relic that he had stolen to the church in which it belonged.[2]

Passengers in fear of their lives during storms at sea commonly commended themselves to God, and begged for aid from above, since the situation was clearly beyond human control. If the ship survived, and its passengers lived to tell the tale, medieval Christians gave credit to God, Christ, the Virgin Mary, or various saints for their rescue. The impression of piety may be enhanced by the fact that most medieval Latin travel narratives were written by pilgrims, who might be expected to credit divine aid more readily, but their beliefs probably also found reflection in the practices of merchants, sailors, and other seafarers. In this passage, the Adornos ascribe their survival to St. Catherine of Alexandria. John and Anselm had called on her during their peril, and later interpreted the fact that they reached safe haven in Brindisi on her saint's day, 25 November, as a sign of her role in their deliverance. Other travelers appealed to different saints. St. Nicholas, particularly, was considered a patron of sailors and those in distress at sea, and medieval paintings of shipwrecks often show him hovering in the clouds above the vessel.

Appeal to the Divine during storms at sea was by no means restricted to medieval Christian practice. Contemporary Jews and Muslims likewise placed themselves in the hands of God during storms at sea. In the 1180s, a Muslim pilgrim named Ibn Jubayr wrote of an experience not unlike that of Anselm Adorno, in which both Muslim and Christian passengers appealed to God during a storm off the coast of Sicily:

> When it came to midnight on Sunday the third of the blessed month [of Ramadan], and we were overlooking the city of Messina, the sudden cries of the sailors gave us the grievous knowledge that the ship had been driven by force of the wind toward one of the shore lines and had struck it . . . Dreadful cries were raised on the ship, and the Last Judgement had come, the break that has no mending and the great calamity which allows us no fortitude. The Christians gave themselves over to grief, and the Muslims submitted themselves to the decree of their Lord, finding only and clinging and holding fast to the rope of hopefulness in the life to come . . . When we were sure that [our time] had come, we braced ourselves to meet death, and, summoning our resolution to show goodly patience, awaited the morn or the time of destiny. Cries and shrieks arose from the women and infants of the Rûm [Christians]. All with humbleness submitted themselves (to the will of God) . . . [3]

Anselm Adorno was born in Bruges in 1424 to a Genoese merchant family long resident in Flanders. During his life, he served as merchant, administrator, and politician in his home city, as well as pursuing diplomatic missions to Scotland and serving at the court of the Scottish king James III. In 1477, he was exiled from Bruges and died in Scotland in 1483, the victim of an assassin. In 1470/71, Anselm Adorno made a pilgrimage to the Holy Land in the company of his son John (born in 1444 and the eldest of his twelve children), who took notes during the voyage and wrote down an account of the trip after their return to Bruges. Their route took them overland from Flanders to Genoa, where they took ship for Tunis, then continued by sea to Alexandria. From there, they journeyed overland to Jerusalem by an indirect route that included visits to Cairo and St. Catherine's monastery in Sinai. Perhaps it was their visit to St. Catherine's shrine that later led the travelers to call on this saint in time of need. After staying in Jerusalem for a period, the Adornos went north to Beirut to catch a ship to the island of Rhodes. The passage presented here picks up at this point, when Anselm and John boarded a boat to travel from Rhodes to Cythera, Brindisi, then up the Adriatic to Venice, from where they would travel home by overland routes.

The Adornos' journey began in April of 1470, and they arrived in Rhodes in early November, after the end of the traditional sailing season in the Mediterranean. The narrative remarks on this, stating that "thanks to God, we arrived safe and sound on Rhodes without encountering any misfortune or storms at sea, which was miraculous since it was already winter."[4] Later, the author openly regretted the fact that they had to travel in this season, when nights were long and dark and storms were common. Most Mediterranean voyages were undertaken between the late spring and early autumn, when the weather was most likely to be calm. Long hours of summer daylight and clear night skies were also an advantage to captains who piloted their vessels by coastal landmarks and by the stars. However, some travelers, like the Adornos, found that their itinerary demanded a sea voyage in winter, and most were able to find a ship to give them passage to their destination.

In this case, the ship was a Spanish vessel, sent to collect grain from southern Italy, and armed by the Knights of Rhodes against Ottoman naval attack. The ship was also prepared for offensive action, should it encounter a Turkish vessel. Since the conquest of Constantinople by the Ottomans in 1453, a couple of decades before the Adornos' voyage, the eastern Mediterranean had become a zone of naval warfare between Christian and Muslim fleets. However, the boat in question was probably a sailing ship, with rounded hull and carrying capacity for passengers and its cargo of grain, not the sleek low-lying galleys, powered by both oars and sails, that were the traditional vessels of naval action in the eastern Mediterranean. In a storm, a round ship was a safer vehicle, especially when high winds and seas forced the vessel to heel over onto its side, as in this passage.

Notes

1. Margery Kempe, *The Book of Margery Kempe*, trans. B. A. Windeatt (Harmondsworth, 1985), 147.
2. Roland A. Browne, trans., *The Holy Jerusalem Voyage of Ogier VIII, Seigneur d'Anglure* (Gainesville, FL, 1975), 69.
3. Ibn Jubayr, *The Travels of Ibn Jubayr*, trans. R.J.C. Broadhurst (London, 1952), 336–37.
4. Jacques Heers, ed., *Itineraire d'Anselme Adorne* (Paris, 1978), 358.

Further Reading

Norbert Ohler, *The Medieval Traveller*, trans. Caroline Hillier (Woodbridge, 1989).
Marjorie Rowling, *Everyday Life of Medieval Travellers* (New York, 1971).
Jean Verdon, *Travel in the Middle Ages*, trans. George Holoch (Notre Dame, IN, 2003).

—12—

Rules and Ritual on the Second Crusade
Campaign to Lisbon, 1147

Susanna A. Throop

CODE OF CONDUCT AGREED AT DARTMOUTH IN MAY 1147

All these people with so many different languages [who made up the crusading armies] swore firm pledges of unity and friendship. Moreover, they decreed strict laws, to be punished [in kind]: a death for a death, a tooth for a tooth. They forbade altogether magnificent, expensive clothing. Also, women should not go about [alone] in public. The peace should be preserved by all, except for when judgment was handed down for an injury. Each week religious services should be held, one for the laity and another separate for the clergy, unless by some great chance one group should require the other to be present. Each ship should have its own priest, and these priests should conduct their religious affairs just as they had in their parishes. No man of status should retain a sailor or servant in his household. Every week every person should confess and take communion on Sunday. . . . Meanwhile, two out of every thousand men were chosen and were called judges and jurors, who, by ordering the constables, could end disputes and distribute money. Then, with these rules thus established, on the sixth morning before the ascension of the Lord [23 May 1147] we set sail . . .

BEFORE THE FINAL CRUSADER ASSAULT ON
THE CITY OF LISBON IN OCTOBER 1147

. . . Finally our siege tower was complete, covered on all sides with wicker and leather spikes, lest it be wounded by fire or thrown rocks. Everyone was ordered, throughout the ships, to place over these [spikes] wicker roofs and coverings made from woven branches. On the following Sunday, when all necessary preparations for defending [the tower] were complete, the archbishop was summoned to bless it before it was moved. After prayer [was said] and the tower was sprinkled with holy water, a certain priest, holding a sacred piece of the Lord's cross in his hands, gave a sermon:

". . . Behold, brothers, behold the wood of the Lord's cross. Bend your knees, lie down flat on the ground, beat your chests, beg the Lord for help. For it will come, it will come. You will see the help of the Lord above you. Adore the Lord Christ, who on this life-giving wood spread his hands and feet for your health and glory. . . . Amen."

At this all present cried out, moaning and weeping, lying flat on their stomachs. Then, at the order of the priest, everyone stood up and they were all marked with the venerable sign of the lordly cross in the name of the Father and the Son and the Holy Spirit. And thus they began to move the [siege tower] almost fifteen cubits toward the [city] wall, unanimously crying out and begging for God to help them . . .

THE SURRENDER OF THE CITY AND THE TRIUMPHAL ENTRY OF THE CRUSADING ARMIES

. . . It was then decreed among us that four thousand armed men from our [Anglo-Norman] contingent, and six thousand from the Germans and Flemish, should enter the city before all the rest, and should take charge of the greater castle [and hold it] securely, so that in that place the enemies [citizens of Lisbon] could hand over their money and all their property, proving their hearts [to be true] by swearing oaths to us. With these men [of Lisbon] thus assembled, the city could thereafter be inspected by our men, and if any more belongings were found in someone's possession than what he had already surrendered, the head of that household should be punished by death, and in this way all spoils should be sent peacefully out of the city . . .

. . . Therefore, with the archbishop and his fellow bishops leading the way with the lordly banner of the cross, our leaders entered [the city] as one with the king, and with those who had been chosen for that matter. O what happiness of all! O what special glory! O what abundant tears [were shed] for joy and piety, when to the praise and honor of God and the holiest virgin Mary the banner of the life-giving cross was seen by all to be placed on the highest arch in the administrative district of the city, while the archbishop and the bishops, along with the clergy and everyone else, not without tears but with admirable joy, preached the *Te Deum laudamus* along with the *Asperges me* and [other] devout prayers. At the same time, the king walked around the high walls of the castle on foot.

Source: C. W. David, ed., *De expugnatione Lyxbonensi*, with a new foreword and bibliography by Jonathan Phillips (New York, 2001).

The Second Crusade (1145–49) was doomed to be viewed as an almost unmiti-gated failure by historians for centuries after it ended. In December 1144, Mus-lims from northern Syria had invaded and captured the Frankish city of Edessa in the Latin East. A year later, in December 1145, Pope Eugenius III issued the fa-mous papal crusading bull *Quantum praedecessores*, urging Christians in western Europe to follow the example of their predecessors on the First Crusade and recapture Edessa. By 1147, King Louis VII of France and King Conrad III of Germany were on their way east, marching overland with two separate crusading armies. Both armies met with disaster, and the remnants made their way home in shame and confusion, forcing Christian theologians to struggle with why God had, apparently, deserted the Christians' military endeavor in the Holy Land.

But not all campaigns of the Second Crusade failed. In particular, efforts to take territory from the Muslims on the Iberian Peninsula were successful. One such success was the campaign to conquer the city of Lisbon in 1147. The campaign was led by King Alfonso Henriques of the newly established kingdom of Portugal, along with an assorted cast of crusaders, featuring contingents from the Anglo-Norman world, Flanders, and Cologne and the Rhineland.

A narrative account of the Lisbon campaign, beginning with the sailing of the fleet from Dartmouth in May 1147 and ending after the conquest of Lisbon in October 1147, has survived to this day in the form of the *Conquest of Lisbon*, in Latin the *De expugnatione Lyxbonensi*. Not only was this detailed account written very shortly after the events it described, it was also composed by a clerical eyewitness of the cam-paign. This priest has been convincingly identified as Raol, an Anglo-Norman cler-gyman who seems to have had some social standing of his own. He composed his ac-count in the form of a letter to a colleague, Osbert of Bawdsey, a cleric connected with the Glanville family of East Anglia—Hervey de Glanville led the Anglo-Norman contingent at Lisbon himself.

The *Conquest of Lisbon* contains fascinating glimpses of crusading religious practice, seen from the perspective of a clerical participant. In the first excerpt above, Raol recorded the details of generally agreed upon rules of conduct among the crusaders, details that are not seen again in a text until a Third Crusade ac-count of Richard I setting forth from Cyprus.[1] Coordinating the movement and behavior of various independent contingents from different regions of western Europe must indeed have posed a severe logistical problem, and the crusaders settled their terms before they set sail in a remarkably democratic manner. It was key that the crusading armies hold together, and so above all the need for unity was stressed. This was to be achieved by downplaying social tensions—for example, by forbidding expensive clothing and preventing men of greater power from tak-ing on lesser individuals as servants. But alongside these sorts of practical social regulations, the crusaders set standards for religious practice, and these duties, in-terestingly enough, relied upon division and delegation, rather than unity and unanimity. Each ship of crusaders was to have its own priest, who was to officiate just as he had done in his parish back home. Clergy and laity alike were to have religious services, but separately from each other. Ultimately, despite appearances, these divisions on the crusade served to ensure religious uniformity in the same way that the parochial structure of the Church at a local level supported overall institutional coherence. The crusaders sought to achieve unity not by prescribing universal religious *practices* for all, but by utilizing individual, local religious customs to buttress the universal *faith* of the overarching, all-encompassing Mother Church.

In the second excerpt, Raol gave detailed descriptions of the rituals that directly preceded pitched battle. The general outline he provided was often similarly described in other crusading accounts: the preparation of military technology, the recitation of blessings, prayer, the delivery of an inspiring sermon emphasizing God's omnipotence and the Christian's hope of everlasting life in Christ, and finally a crusader assault accompanied by requests for God's aid. What is interesting about the second passage from the *Conquest of Lisbon* is the detail Raol provided, along with the fact that many historians now believe that Raol himself may have been the "certain priest" who delivered the rousing sermon to the armies.

In his account, although the crusaders worked in different groups to prepare for the battle, they came together to pray—again, the unity of the crusading forces was paramount, and from this point in the text until the battle they moved and spoke as one. The siege tower was blessed and sprinkled with holy water, a common Christian step when preparing for battle that was intended as an act of purification symbolically similar to the sacrament of baptism. Then a priest (presumably Raol), holding aloft in his hand a portion of the relic of the cross of Christ, delivered an inspiring sermon. Relics were seen as conduits to holy power, and Raol called upon his listeners to abase themselves before the relic of the cross—it was generally believed that the proper way to solicit help from a relic was through humility and humiliation. The crusaders believed that they could only conquer through God's will and with his help, making it extremely important to beg for his aid and express their own awareness of his power and their comparative impotence.

In this section of the passage we see the crusading armies shifting context, moving from the practical, individual preparations for war (such as constructing and shielding a siege tower) to the unanimous spiritual experience of utter reliance on God's will and their faith in divine aid. In addition, the correlation between Raol's instructions ("bend your knees . . . beg the Lord for help") and the behavior of his audience emphasizes that this ritual was one in which all present understood their roles and fully participated, rather than simply a clerical performance to a passive lay audience. Finally, through this act the practical and the spiritual were united, just as the different crusading contingents were united, moving forward with determination to the city walls.

In the third excerpt, Raol described the surrender of the city of Lisbon and the triumphal entry of the crusading forces into the city. Here, again, we see the tension between the individual and the universal, the crusaders as different contingents and the crusaders as a unified whole. By this point in time the different crusading contingents had virtually come to blows, and as a last effort to ensure peace among themselves, they all swore fidelity to King Alfonso Henriques for as long as they were in his territory. In order to prevent personal looting and gratuitous violence, the crusaders decided that only a certain number of armed men would enter the city at first, in order to guarantee an orderly and secure collection of booty and prisoners. However, as Raol related in a section not here translated, this agreement was broken when the Flemings and Rhinelanders snuck extra men into the city and (according to Raol, whose identity as Anglo-Norman should not be forgotten) embarked upon rape, murder, and personal looting.

But although Raol reported the divisions among the crusaders, by the time the formal ceremonial entry into the city took place, all were again united in his account. His description of the triumphal entry of the crusaders into Lisbon recalled many different traditions, and the ritual actions of participants are deeply symbolic. Based upon the Roman triumph, there were two main forms of ritual entry seen in the medieval Christian West. First, there was a type commonly known as "penetration," in which the procession passed through the city directly to its center and laid claim to it—the direct movement of the archbishops and their train to the "highest arch" of the city of Lisbon and the placing of their banners there was this kind of entry. The other type of triumphal entry was based upon the Roman *circuitus murorum*, or circuit of the city walls before entrance. In the Middle Ages this ritual was incorporated into church dedication ceremonies and the *adventus novi episcopi*, "the coming of a new bishop" into his city. Raol's description of the king walking around the high walls of the castle would seem to have been a form of *circuitus murorum*. The circuit of the walls was not only a territorial proclamation but also and more importantly a ritual cleansing, a purification of space. The fact that the king circled the walls on foot and not in a chariot as the Romans had done may well have been intended as a sign of his humility before God, and of his reluctance to claim a victory that was seen as God's alone.

The liturgy Raol reported supports the idea that the crusaders' actions upon entering Lisbon were aimed at purification and attributing victory to God. The *Te Deum laudamus* ("We praise you God") was an anonymous Latin hymn of praise, traditionally sung during Matins but also commonly used on other occasions to express humble worship. The *Asperges me* was an antiphon, a short prayer usually sung at mass. It was one of seven traditional penitential antiphons long associated with atonement, and was derived from Psalms 50:9 (Vulgate): "purge me with hyssop and I will be clean, wash me and I will be whiter than snow."[2] The performance of this antiphon directly after entering Lisbon emphasizes the need to purify not only the city itself, which many had seen as "defiled" by the presence of the unfaithful, but also the need to again cleanse the crusaders of their sins and thereby underline their belief in their own resurrection through baptism and faith in Christ.

Notes

1. W. Stubbs, ed., *Gesta regis Henrici Secundi* II, Rolls Series 49 (London, 1867), 110–11.

2. *Asparges me hysopo et mundabor lavabis me et super nivem dealbabor.* Available at www.lib
.uchicago.edu/efts/ARTFL/public/bibles/vulgate.html.

Further Reading

S. Bertelli, *The King's Body*, trans. R. B. Litchfield (University Park, PA, 2001).

Giles Constable, "The Second Crusade as Seen by Contemporaries," *Traditio* 9 (1953):
213–79.

T. J. Heffernan and E. A. Matter, eds., *The Liturgy of the Medieval Church* (Kalamazoo, MI,
2001).

Jonathan Phillips, "Ideas of Crusade and Holy War in *De expugnatione Lyxbonensi*," *Studies
in Church History* 36 (2000): 123–41.

Churches, Parishes, and Daily Life:
Consecration

— 13 —

The Consecration of Church Space

Dominique Iogna-Prat

Thus, when, with wise counsel and under the dictation of the Holy Ghost Whose unction instructs us in all things, that which we proposed to carry out had been designed with perspicuous order, we brought together an assembly of illustrious men, both bishops and abbots, and also requested the presence of our Lord, the Most Serene King of the Franks, Louis. On Sunday, the day before the Ides of July, we arranged a procession beautiful by its ornaments and notable by its personages. Carrying before ourselves, in the hands of the bishops and the abbots, the insignia of Our Lord's Passion, viz., the Nail and the Crown of the Lord, also the arm of the Aged St. Simeon and the tutelage of other holy relics, we descended with humble devotion to the excavations made ready for the foundations. Then, when the consolation of the Comforter, the Holy Spirit, had been invoked so that He might crown the good beginning of the house of God with a good end, the bishops—having prepared, with their own hands, the mortar and the blessed water from the dedication of the previous fifth day after the Ides of June—laid the first stones, singing a hymn to God and solemnly chanting the *Fundamenta ejus*,[1] to the end of the Psalm. The Most Serene King himself stepped down [into the excavation] and with his own hands laid his [stone]. Also we and many others, both abbots and monks, laid their stones. Certain persons also [deposited] gems out of love and reverence for Jesus Christ, chanting: *Lapides preciosi omnes muri tui*.[2]

Source: *Abbot Suger on the Abbey Church of St.-Denis and Its Treasures*, ed.
Erwin Panofsky and Gerda Panofsky-Soergel, second ed.
(Princeton, NJ, 1948), 101, 103.

In these words Abbot Suger (1081–1151) recorded the ceremony of the laying of the first stone of the chevet of the abbey church of St. Denis, on 14 July 1140. This extract forms part of the discussion of the building of the chevet, which also considers the earlier church, dedicated to Christ, and the stones of which were treated as if they were relics; the early, Carolingian, part of the church thus became a reliquary-crypt of sorts. After detailing the complexities of new construction upon a site deemed to be sacred (the crypt), Suger describes the ceremony of laying the foundation stone, on Sunday 14 July 1140, in the presence of King Louis VII and several prelates, bishops, and abbots, whose identity he does not disclose. The ritual described unfolds in three stages: a procession with the relics of St. Denis (the Christ-related nail and Crown of Thorns), up to the foundations of the structure; the bishops then prepare the container with water already blessed for the consecration of an earlier stage of the construction (this ceremony was also described by Suger); finally the assembled solemnly chanted Psalm 86[87], and the King, Abbot Suger, abbots, and other religious (perhaps including monks of St. Denis), queued up to pose the first stones of the chevet, upon the foundations, while intoning the hymn *Lapides preciosi omnes muri tui*.

To what extent does Suger's description correspond with our knowledge of similar rituals? Is the ceremony described by him a long-established one? What was its aim, and what role did the presence of the king play in this religious ceremony? These questions will guide the following commentary.

The Interface between Church-Founders and Priest-Consecrators

The Latin west possessed no description of the construction of a church to rival that provided by the Greek *Life of Porphyry of Gaza* (who died in 420), a text of circa 600.[3] The first mention of the laying of the first stone in the west is attested in the Romano-Germanic Pontifical, a bishop's liturgical book, composed in Mainz in the 960s. This book reached usage in the whole of Europe in the course of the eleventh century through papal dissemination. The passage in the Pontifical that interests us here is part of a section on the dedication of a church and the consecration of its associated parts (cemetery and baptistery).[4] Chapter 36 is entitled *canon*, a juridical term, since at this point it incorporates a legal text *Nemo ecclesaim aedificat* (*No one should build a church*), which originates in the Emperor Justinian's (d. 565) law code, the *Novellae*.[5] This law code was transmitted throughout the sixth century in the west through a Latin truncated edition, and later appeared regularly in collections of church law, canon law. According to the *canon*, based on imperial decree, the founder of a church was obliged to seek authorization by the local bishop for the construction. The editor of the Romano-Germanic Pontifical integrated this canon and also provided the relevant ritual, which enacted the bishop's control over the building of a new church. The ceremony began when the bishop, or his representative, posed a cross over the building site. He then designated the space for eucharistic celebration, the place of the altar, by planting a cross on the spot. Finally, he pronounced a lengthy blessing that placed the purification of the founder, who was placed in the hallowed line of David and Solomon, and chased out any demonic powers, seeking the resurrection of the church in the "simplicity and candor of innocence."

No stone is laid in the course of this ritual. The liturgical logic of the Pontifical, the passage from cross to foundation stone, is made by the fact that Christ, foundation of all foundations, is considered to be that cornerstone—first stone—of the ecclesiastical edifice. This understanding, which is so hard to date with precision, is attested in thirteenth-century Pontificals—that of the Roman papal court and that of William Durand (1230–1296)—and is abundantly documented from the second half of the twelfth century by commentators on the liturgy. The Roman Pontifical of the twelfth century does not include this ritual, while that of the thirteenth only comments rather soberly on the blessing of the laying of the foundation stone. It is only with William Durand that the ceremony receives full treatment.[6] Twelfth-century commentators on the whole followed the lead of John Beleth (ca. 1150) who described three stages in the ritual of the laying of the foundation stone: preparation of the foundation, exorcism and purification of the building space, and the laying of the stone with a cross on top.[7] These texts are describing a well-established ritual. Archaeological findings have discovered foundation stones with a cross, datable to the early eleventh century, such as that placed by Bishop Bernward in 1010 at the foundations of the church of St. Michael in Hildesheim. It is worth noting that the materialization of the Christ-related signs occurred in a period of great controversies over the Eucharist, raised by Berengar of Tours (ca. 1010–1088), around the nature of the presence of Christ's body—real or symbolic—in the consecrated bread and wine of the mass. Berengar, who was an expert in grammar, emphasized the meaning of the words pronounced by Christ at the Last Supper—*hoc est corpus meum* (this is my body)—questioning the nature of the predicate *hoc* (this). He also referred to the image of Christ as a cornerstone, by which he easily demonstrated that such an image ought not to be taken too literally. The detail of the debate is of little importance here; for our purposes it suffices to note that during these years the authority of the papacy in Rome was used to impose a more "realist"—even hyperrealist, interpretation of the eucharistic transformation. The placing of the first stone, marked with a cross, contributed, in its way, to the process of materialization of Christ's body.

It is also useful to point out the absence of any reference to a ritual of placing the first stone in the Roman Pontifical of the twelfth century, and to note the restraint with which the thirteenth-century Pontifical mentions it. Quite unlike the ritual of consecration of a church, a rite that was lavish and the elaboration of which was an object for episcopal and papal control, the ceremony of placing the first stone for an edifice drew for very long only minimal clerical attention. The ritual that ultimately emerged was merely a frame of reference for the development of the most varied and even exuberant practices. Hence we have evidence of a great variety of ceremonial practices, from around the year 1010, in hagiographical accounts, which seem—as Suger's own text exemplifies—each to elaborate without restraint a shared experience, turning it into a unique local occasion, fixed in memory, of the building of the local church. This very freedom of movement renders the development of the ritual particularly interesting, since the beginning of construction of a church created an occasion on which actors other than clerics-consecrators could express themselves: donors, founders, builders, and simple believers, too.

So it was that the interplay of power in the work of construction is binary in nature. It essentially involved, on the one hand, powerful laypersons seeking to build a place for themselves in the hereafter, and on the other, clerics, whose function it was to consecrate pious acts of personal edification. It is all a matter of the dialectic of power and of a balance between the two socially dominant forces. The ritual of placing the first stone is a gesture of affirmation of clerical prerogatives, which nonetheless recognize the place allocated to laypeople. The frequent reference to the Emperor Constantine—working with his own hands in the construction of the Lateran church in Rome—is part of the subtle set of relations between spiritual authority *auctoritas* and temporal *potestas*; the former activating the latter.

From here arises the political-ecclesiastical context within which the ritual is fully realized. In the case of St. Denis, it is clear that the participation of King Louis VII at the placing of the first stone (and then, four years later, at the dedication of the chevet) makes no sense except as part of a set of idealized relations between the kingdom and abbey, of which Suger dreamed, and in which the sovereign is a vassal of the patron saint, and St. Denis becomes the epicenter of Capetian building. But there are also other grounds for expression and other stagings of great lay builders of churches that were not mediated by clerics. One may think of the representation, in the *Grandes Chroniques de France* (*The Great Chronicles of France*), which were illustrated by Jean Fouquet (d. 1477/81), of Charlemagne surveying the construction of palaces and churches, where the sovereign—a real clerk of works—occupies a central, even unique, position.[8] It is true that the two actors—building cleric and building sovereign—do not seem at this late date to be situated on the same level. Their bodily separation had taken place some while earlier. In fact, the *ordo* of blessing and of placing the first stone elaborated in the thirteenth century never mentions the founder. While in the Pontifical of William Durand the ceremony takes its full shape and has its real rise, the absence of even the slightest mention of a donor, a founder, or a builder is particularly striking. While the Roman-Germanic Pontifical made place for "him whom God had chosen to build" the edifice, the ritual put in place by William Durand displays an interest only in the great characters of divine theater: the Virgin and the patron saint of the church; the apostle Peter and Christ, the former as a stone (*Petrus/petra*) placed upon the latter (cornerstone=Christ); David and Solomon. The last two do not shift the shadow cast over the founder.

Why is this the case? The reason for the silence is obvious. Elaborated within a context of an imperial church in which the sovereign is also, in a way, the master of ceremonies, the Romano-Germanic Pontifical provides the suitable place, at the side of the celebrant, for the one who has provided the resources for the edification of a dwelling for the Lord, and a place where His memory was fixed.

Three centuries later, William Durand, as bishop, was part of an ecclesiastical system that claimed to be the sole representative of the City of God on earth, had no reason to cede space in this vital ritual to laypeople. They sought after all, in the building stones, some treasure in a future life.

Translated from the French by Miri Rubin

Notes

1. Psalm 86: "The Foundations thereof are in the holy mountain."

2. Fifth antiphon of the service of dedication of a church: "All thy walls are precious stones, and the towers of Jerusalem shall be built of gems."

3. Marc the Deacon, *Vie de Porphyre de Gaza*, ed. and trans. H. Grégoire and M. H. Kugener (Paris, 1930).

4. R. Elze and C. Vogel, eds., *Le pontifical romano-germanique du dixième siècle* I, (Vatican, 1963), 122–23.

5. Novella 67.

6. M. Andrieu, ed., *Le pontifical Romain de moyen-âge* III (Vatican, 1940), 451–55.

7. John Beleth, *Summa de ecclesiasticis officiis*, ed. H. Douteil, CCCM 41A (Turnhout, 1976), 6–7.

8. The image and an analysis appear in C. Beaune, *Le miroir de pouvoir: Les manuscrits des rois de France au Moyen Age* (Paris, 1997), 48–49.

Further Reading

P. L. Gerson, ed., *Abbot Suger and Saint-Denis: A Symposium* (New York, 1986).

L. Grant, *Abbot Suger of St-Denis: Church and State in Early Twelfth-Century France* (London and New York, 1998).

M. Wyss, ed., *Atlas historique de Saint-Denis: Des origines au XIIIe siècle* (Paris, 1996).

Churches, Parishes, and Daily Life:
Pastoral Care

—14—

Fourteenth-Century Instructions
for Bedside Pastoral Care

Joseph Ziegler

On the causes of disease, that is, why people become ill, I reply, there are three [*sic*] reasons: the first is spiritual, that is, sin. The other is corporeal. But in this chapter we shall speak of the first cause. Therefore I say that had man never sinned, and had he remained in the state of innocence, he would never have had any diseases, neither would he have ever died.[1] But when he had stayed in this world for twenty or thirty years, or as long as would please God, then without the death of the body he would go to paradise in soul and in body. Pay attention that in the treatise on the nature of serpents (*Tractatus de natura serpentium*) it says that before sin no animal was poisonous, and if it were, it would not harm man. In the same way I believe that if man had been in fire or in water or among a thousand swords, none of this would have harmed him had he not sinned. The reason for all this is that the first man was created by God, was placed in terrestrial paradise in a state of innocence, and had [in him] original justice. Nothing would have protected him except for the fact that the soul was entirely obedient to God and that the powers of the lower senses entirely obeyed the soul. He was righteous and consequently God bestowed his grace upon the human being, so that he took care of the active and passive qualities of the elements that were in the human body lest they act against each other, and caused them to remain in due proportion. Consequently the human body would never be ill, neither would it ever die. But because Adam was disobedient to God, eating from the tree against the precept, immediately the lower powers of the soul rebelled against the sensibility of the soul. This is experienced individually by everyone, seeing that the flesh desires things that counter the spirit. And this was original sin, that is, the lack of original justice. And then immediately God permitted the elemental qualities to act against each other and consequently the body becomes ill and dies. And God explicitly said to the first parents in Genesis [2:17], namely "On the day you eat of it you shall most surely die." Hence sin is the cause why man becomes ill and dies. And the Apostle says [Romans 6:23]: "For the wage paid by sin is death," and because of this in a certain general council (in the canon titled *Cum infirmitas*)[2] all physicians of the body are ordered that whenever they are called to treat ill people they should first admonish and induce the sick to call in the assistance of physicians of the soul, that is confessors, since sometime illness originates from sin. Also, the same canon by imposing anathema prohibits physicians to give, for procuring bodily health, medical advice that is sinful. Therefore confess, if you do not wish to be ill.

On the formal cause[3] of disease, which is manifold according to the elemental qualities, namely hot and cold, humid and dry. Whence Hippocrates said that [humoral] changes (*mutationes*) generate diseases. For a sudden change from warm to cold alters and changes the body and consequently generates disease. Similarly another cause of bodily illness is excess or highly inadequate eating and drinking, or foul air.[4]

On fever; how are they generated? Superfluous food is discharged by nausea and vomiting and then the person is cured. Aristotle in *Secret of Secrets* says that once or twice a month and especially in the summer one should practice vomiting, since vomiting cleanses the body and purges the stomach of the worst and most putrid humors. If such humors are sparse in the stomach it fortifies the digestive heat. But if the superfluous food remains inside then it either corrupts the sanguine humor and thus continuous fever (*febris continua*) is caused, or the melancholy and thus quartan fever (*febris quartana*) is caused, or the phlegm and thus quotidian fever (*febris cotidiana*) is caused, or the choleric humor and thus tertian fever (*febris tertiana*) is caused.[5] But pay attention to the fact that the lion has all the time quartan fever in order to tame his ferocity, that is, to temper his irascibility and pride. As Galen says similarly, among diverse people diseases correct diverse faults (*vicia*) . . .

Pay attention that as the disease follows its course, it intensifies as the sun gets farther, as is clear in quartan fever, and when the sun gets nearer the disease abates. For it follows the movement of the sun, so that notwithstanding the different hours of the day, it affects the sick person rather at sunset than at sunrise. This is similarly the case in other diseases that follow the course of the sun. In the same manner it is possible to speak of the ruinous sickness of the soul that [is affected] by the sun of justice, that is Christ . . .

Source: Liber de introductione loquendi, book iv, Munich,
Staatsbibliothek, MS Clm 16126, fols. 1r–92v (at fols. 62rb–63vb).

The text reproduces the first two chapters of book four of what is known as *Liber de introductione loquendi* (sometimes known as *Liber introductorius loquendi in mensa* or simply as *Liber mensalis*, since almost half of the text is dedicated to table talk) compiled by the Dominican Philip of Ferrara (d. ca. 1350). Little is known of the author. Philip probably was born in Ferrara in the second half of the thirteenth century. He joined the Dominican Order in Faenza and had some theological training in the Dominican convent in Venice in 1307/08. In 1313 he was in the Dominican convent in Bologna, but never became a lector there. Subsequently, at an unknown time he was in Bergamo. Beyond the *Liber de introductione loquendi* he left behind a commentary on Peter of Spain's *Summulae logicales*. In the short prologue to the *Liber de introductione loquendi* Philip explains that he composed the book because of a deficiency he detected among clerics: their inability to utter edifying words (*edificatoria*) when they encountered other people. From Philip's text it is clear that he was obviously convinced that the duty of all clerics, friars and preachers in particular, was to use every encounter, whenever it happened, to educate the people, to forming the religious identity of the faithful, and consequently to furthering the cause of the Church and the Dominican Order. He divided his treatise into eight books, which instructed the friar what to say in different social circumstances. These were when the friar was invited to a meal; when he sat with other people at the fireplace after the meal; when he traveled with other people; when he visited a sick person, mourners, or a person vexed by any other tribulation; and when he wished to arouse friendly attitudes toward himself or to say something about vices or virtues. Most of the material Philip uses is unoriginal. His sources, sometimes explicitly stated, but very often hidden, include Gerard of Fracheto's *Vite fratrum predicatorum*, Humbert of Roman's *Liber de dono timoris*, Arnold of Liège's *Alphabetum narrationum*, *Golden Legend* of James of Voragine, Pseudo-Aristotle's *Secret of Secrets*, Giovanni da San Gimignano's *Liber de exemplis et similitudinibus rerum*, and Marco Polo's account of his voyage to the East. Philip of Ferrara's book reflects the growing tendency from the mid-thirteenth century onward, and especially among the mendicants, to set out knowledge and religious practices into practical manuals. From these largely formulaic collections the historian can sieve invaluable insights about attitudes, beliefs, sources of knowledge, and religious practices. By cautiously using these manuals, she or he can reconstruct parts of those precious moments of encounter between the learned cleric and the ordinary believer. The text translated above gives the first two chapters of *Liber introductorius loquendi* book 4, which is devoted to topics of conversation when the friar visited the sick. In addition to the specific humane encounter between the sick person and the friar, it illuminates the intricate relationship between religion and medicine in the first half of the fourteenth century. The translation is based on one of ten known manuscripts containing the text, which is unique in its kind in mendicant writing and which, on the whole, did not exert a great impact on ethical and courtesy literature in the author's time and beyond.

The churches, the convents, the papal court, and the teaching halls of the universities were not the only arenas of mendicant activity that imposed order on the society. The urban environment in which the friars lived together with their pastoral ideology inevitably created many opportunities for encounters with the faithful beyond the Sunday sermon or the schoolroom. Friars were invited to dine, traveled in the company of others, sat by the hearth; and they socialized with the faithful, visited the sick, and comforted the dying and the mourners. This was part of practical religion, which mingled teaching, preaching, and humane relationship. On all these occasions the friar appears as a teacher and a spiritual guide, but also as a compassionate friend and even as an entertainer; not as an austere guardian of moral mores remote from the society in which he lived.

How could the friar react and respond to the needs of the sick, suffering, faithful in search of answers and spiritual guidance? The first strategy that Philip adopts is to highlight the direct relationship between sin and disease, and to induce the sick man to confess his sins. There is nothing original about this strategy. The association between disease and personal sin had been a commonplace among Christians and Jews since late antiquity. The analogies between corporeal and spiritual diseases would naturally encourage the cleric to use the situation of visiting the sick in order to discuss the really important issue, namely the state of the patient's soul and the ways to cure it. The spiritualization of the disease situation made it possible to render the Christian history and way of life as more intelligible, concrete, and real. According to Philip in later chapters of the book, after the disease is over the body is weak and needs protection from the wind, for example. Similarly, Philip preaches extensively on how the soul should avoid the "winds of temptation." A disoriented sense of taste is a sign of disease (when one misperceives sweet for sour and vice versa); similarly, says Peter, when one is unable to distinguish between the taste of vice and of virtue, one is spiritually sick. The timing of therapy is as important for the success of the cure as the timing of penance for the success of spiritual cure. And in our text, the closeness or distance of the sinful believer from Christ and his grace has the same effect on the soul as the changing course of the sun has on the body of a fever-stricken patient.

But by preceding the suggestion that the patient confess his sins with the story of the real cause of disease, Adam's original sin, Philip introduces a unique touch to the conventional discourse about sin and disease. Canon 22 of Lateran IV was a locus classicus for this debate. Confessors' manuals of the late thirteenth and fourteenth centuries cited abbreviated parts of this canon or quoted it verbatim. These were important channels of communication through which the ideas and practices that were formulated in Church councils were being disseminated among the faithful. But there is something peculiar in the way Philip cites Canon 22. By the long introduction, which describes how disease entered our life as a direct result of original sin, Philip forces down the real message of the Canon, namely that individual sin is responsible for one's disease. The traditional nexus of sin-disease is elegantly turned into the nexus of original sin-disease, and the immediate cause of disease (the "formal cause" as he calls it) is natural, not spiritual.

Confession was only one of several religious practices that the friar could advertise on such an occasion. The encounter between the mendicant friar and the sick person could produce an attractive opportunity for the cleric to convince the sick person and his entourage to join the Christian cause of humility, repentance, confession, conversion, and the fortification of the Church's authority. The friar could use the opportunity to make new recruits for the Dominican order and guide the people to give alimony while alive rather than relying on mass and alimony that may or may not be given by relatives later. On all these topics Philip has suggestions to make, which he introduces in the form of short and sometimes entertaining exempla.

But the friar could enhance another aspect of his religious vocation, namely teaching the faithful the Holy Scripture and key doctrinal points. In this case concepts such as original sin, Adam's state of innocence before sin and his immortality, or key and complex questions debated by theologians and exegetes particularly in the thirteenth century, such as how Adam's body functioned before sin, or what would have happened to his body had he not sinned, could be simplified and summarized for the faithful.

Visiting the sick was therefore a locus of education and moral guidance, but it was also a platform for the cleric to show off his practical medical knowledge. From Philip's book it is clear that the friar was expected to speak about the physical causes of disease (fevers in the translated section), the signs of approaching death or recovery, and even to make prognosis according to the urine picture, the patient's sweating habits, his appetite and digestion, and his sleeping patterns. He could express his views about the importance of considering the state of the moon (in conjunction with the stars) for proper therapeutics and prognosis and then add to it a Christian interpretation. The *subtilis/perfectus medicus* who must be also a skillful astrologer is similar to God the father, who, like a wise phlebotomer, phlebotomizes and injures the right arm for curing the left arm (that is, introducing the healing powers of Christ's passion). All this was done with occasional references to medical authorities (Galen and Hippocrates) in order to enhance the credibility of his words. This name dropping did not reflect direct exposure to medical books, but an indirect knowledge through encyclopedias and other manuals.

The friar seasoned this activity with biblical quotations and canon law, with medical allusions, and excerpts from Pseudo Aristotle's *Secret of Secrets*. Hippocrates, Aristotle, the Bible, and canon law are all part of one body of knowledge, which if properly used can assure the person's health in body and in soul. Despite the opportunity that was offered to him, Philip neither rejected nor ridiculed conventional medicine, which is elegantly fused with spiritual medicine. There is no indication that the friar intervened in the actual physical cure, which certainly was not loathed by the cleric, and which certainly could be enhanced by the spiritual medicine that the cleric was willing to supply. Together with the overwhelming scarcity of official healing liturgies in the medieval Church, this suggests perhaps a clerical reluctance to interfere with the physical therapeutic activity, and a cooperative division of labor between clerics and medical practitioners at the patient's bed.

Notes

1. For the scholastic debate about Adam's immortality in terrestrial paradise, see J. Ziegler, "Medicine and Immortality in Terrestrial Paradise," in *Religion and Medicine in the Middle Ages*, ed. P. Biller and J. Ziegler, York Studies in Medieval Theology 3 (Woodbridge and Rochester, 2001), 201–42.

2. Canon 22 from the Ecumenical Council Lateran IV (1214), in *Constitutiones concilii quarti lateranensis una cum commentariis glossatorum*, ed. A. García y García (Vatican City, 1981), 68–69. English

version in *Decrees of the Ecumenical Councils*, ed. N. Tanner, 2 vols. (London and New York, 1990), 1: 245–46. See also Joseph Ziegler, *Medicine and Religion c. 1300: The Case of Arnau de Vilanova* (Oxford, 1998), 9–10, 212–13.

3. It would make more sense to speak here of "material cause," but the author insists that he is speaking of "formal cause."

4. Using Hippocratic and Galenic medical theory, the medieval physician thought that diseases, and most especially fevers, were caused by a disturbance in the four humors of the body: blood, melancholy, choler, and phlegm. The humors might become disturbed by a faulty regimen or a badly regulated life, that is, for example, by excess or deficit in eating or drinking, exposure to heat or cold, violent exercise, or unusual fatigue. In addition, humoral fluctuations might be associated with environmental and climatic variables. Philip's concise discussion conforms to the common medical notion of disease.

5. The four categories of fevers (which could be possibly identified with what we know today as malaria) are ordered according to the frequency in which the patient suffers from the attack of fever (normally lasting six to twelve hours). Quotidian fever is a daily attack of fever, while tertian and quartan fevers are attacks that take place every third or fourth day respectively. Philip accurately conveys here the Galenic belief that fevers might result from excess of blood, choler, melancholy, or phlegm, and fuses it with the insight from *Secret of Secrets* about the importance of occasional vomiting.

Further Reading

Michael R. McVaugh, "Bedside Manners in the Middle Ages," *Bulletin of the History of Medicine* 71 (1997): 201–23.

Nancy Siraisi, *Medieval and Early Renaissance Medicine* (Chicago, 1990).

Joseph Ziegler, "Medicine and the Body at the Table in Fourteenth-Century Italy: Book 1 of Philip of Ferrara's *Liber de introductione loquendi*," in *Between Text and Patient: The Medical Enterprise in Medieval and Early Modern Europe*, ed. Brian Nance and Florence Eliza Glaze (Florence, 2009).

─── 15 ───

How to Behave in Church and
How to Become a Priest

Daniel Bornstein

HOW TO BEHAVE IN CHURCH

ABOUT THE WAY ONE SHOULD ACT IN CHURCH, AND WHAT CHURCH ONE SHOULD
GO TO, AND IN WHOSE COMPANY, AND IN WHAT MANNER ONE SHOULD GO, AND
WHAT PRAYERS ONE SHOULD SAY AT VARIOUS TIMES

In the morning when you leave the house, you should first make the sign of the
holy cross and commend yourself to God, and then set out with measured steps
and proper bearing, saying this psalm, which is the 27th: "The Lord is my light
and my salvation; whom shall I fear?" [Ps. 27:1]. Entering the church, you shall
say that verse which is in my word, that is: "I will enter thy house, I will worship
toward thy holy temple in the fear of thee" [Ps. 5:7]. Then throw some holy water on your head and say: "My Lord and my Savior, by your grace may this water
be an ablution cleansing me of my every vice and sin." And thus you shall remain in your usual place in silence and dread, without looking here and there or
making any uproar the way some people do, beating their breast and sighing dis-
gracefully. You should direct your devotion to those churches in which you see
God praised most frequently and with the greatest devotion, and where there are
the most devout and wisest people. Then go with great devotion to hear the
mass; and when the priest approaches the altar, say: "My Lord, God most kind,
grant this priest the grace that he might perform to your praise and glory that
which he has come to do." And when he says the confession, you shall say it quietly with him. Then pray God for him, since he enters upon his office as our representative, to offer sacrifice to God for himself and for the entire people and to
implore for us his kindly mercy and grace.

When he says the *Kyrie eleison* (which is said nine times in memory of the
nine orders of angels which are in heaven, so that some from every order may

descend to that mass and enliven and move our hearts to devotion for the blessed Christ who is to come), then say that sweet song the angels sang when Christ was born to the Virgin Mary, that is: "Glory to God on high, and on earth peace, good will among men" [Lk 2:14]. And then you should fervently praise and bless the name of the Savior, and pray him that he may send peace from heaven to earth and maintain our hearts in his holy love and fear.

Next the epistle is read, and this is like a messenger who brings us the news that the Lord will soon come. Then you should say: "Lord God, you have sent to tell us that you shall come to us. I pray you that in your charity you might make me, an unworthy sinner, see with devotion your holy coming and desired presence."

When he says the gospel, which is the true messenger who cannot lie, stand up straight and sign yourself [with the cross] on the forehead, mouth, and chest, and remain with your head somewhat bowed. You should devoutly hear and understand what the gospel says because these are the words that our Lord said with his own mouth, that teach how we should lead our lives if we want to be in the mansion of so great a Lord. And with great reverence pray God that he may make you understand well and put into practice even better all that which he commands.

Then he says the Credo, which contains all the articles of the faith. And you should say it with him with complete devotion and reverence, and pray God that he may keep you in his holy faith, hope, and charity.

And then he turns and says: "Pray for me, brothers and sisters." And he says this because he goes on everyone's behalf to present our prayers and our needs to God. And we should pray for him and say: "May God, our most holy Savior, illuminate your heart and your soul. And may he give you the grace to be able to celebrate his most holy body to his praise, glory, and reverence, and for the salvation of your soul, and may he also give you the grace to be able to pray for us all."

Next the priest says: "Per omnia secula seculorum, sursum corda." That is, lift up your hearts to God on high. And we respond: "We have [lifted] them unto the Lord." We must so prepare our hearts that we do not lie to God. And then lift up your mind on high, and contemplate that blessed court with great humility, and pray God that he may send his son from heaven to earth on the altar so that we can see him with the mind's eyes, and that he may send his eternal blessing upon us all.

And then the priest says that sweetest song that the angels sing continuously and endlessly in God's presence, that is: "Sanctus, sanctus, sanctus." You should know that the angels then descend upon the altar to prepare the place where our lord Jesus Christ will come in the holy sacrifice. And then so dispose your heart that it burns with his love and say: "My most holy savior, I pray that you may not scorn to descend in the presence of me, a wicked sinner who has offended you so much."

When you see the priest bow before the consecrated host, you shall first say this verse: "Thou hast loosed my bonds. I will offer to thee the sacrifice of thanksgiving and call on the name of the Lord" [Ps. 116:16–17]. Then say: "I offer infinite thanks to you, most holy Savior, because you have deigned to make manifest your most holy son Jesus Christ, in pure and true flesh." And call upon his mercy and say: "I am guilty, my Lord, of that which I have done to offend you in the days of my life." And pray to him for all Christian souls, and especially for those who are in mortal sin and those who are in the pains of Purgatory.

And when the Lord is placed down and the chalice raised, you should say: "O most holy blood, which redeemed us sinners by that holy fluid, make me drunk with the love and fear of our most holy lord Jesus Christ, me and every creature that believes in you."

Source: Agostino Contò and Caterina Crestani, "Un testo quattrocentesco inedito: 'Del modo che si die tenere in chiexia,'" *Fedeli in chiesa*, Quaderni di storia religiosa 6 (Verona, 1999), 223–35; text translated on pp. 228–31.

HOW TO BECOME A PRIEST

THE CLERIC GIOVANNI DI NASCIMBENE OF MONSELICE STIPULATES A CONTRACT OF FIVE YEARS DURATION WITH THE PRIEST PIETRO, RECTOR OF SAN LORENZO, TO LEARN TO EXERCISE THE *ARS ET OFFICIUM CLERICATUS*

In the year 1340, eighth indiction, on Tuesday, August 8, in Treviso, in front of the town hall, in the presence of the priest Dom [. . .], rector of the church of San Leonardo of Treviso, the priest Dom Enrico, rector of the church of Sant'-Agnese, Niccolò, son of the late notary Ser Enrico of Selerio, and others.[1] The cleric Giovanni, son of the late Nascimbene, who was from Monselice and presently resides in Treviso, declaring and affirming that he is older than eighteen years and acts on his own behalf like a public merchant,[2] even swore on the holy gospels of God that there is no impediment to this opportunity by reason of being under age, et cetera, and undertook to stay, remain, reside, and enter upon the profession and duties of a cleric of God omnipotent, Father, Son, and Holy Spirit, with the priest Dom Pietro, rector of the church of San Lorenzo in Treviso, beginning on the said day and continuing for the next five years to come. The said priest Pietro promised to teach him the said worthy office of a cleric well, faithfully, and legally during that term, and before that term if he can, and to give and provide him with adequate food and clothing in his home so far as he is able. The cleric Giovanni promised, along with pledging his own goods for expenses and surety, to actually and personally stay and reside with him, and to observe and obey his instructions day and night, and not to separate from him nor steal any of his goods or furnishings, but rather to conserve and keep whatever of those goods come into his hands; and if he contravenes this, he promised to make good any damage which may be sustained et cetera.

As guarantor for the said cleric Giovanni in each and all of the aforesaid, Niccolò, son of Ser Ventura of Florence, who resides in Padua, promised to fulfill out of his own goods up to the sum of 100 small pounds et cetera, the which penalty et cetera.

Source: Giampaolo Cagnin, "'Ad adiscendum artem et officium clericatus':
Note sul reclutamento e sulla formazione del clero a Treviso (sec. XIV)," *Preti
nel medioevo*, Quaderni di storia religiosa 4 (Verona: Cierre Edizioni, 1997),
93–124; text translated on pp. 113–14.

Priest and parishioner alike had to learn how to conduct themselves in church, whether their place there was the altar or the nave. Laypeople normally learned what was expected of them by accompanying their parents and by observing and imitating others. However, various writers of devotional handbooks realized that there might be a market for advice on this subject as well, and so produced explicit written guidelines. One of the best-known examples is *Il soccorso dei poveri* (*Help for the Poor*), by the fourteenth-century Tuscan author Girolamo da Siena.[3] The forty brief chapters of Girolamo's handbook include several that treat the same topics covered in the text translated here, and in much the same terms: what church one should go to and how one should go there; how one should act in church; and what prayers one should say during the mass.

The instructions on how to behave in church are found in a manuscript that belonged to the convent of the Santo Spirito in Verona, but they clearly were not intended for cloistered nuns since they begin with setting out from one's home to go to church. Once there, the layman—the language of the text is gendered male—should observe respectful silence and refrain from looking around. In a society that kept women of good family confined at home, churches offered opportunities for social mingling, and moralists like this author constantly sought to keep the focus of attention (and of the worshipers' eyes) on matters of faith, often with little success. In one of her letters, the Florentine patrician Alessandra Macinghi Strozzi described looking over a prospective bride for her oldest son in church, an opportunity for careful inspection that Alessandra attributed to God's providential arranging.

Like Alessandra, the author of these brief guidelines is much concerned with comportment. He conceives of the good Christian in terms of worship rather than belief. He wants his reader to adopt and maintain a proper bearing: composed, restrained, and respectful. He assumes that conducting oneself in a properly reverential manner will put the worshiper in a properly devout frame of mind. Belief is expressed publicly, through words and gestures: going to church to witness the mass, join in collective prayers, and recite the Creed, and, when one leaves church, putting into practice the gospel message of peace and charity.

This handbook guides the reader step by step through the central act of Christian worship, the mass. It emphasizes the priest's role as intercessor between God and humankind and explains the significance of the prayers he offers on behalf of all the faithful. It calls upon the worshiper to pray as well, reciting the Creed along with the priest, psalms in response to the priest, and prayers in support of the priest. The worshiper's prayers and responses are given in the vernacular, whereas all scriptural citations, from the psalms and from the gospels, are in Latin. Our author, writing in the vernacular for a lay audience, assumes that his readers will be able to recite from memory, and perhaps even with a fair bit of understanding, short passages from the Latin Bible. By the same token, we might presume that regular worshipers, aided by long familiarity and a few words of guidance from handbooks such as this, would have understood much of the Latin liturgy—especially when they spoke a language derived from Latin. In regions that spoke Germanic languages, the Latin mass would have seemed far more opaque; and it is surely no coincidence that those were the regions that welcomed the Protestant Reformation, with its adoption of the spoken vernacular as the language of formal worship.

In this period before the Counter-Reformation creation of diocesan seminaries, priests, like parishioners, learned how to behave in church by observation and imitation. Priestcraft was a trade, learned like other trades by apprenticeship under an experienced master. The second document translated here is an apprenticeship contract from the diocese of Treviso, in northeastern Italy. In the standard fashion of such contracts, it specifies the reciprocal obligations of master and apprentice: the young aspirant to the priesthood promises to reside with his priest-mentor for five years, during which time this priest will provide him with food and clothing and instruct him in the profession and duties of a priest. The contract was drawn up by a notary, using the standard legal formulas (heavily abbreviated in this working draft) in standard legal Latin, and witnessed by a guarantor who stood surety for the priest in training. The only unusual feature about this contract for an apprenticeship at the altar is that it is signed by the apprentice himself, who has already reached the age of eighteen and entered minor orders; more typically, apprenticeship contracts were signed by the father who placed his son with a master artisan, such as a cobbler or mason, who would impart the skills of that craft to the young apprentice. The apprentice priest offers his services freely, expecting no compensation for five years of assisting at the altar other than his room, board, and clothing. In those five years, however, he expects to acquire sufficient familiarity with the solemn words and gestures of the liturgy to be able to repeat them correctly and appropriately, and with adequate understanding of their meaning. He can also expect to embark on his own ministry with a solid practical grasp of his other duties, gained from close and constant observations of the interactions between priest and parishioners in his master's "workshop," the church of San Lorenzo.

Notes

1. Dominus (translated here as Dom) was the honorific title given to a priest, as Ser was the title of a notary. The name of the first witness, the rector of San Leonardo, was left blank in the manuscript. The indiction refers to a fifteen-year cycle used in dating documents.

2. Canceled: and father of a family—a standard formula that the notary employed automatically before realizing the impropriety of applying it to a cleric vowed to celibacy.

3. Giuseppe De Luca, ed., *Scrittori di religione del Trecento* (Turin, 1977), 275–328, esp. 312–14. Originally published in Vol. 1 of *Prosatori minori del Trecento* (Milan and Naples, 1954).

Further Reading

Daniel Bornstein, "Priests and Villagers in the Diocese of Cortona," *Ricerche Storiche* 27 (1997): 93–106.

Katherine L. French, *The People of the Parish: Community Life in a Late Medieval English Diocese* (Philadelphia, 2001).

Peter Heath, *The English Parish Clergy on the Eve of the Reformation* (London, 1969).

R. N. Swanson, *Religion and Devotion in Europe, c. 1215–c. 1515* (Cambridge, 1995).

Churches, Parishes, and Daily Life:
Preaching

—16—

A Sermon on the Virtues of the Contemplative Life

Katherine L. Jansen

A SERMON OF FRIAR GIORDANO DA PISA, LECTOR OF THE
PREACHING FRIARS OF SANTA MARIA NOVELLA ON JULY 22, 1305,
THURSDAY MORNING ON THE FEAST DAY OF ST. MARY MAGDALEN
AT THE CHURCH OF ST. MARY MAGDALEN ON THE RIVERBANK

"MARY HAS CHOSEN THE BETTER PART"

One of the ways through which God is shown to be ruler and steward of the
world is that he gives to creation what it needs for self-fulfillment. That is, not
only did God become a man and come into the world, and undergo suffering and
death for the salvation of the human race, but he also gave the world his example and his saints, who are the way to approach salvation. Nonetheless, all examples of perfection are to be found first and foremost in Christ himself and in him
they were perfect. This is the first rule and the truth. There is one thing, however,
of which the Lord was not a perfect example (though indeed he was of everything else)—that which is the most necessary thing in the world. And what is
that? Penitence. Christ never made himself an example of penance because he
was unable to do so, although he gave himself as a true example of every other
perfection. The reason is this: What is penitence? To repent sin. In Christ there
was never any sin, so therefore he never repented sinning. Thus he was unable to
give you an example. Nor could his mother give you an example of penitence,
since like Christ she was pure. He also wanted his mother to be pure without sin.
Thus from them you have no examples, but in doctrine and in teachings you
have many as the Lord says and commands in the gospels. Do not think that
penitence is fasting, or wearing a hair shirt. These are the works of penance. For
if you call this penitence then Christ performed penance, as did his mother since
they were poor, they shunned the delights of the world, and they suffered many
trials and tribulations in this life. But he gave us a great abundance of penitential
examples in his saints, both men and women, such as in St. Peter, apostle, who

deeply repented his sins, and St. Mary Magdalen in whom Christ showed the
world the highest example of penitence, more perfectly than in any other saint.
Indeed one reads about St. Mary of Egypt and St. Paul, the first hermit, who were
great examples of penitence, but they ate herbs and other things. Whereas Mary
Magdalen for thirty-two years neither ate nor drank anything, except for angelic
food.[1]

True penance consists of three things: the first is to avoid evil; the second is
to choose good; the third is to persist until the end. There are two ways in
which the greatest wisdom of this life and true knowledge are demonstrated:
the first is knowing how to choose the lesser of two evils. Although every evil
must be avoided, nevertheless there are cases where it is necessary to accept
one. Wisdom is knowing how to admit the less wicked. The other is knowing
how to choose the better of two goods, the best one. Here is the sum of knowledge: that is, to know how to choose only the best, which is the flower of all
wisdom. I say, then, that you must choose the lesser of two evils when necessity constrains you to endure one. Such is the case of the sailor who, having
good fortune, perceives that death is at hand. He is compelled to throw his
merchandise overboard in order to survive; otherwise he would perish with it.
The wise sailor therefore chooses the lesser evil, and throws his merchandise of
silk, gold, and silver overboard and survives. This is true also for the doctor who
has a patient in his care with a gangrenous limb. So that the body does not die,
he will choose to amputate it.

Such was the case with Mary Magdalen, this wise and precious saint, who acted
in the same way as the sailor. She perceived fortune to be a danger of the world
and threw the merchandise and goods overboard; that is, her riches, which she
gave to the poor in order to be poor and to follow Christ.[2] She wanted to abandon
the world in order not to perish, since the things of the world are very harmful. As
she was a very rich woman, of royal stock who had lived a too comfortable and
delicate life, she fell into sin. You see then that the comforts of the world and too
much indulgence are the instrument and reason for much sin. Therefore, she
abandoned them and followed Christ so that by forsaking these things she saved
her soul, not forsaking the true goods. It would be better for the usurer to return
interest in order to save himself, so that at least before death he is rid of it and then
perishes. You see how Mary Magdalen knew how to choose the less harmful of
two evils. She chose the shame of this world, which is small. To escape the greater
shame and cruelty without end [damnation] she chose the smaller shame that
lasts only a little while, while you are confessing to the priest. This shame of the
world is small: you tell it to the priest, it goes away, and no one knows. But all the
world will know the other one since the shame of the sinner in the next life will
be endless. And since it will be great, and because it will be eternal, everyone will
know about it.

I would like to talk about this theme at greater length at another time; however, let me say just a few things about it now. Everyone has a lot of questions

on this subject: those who are good and those who want to be better, but they do not have the answers. Which way is the better way? Which way is more useful? And so everyone comes to me demanding, "Fra Giordano, which is better: fasting, saying the *Lord's Prayer*, or going on pilgrimage for indulgences? Or which is better: either giving alms, or serving in hospitals?" Everyone asks this sort of question. All these questions are answered by the teachings of Jesus Christ—by example and by word—that is his answer. By example he shows us in two sisters Martha and Mary Magdalen . . . two good lives, one is the active life, and the other is the contemplative life. Which of these two lives is it better to follow? The wise of the world before the coming of Christ also looked for the answer. What a wonderful thing that they had the same teachings as Christ! You see a great thing! So says the great philosopher Aristotle in his *Ethics*, which proposes these two lives: the life we call the active life he calls the civil and political life, the other he calls the intellectual life or the noble life. . . . All the works of all people even though they are diverse and variously discharged all boil down to these two: that is, the active life and the contemplative lives.

What is the active life? This life has many parts. The active life is giving food to the poor, clothing them, quenching their thirst, visiting the sick and those who are incarcerated and such things, and giving alms, and fasting, and wearing a hair shirt, and practicing abstinence, and such works. The active life is also marriage, managing the family, governing it, and teaching the commandments of God, ruling and directing people, and also defending the faith against infidels. The active life is also building bridges and hospitals, serving the sick and going on pilgrimage. Preaching is also part of the active life.

The contemplative life does not have so many parts, rather it has only one and that is to think about the creator, and to delight in him, either in beautiful songs, or thoughts, or in whatever way you can, and to study always how to grow in his love. This is the contemplative life. Which of these is it better to choose? Here is how the Lord answered this question. He said: "Mary has chosen the better part." If the Lord had not answered the question there would be much debate, but even today it is still a question for those who do not know Christ's teachings. And why does this question arise? It is born in the active life that seems to bear more fruit in comparison to the contemplative life. And therefore a heretic condemned virginity saying that marriage was better on account of the fruit which comes from it, and that many souls destined for heaven are born from it. He did not know what he was saying because even if I had a thousand children, they would do nothing for my salvation. Only my works and the love I have for God will save me. He still says: "but those children are born for God." But perhaps they are born for hell. This is clear and thus the heretic erred gravely.

It is true that active works are more visible. It appears to be a great thing to go on pilgrimage, or to serve in a hospital. Preaching seems to be a great thing and of great utility, and so it is as it is more visible than other things. The contemplative life does not seem to be so, but it is much better. The same philosophers judge it to be the case, as does Christ for the same reasons. And what are these reasons? The active life has four defects the contemplative life does not have and the Lord shows all these shortcomings when he says: "Martha, Martha, you worry and fret about so many things and yet few are needed, indeed only one. It is Mary who has chosen the better part, and it is not to be taken from her." All of them are shown here. . . . There is no active life without anxiety. How much difficulty will a man have who serves the sick? How much anxiety is there in going on pilgrimage? How much worry and turbulence in marriage and in running a household and in other things? You could not even begin to say. And you should know that there can be no anxiety without sin, or at least venial sin. But the contemplative life is not like that, there is no anxiety. Therefore, in the first place, the active life is less pure.

Another defect is the doubt that exists in the active life. In the active life one can sin mortally in many ways: A man comes and goes on pilgrimage to Santiago [de Compostela] and before getting there he falls into a mortal sin, or perhaps two, and then from there perhaps three mortal sins, or perhaps even more. Now what kind of pilgrimage is this? . . . What you ought to know is this: that whoever wants to receive indulgences, it behooves him to be pure, as if he were going to receive the sacrament of the body of Christ. . . . Now I counsel no one to go to St. Gall or on any other pilgrimages because they cause more damage than good. People go here and there mistaking their feet for God. You are deceived: this is not the way. Better to pull yourself together, think of the creator, and weep about your sins and the misery of your future than all the comings and goings you could make. Therefore another defect of the active life is the impurity, the doubt, and the anxiety.

The active life is not indeed necessary. This is what the Lord says: "few are needed, indeed only one." Do you want to see how one is necessary and the other is not? Does life require food, a horse, and such things? A horse is not really necessary because you can live without a horse, nor will you die since you can go on foot, but life would be more enjoyable with one. But without food I cannot live. Therefore I say to you these works of the active life are not necessary for my salvation, as a horse is not necessary for life. Thus fasting is not necessary to eternal life, nor hair shirts, traveling, nursing, or marriage, since without doing all these things I can have eternal life. This is shown by the example of the Magdalen. Without the contemplative life one cannot live; it is of necessity, otherwise you cannot attain salvation.

The other shortcoming of the active life is that it ends at death. Since in the eternal life there will be no need for hospital attendants, there will be no sick people, or poor, nor will there be preaching, nor will you fast, or anything else since it will amount to less. But the contemplative life does not amount to less because in the eternal life you contemplate God perfectly, much better than you do in this life. Rather it is the art of the angels who in heaven have no other art. This angelic art begins in this life and accordingly the holy men in this life have already had a foretaste, a part, a piece of that glory.

And therefore the height of the contemplative life is shown in that it is pure and free of doubt except in very rare cases as happened to a heretic, but this is very rare, it hardly ever happens. This life indeed is necessary, and without anxiety, and it does not amount to less, and it is not lost, as we have shown. Since the time that the Lord uttered those words it has not been necessary to defend it, but nonetheless to make it clear one tries to demonstrate it through these reasons and many others. Of the third shortcoming of the active life—that it comes to an end—we will say no more because we have already talked enough. It is best now to tell the story of Mary Magdalen.[3] Thanks be to God.

Source: Giordano da Pisa, *La vita attiva e contemplativa,*
ed. P. Zanitti (Verona, 1831).

The Thirteenth-Century Preaching Revolution

Alain of Lille (d. 1202), author of a celebrated manual on the art of composing sermons, defined preaching as "an open and public instruction in faith and morals, at the service of the faithful, which derives from reason and authoritative sources."[4] In the medieval world preaching was delivered in the form of a sermon, the primary vehicle for transmitting the basic tenets of the Christian faith, particularly religious and moral instruction. Moreover, in a society in which few people were literate, it was the predominant medium through which the Bible was imparted and explicated to the Christian faithful.

Although preaching to the lay community had always been a part of the Christian tradition, a duty that traditionally had fallen within the purview of the bishop, beginning in the twelfth century a circle of theologians and students who gathered around Peter the Chanter at the University of Paris began to reformulate the Church's pastoral mission in the world. One practical result of their work was to center moral instruction in the form of preaching at the heart of the Church's ministry to the laity, and one of the first to do so in this new spirit was the great preacher, Jacques de Vitry (d. 1240), whose sermons to the laity were gathered together and preserved in *ad status* collections, groups of sermons directed at specific audiences such as "merchants," "widows," "virgins," "peasants," "soldiers," and the like. They were, in essence, sermons tailored to the expectations of diverse demographic audiences.

The sermons of Jacques de Vitry stand at the beginning of what would become a preaching revolution due in no small part to Pope Innocent III (1198–1216), who had possibly studied with Peter the Chanter in Paris. Innocent's Parisian training marks his pontificate; it is characterized by an unprecedented concern for the spiritual welfare of the laity, which, he believed, could both be improved and protected through preaching. In his view, capable preachers could edify by teaching the word of God, but equally important they could counter the powerful oratory of the Cathars and Waldensians whose heretical preachers had begun to compete with the Church for the hearts and minds of the laity. The pope's belief that preaching could and should be used to combat heresy certainly informed his decision to approve Dominic Guzmán's new order, the Order of Preachers (commonly known as the Dominicans), while his oral approbation of Francis of Assisi's mission to exhort repentance was yet another use to which preaching could be put in the service of the laity. Significantly, Innocent III's solicitude about preaching eventually found legislative expression in Canon 10 of the Fourth Lateran Council (1215), which ordained that bishops must "recruit persons mighty in word and work, capable of fulfilling . . . the duty of holy preaching."[5] Although this decree did not directly license the new mendicant orders—the Dominicans and the Franciscans—it was certainly they who in subsequent years rose to the challenge of fulfilling the Lateran mandate by undertaking preaching missions not only throughout Europe, in North Africa, and the Holy Land, but also in the Far East.

But it soon became clear, especially to Dominic, that in order to preach persuasively systematic training was required. To that end, he endeavored to ensure that his nascent community be well trained in theology, a concern that eventually resulted in the emergence of a Dominican conventual school system dedicated to educating preachers. A minimum of three years of study under a lector was required of a friar before he was allowed to engage in any sort of preaching mission. Those who were being groomed for teaching careers were sent off to the *studia generalia* at the universities of Paris and Oxford to further their education. As our sermon demonstrates, the curriculum was firmly anchored in Aristotelian thought, the "new learning" that had been incorporated into the Dominican theology syllabus by the middle of the thirteenth century.

Giordano da Pisa

Our preacher, Giordano da Pisa (d. 1310), born in a little town called Rivalto just outside of Pisa, was brought up through the ranks in just this Dominican system. Giordano entered the Order of Preachers in 1279 at the convent of Santa Catarina in Pisa where he received his primary training. Between 1284 and 1286 he was sent to the university of Bologna, then Paris to continue his studies, and by 1287 when he returned to Italy he was commissioned as a conventual teacher, or lecturer. He discharged this duty in various convents in central Italy before arriving in Florence in 1303 to teach at the convent of Santa Maria Novella. In that same year he was named preacher-general, a preacher whom the Order had permanently licensed to preach. Giordano seems to have been a born charismatic preacher. According to one source, his eloquent orations—sometimes four or five times a day—would soften the hearts of men who then "turned their enmities into friendships, and many having renounced all vices, baptized themselves in this same Jordan [Giordano] with their tears, [and] changed their lives for the better."[6]

Giordano's Sermons: Audience, Language, Date, and Venue

From Giordano's sojourns in Florence and Pisa in the first decade of the fourteenth century we are extremely fortunate that over seven hundred of his sermons have survived, thanks mainly to a group of laymen, probably confraternity members, associated with the Dominican convent of Santa Maria Novella. They made reports (*reportationes*) or transcripts of the contents of his sermons as delivered "live," which accounts for their raw and unpolished quality but nevertheless has the advantage of bringing us into the proximity of the actual oral preaching event. Whereas most extant medieval sermons were preached in the vernacular but preserved in Latin—the language of the clerical elite who composed the sermons—Giordano's sermons were delivered and transcribed in the Italian vernacular, the spoken language of the laity. Significantly, Giordano's language and indeed the language of the preacher's recorders is the idiom of their great contemporary, the vernacular poet, Dante Alighieri.

As distinct from most medieval sermons, Giordano da Pisa's are unusual inasmuch as they are very firmly anchored in time and place. Thanks to his scrupulous recorders, we know where and when most of his sermons were delivered. For instance, we know that our sermon was delivered on 22 July 1305, on the feast day of St. Mary Magdalen, the most popular female saint of the medieval period (notwithstanding the Virgin Mary). For the feast days of those saints included in the Roman liturgical calendar, Giordano normally preached at their titular churches, although on most other occasions he usually preached at his own convent or confraternities associated with his convent. In the case of this sermon, the recorder tells us that Giordano preached it at the church of St. Mary Magdalen, a church that was once located on the bank of the Arno River in Florence, but now sadly no longer exists.

The Sermon: Form and Content

By the fourteenth century, preachers had begun to favor a new type of sermon called simply the *sermo modernus*, or modern sermon. In contrast to the old style of preaching (*modus antiquus*) which explicated the liturgical reading in its entirety, the modern sermon was an extended elaboration of one single *thema*, an individual line of scripture, drawn from the prescribed gospel reading for the day. Giordano's sermon on St. Mary Magdalen is decisively a modern sermon on the theme "Mary has chosen the better part" (Luke 10:42), one of the suggested readings for her feast day celebration.

Ostensibly Giordano's sermon is concerned with extolling the virtues of the contemplative life, a life characterized by solitary study and meditative prayer, the life normally associated with members of religious orders. A sermon on St. Mary Magdalen was the appropriate means for such teaching, as medieval thought interpreted the sisters Martha and Mary of Bethany as symbols of the active and contemplative lives, respectively.[7] Such an interpretation was based on the gospel passage in which the evangelist Luke (10:38–42) recounts Jesus's visit to Bethany. Whereas Mary sat at Jesus's feet rapt in contemplation hanging on his every word, Martha, preoccupied with domestic matters, busied herself at the hearth. While the contemplative life of Mary earned Jesus's praise: "Mary has chosen the better part which shall not be taken from her," Martha's active life elicited only a gentle rebuke from the Lord: "Martha, Martha, you are worried and distracted by many things." By associating contemplation with eternity (it "shall not be taken from her"), medieval commentators used Jesus's praise of Mary to glorify the contemplative over the active life. The active life, then, was deemed to be a lesser life, mired as it was in the mundane concerns of the secular world.

Nonetheless, it was the active life of Martha, her busy engagement with the world, not the contemplative life of Mary that most typified the lives led by Giordano's most faithful auditors, members of the *popolo grasso*, the wealthy merchant class that had dominated both Florentine politics and society since the late thirteenth century. This was a fact, of course, not lost on the preacher whose examples of the merchant-sailor, the doctor, and the usurer were calculated to appeal to his listeners' social experience. But more than Giordano's edifying stories, it is his vivid and detailed descriptions of the active life of the laity, particularly their devotional practices, that provides a window onto this urban bourgeois society. Although Giordano refers to this way of life as the active life, we might more readily identify it as a life devoted to good works and charity, characterized by works of penance. For the Dominican preacher the works of penance such as fasting, practicing abstinence, mortifying the flesh by wearing a hair shirt, and pilgrimage—all penitential devotions clearly practiced by Giordano's audience—were to be distinguished from true penitence, or repentance of one sins, for which St. Mary Magdalen provided the paradigmatic example. In Giordano's opinion, such devotional practices, while good, constituted a "lesser good" in comparison to true repentance, that exemplified by St. Mary Magdalen.

No less interesting than the penitential practices of the urban laity were their social and charitable activities, enumerated at length in this sermon. Laypeople were clearly carrying out the corporal works of mercy, which in this period stabilized at seven.[8] But they were also defending the faith, teaching the Ten Commandments, nursing the sick, governing the city and their families, and building hospitals and bridges, all activities that illuminate a strong tradition of civic religious spirit as it was lived and practiced by the laity in early fourteenth-century Florence.

Notes

1. Giordano is referring to legendary material about Mary Magdalen such as that included in Jacobus de Voragine's *Golden Legend* (trans. William Granger Ryan, 2 vols. [Princeton, 1993]), the most widely used compendium of saints' lives compiled circa 1260. See vol. 1: 374–83. The lives of St. Mary

of Egypt and St. Paul of Thebes, two important "desert saints," are also contained in the *Golden Legend*. See vol. 1: 84–85 and 227–29.

2. Jesus's injunction to his followers to renounce riches (Matt. 19:21) was a text of paramount importance for those religious orders such as the Franciscans, Dominicans, and Humiliati, who were committed to living in voluntary poverty.

3. To conclude the sermon, Giordano recounts the legend of Mary Magdalen, most likely drawn from Jacobus de Voragine's *Golden Legend*.

4. *Ars praedicandi*, in J.-P. Migne, *Patrologiae Cursus Completus: Series Latina* CCX, cols. 109–35; translated as *The Art of Preaching*, trans. Gillian R. Evans (Kalamazoo, MI, 1981).

5. Norman Tanner, ed., *Decrees of the Ecumenical Councils* I (Washington, DC, 1990), 239–40.

6. *Cronica antiqua conventus Sanctae Catharinae de Pisis*, in *Archivio Storico Italiano*, ser. 1. vol. 6, pt. 2 (Florence, 1845), 451.

7. Giordano is assuming that the scriptural figures Mary Magdalen and Mary of Bethany are one and the same person, an assumption that had its origins with Pope Gregory the Great in the sixth century and that continued to structure thinking about the saint throughout the Middle Ages. For a more extended treatment of this subject, see Katherine Ludwig Jansen, *The Making of the Magdalen: Preaching and Popular Devotion in the Later Middle Ages* (Princeton, 2000).

8. Giordano lists "giving food to the poor, clothing them, quenching their thirst, visiting the sick and those who are incarcerated." The other two unmentioned corporal works of mercy are giving hospitality and burying the dead.

Further Reading

John Baldwin, *Masters, Preachers, and Merchants: The Social Views of Peter the Chanter and his Circle*, 2 vols. (Princeton, 1970).

Brenda Bolton, *Innocent III: Studies on Papal Authority and Pastoral Care* (Aldershot, 1995).

Carlo Delcorno, "Medieval Preaching in Italy (1200–1500)," trans. Benjamin Westervelt, in *The Sermon*, ed. Beverly Mayne Kienzle, Typologie des sources du moyen âge occidental, fasc. 81-83 (Turnhout, 2000), 449–560.

John Henderson, *Piety and Charity in Late Medieval Florence* (Oxford, 1994).

Daniel R. Lesnick, *Preaching in Medieval Florence: The Social World of Franciscan and Dominican Spirituality* (Athens, GA, 1989).

Lester K. Little, *Religious Poverty and the Profit Economy in Medieval Europe* (Ithaca, NY, 1978).

M. Michèle Mulchahey, *"First the Bow Is Bent in Study . . .": Dominican Education before 1350* (Toronto, 1998).

Augustine Thompson O.P. *Cities of God: The Religion of the Italian Communes, 1125–1325* (University Park, PA, 2005).

André Vauchez, *The Laity in the Middle Ages*, trans. Margery J. Schneider (Notre Dame, IN, 1993).

—— 17 ——

Preaching and Pastoral Care of a Devout Woman
(*deo devota*) in Fifteenth-Century Basel

Hans-Jochen Schiewer

"OF THE HOLY SCHERERIN"

A person was lying there who was in the throes of death. His entire senseless life appeared before him. And the person saw his soul standing there naked and bare and was very sad that it should part and go, and did not know whereto. Thus, body and soul were in great suffering. Thereupon, a voice came in a great light and spoke to him: "May God clothe your soul with his suffering, then it will stand honestly and chivalrously and well dressed before him" [. . .] That person then came to, and our Lord gave him another x years to live [lines 2–7, 27].

She was a married woman who asked her husband for permission to lay down worldly clothes. He agreed and she started to live a pious and spiritual life. Our Lord sentenced her to great sorrow and distress in many different ways, but she overcame all suffering with our Lord's force and fled to him. And our Lord looked at her with pitiable eyes and was full of mercy and sweet comfort. She got a great desire and longing for the holy sacrament and with the permission of her father confessor often partook of it. Her father confessor was told that he permitted her to partake of it too often, since she was a married woman. And her father confessor told her that there was a rumor about, and what he was told. She became very sad and thought that the evil will prevent her from this good deed and blessing, which she got from the holy sacrament. She said: "You know me best from confession and you know my virtuous lifestyle, my peaceful behavior with neighbors and the permission I got from my husband to live a spiritual life, that you are not doing wrong to allow to partake of the holy communion." She said also: "If there are no obstacles, our Lord makes me so eager for the holy communion, that I nearly die." The father confessor said: "You don't die." And he talked to her, that she shouldn't partake that often, and forced her to obey his recommendations which would bless her. The father confessor thought to try out her obedience, and she was convinced that she would be able to follow his advice. But she came into a big heartbreaking suffering and longing. But the desire was so overwhelming that it became life threatening. Her friends became frightened and asked the father confessor to come as quick as possible, because they thought she was going to die. He came and saw her suffering and throes of death, because she had such a longing to the holy sacrament. It gave him quite a fright. He started to comfort her and said: "I don't do it again." He gave her every permission to partake of the holy sacrament as he did before. Immediately she felt much better through the good comfort, and her will to live returned.

From time to time she told her father confessor some beneficial secrets, which our Lord quite often revealed to her. One great goodness and grace happened to her in the church of the Franciscans. Her father confessor was preaching on the steps of the Franciscan church in front of the choir. He said: "When a person would like to receive the holy sacrament and is prevented from doing so and refrains from it obediently, then he shall receive the flesh and the blood of our Lord spiritually." And as he said this in the same moment, she saw a cross floating before her with vj wings that were formed like a cross, which Saint Francis had seen when the five signs of love were impressed into him. And the cross floated before the altar of Saint Paul. And she was sitting across from it before the altar of the martyrs and saw that four rivulets of blood ran from the five signs of love, two from the hands and one from the feet and one from the side. And the four rivulets of blood came together in front and became one trunk or rivulet and ran into her mouth. And she felt the feeling and taste of the blood in her mouth and was very much strengthened from this drink of love and cheered by it [lines 28–73].[. . .]

She was very hungry and desired very strongly to have the holy sacrament and often went there daily and always with the permission of her father confessor. And when she did not go there for reasons of good behavior, she was often most sweetly refreshed and strengthened by our Lord. And this strengthening was thus that she felt in her mouth the worthy blood of our Lord, tasting with the same taste and strength that she had also had during the sermons of her father confessor, when the liquid of the five signs of love had been poured into her mouth. She often felt the blood and the taste of the blood in her mouth, and this also strengthened her. Because the hunger she had for the holy sacrament was so great that had she not had the refreshment and the strengthening, she might have died from the hunger she felt for the holy sacrament. But if there was any obstacle, especially through her husband, that forced her not to go to the Holy Communion, she didn't get the vital refreshment and strengthening. Then she became very hungry, and if the father confessor didn't allow her to have the holy sacrament, she felt deadly sad and injured [lines 86–101].[. . .]

Once at Pentecost she really desired to receive our Lord, but she didn't dare to do so because of obedience and discipline, but our Lord did her a great blessing. She started to pray and to reflect on the holy sacrament with humble calmness. She considered the dignified passion of our Lord and her soul was enraptured and sat beside the crucified Christ close to his heart. And our Lord's heart burst open and the divine blood poured into her mouth. As she regained consciousness, she thought to herself: "Great things have happened to you. You drank from the heart of our Lord. Thus, Saint John has appeared to you. He was refreshed from the heart of our Lord and became very wise. What did you drink? I drank our Lord's will. I immersed on the bottom of his will. What did you drink? I drank that my heart will happily burst open through divine love as our Lord's heart burst open through love. What did you drink? I drank: 'My soul doth magnify the Lord, And my spirit hath rejoiced in God my Saviour'" (Luke 1:46).

During the great dispute going on at the time, she asked our Lord that he should give a sign of what was to be done in order to end it. A voice spoke and said: "This is a difficult matter. But if the clerics and the laymen better themselves, then everything shall be smoothed. Let all those who love the Lord pray to him!" And the same voice also spoke to her: "The rule of the Order shall no longer be harmed. Although one must suffer costs and sorrow, they shall be gratefully received." She told her father confessor. He thereupon told her that they had suffered great sorrow and costs because of the rule, and that Mulberg was one of the reasons. She said, "Lord, I did not know that there was distress or sorrow concerning the rule or that I was to pray for them, and nonetheless, the voice spoke to me of this." Afterward, she thought of how this should be made known to those who love God so that they might pray to our Lord. And for this reason, she was in great distress and asked our Lord that he might give her instructions. And a voice sounded and said, "Tell your father confessor that he make this affair his own." She told this to her father confessor, and he sent her together with two other persons, two distinguished persons, and told her to ask that the clerical women pray, and the priests in the order also, to our Lord for the dispute. And they did this with great diligence, and great prayer was held, and many masses were sung by the order. And the great and harsh dispute was reconciled.[. . .]

Once she went to a spiritual community, which was in great sorrow and distress, and she was going to comfort her "spiritual children." She had great pity on them and our Lord appeared to her and (242v) stood before her bare and naked and was full of wounds all over his body and he looked at her most pitiably. She hid her face and did not dare look at him fully, as she would have liked to, since she thought, that she would feel a great weakness and wouldn't be able to go home, because the way home was remarkably long. And she thought, that our Lord showed mercy with this community, because he appeared to her in a pitiable manner, as if he would be full of mercy showing himself full of wounds and full of pity [lines 262–71].

Source: Staatsbibliothek zu Berlin—Preußischer Kulturbesitz, Haus 2,
Ms. germ. 4° 191, fol. 237v–243v and 276r–277v. The manuscript belonged
to the library of the convent of St. Nikolaus in Undis in Strasbourg.

This passage is a unique devotional biography, written in the third person, about the experiences of a woman who devoted her life to God—*deo devota*—the Schererin, the female appellation for the wife of a man named Scherer. It offers a view rare in its intimacy of the religious life of a highly motivated laywoman, and the many religious experiences—private and public—that contributed to her religious formation. The unique feature of the text is in describing the mundane and daily problems of a *deo devota* in a late medieval town like Basel.

The Schererin's revelations are introduced as phenomena that followed a revelation of God's grace to her as a refulgence of light. This led her to decide to live a life of a devout married woman, like a beguine; she sought her husband's permission to remove her worldly clothes and don a habit. He allowed her to do so. At a later time, we learn that she had a daughter whom she wished to direct toward a life of chastity or enclosure. She received a vision that prophesied that the daughter would not be able to keep her virginity. Her husband appears little in the text, yet the text hints at the conflicts that arose when a married woman chose a chaste life. Her husband is mentioned in connection with the prohibition to partake of communion immediately after sexual intercourse. For life within the family, as *deo devota in matrimonio*, did not release the wife from her sexual marriage debt. Quite the contrary, the father confessor actually admonishes the Schererin to reduce the frequency with which she partakes of the communion in order to not force her husband to go elsewhere.

The Schererin's life is marked by a physical and spiritual dependence on the Eucharist, which led to discussions with her father confessor since this dependency on the communion and the frequency with which she partakes of it lead to public complaints. The fact that the Schererin's Franciscan father confessor satisfied her desire for the sacrament by admitting her to communion frequently was not only criticized within the context of the present text, but was also a point of criticism that was raised against the Franciscans in the context of the Basel beguine dispute (1400–1411). The father confessor reacted to the public outcry the Schererin's behavior had caused by more restrictively handling the communion. He recommended to the Schererin that she partake of communion only on Sundays or at least not more often than thrice weekly. As a substitute, he recommended spiritual communion when the host is raised. The consequence of this behavior is a conflict of obedience that leads to life-endangering physical states in the Schererin, as the deprivation of the host causes her to be weakened and to suffer outbreaks of cold sweat.

The environment of the parish church and the sermons heard in it affected the shape of the Schererin's visions. The imagery of the Franciscan church in Basel may have inspired the vision of stigmatization, following that experienced by St. Francis. The visionary imagery of the Schererin was affected by the furnishings of churches and by private devotional objects such as crucifixes or *martel bilde* (pictures of the crucified Christ), which are repeatedly mentioned.

The eucharistic visions refer exclusively to the blood of Christ and thus complement the communion, which is limited to the host. In the visions the Eucharist was experienced in both kinds. References are frequently made to the taste of Christ's blood, also in the church itself. As a matter of fact, it is a prerequisite for the revivification of the Schererin who cannot leave the church because she feels so weak.

The revelations of the Schererin also contain a political point of view, which refers to the conflict the Dominican Johannes Mulberg initiated with the beguines and members of the Franciscan Third Order in Basel. In the year 1400, Mulberg and the Master of the Basel cathedral school, Johannes Pastoris, preached against beguines and beghards who wrongfully assumed clerical status and lived by demanding alms. These accusations are also valid for those living *secundum tertiam regulam sancti Francisci* according to the Third Rule of St. Francis). The Franciscan lector Rudolf Buchsmann objected to these accusations, also speaking from the pulpit. Christ had given permission to ask for alms, the imitation of Christ entailed the renouncement of all material goods and to take up a life of begging, and the adherents of the "beloved begging poverty" stand for the most noble evangelical truths; this should be compensated in like manner as physical work. The conflict apparently simmered for five years until Mulberg, supported by the Augustinian Hermits, took up preaching against the beguines again, in 1405, in Basel as well as in Strasbourg. After this explosion of religious feeling, events were moved to the papal court, where Mulberg struggled during the following four years in order to enforce his goal. In 1409, the papacy revoked all measures decreed against members of the Third Order and against Franciscans.

The structure of the text is simple: the visions follow one another consecutively, generally joined to each other by *einest* (once). An everyday context is maintained by mentioning the churches of Basel, the intercessions for the husband, worries concerning the only daughter, the yearly donations, self-stylizations as *einer armen hantwerg frowen* (a poor craftswoman). The revelations are rendered even livelier by their dialogic character: the Schererin again and again addresses God and her father confessor directly.

It has not been possible to identify the Schererin; she is not the Basel beguine Adelheid Schererin who is recorded in the year 1376, since that woman already lived in separation from her family, as a candle maker. Any connection to the Guardian of the Basel Franciscans, Johannes Scherer, who was in office between 1405 and 1410, must remain a question still to be solved. Further proof of the family name is to be found for this time in the book of records of the city of Basel, although for the time being no connections can be drawn to our Schererin. However, archaeological studies of the Franciscan church in Basel suggest that an inscription in the square adjacent to the church, noted in an eighteenth-century ledger: "*dies • Grab • ist Iohans • Scherer • und siner Wirtin*" (*this is the grave of John Scherer and his wife*), may indicate that the Schererin was granted her wish of being buried near the church that was so central in her religious formation.

Further Reading

Nancy Caciola, *Discerning Spirits: Divine and Demonic Possession in the Middle Ages* (Ithaca, NY, 2003).

Jeffrey F. Hamburger, *St. John the Divine: The Deified Evangelist in Medieval Art and Theology* (Berkeley, Los Angeles, and London, 2002).

Miri Rubin, *Corpus Christi: The Eucharist in Late Medieval Culture* (Cambridge, 1991).

Hans-Jochen Schiewer, "Auditionen und Visionen einer Begine. Die 'Selige Schererin,' Johannes Mulberg und der Basler Beginenstreit," in *Begegnungen mit dem Mittelalter in Basel*, ed. Simona Slanicka, Basler Beiträge zur Geschichtswissenschaft 171(Basel, 2000), 55–90.

Hans-Jochen Schiewer, "German Sermons in the Middle Ages," in *The Sermon*, ed. Beverly Kienzle, Typologie des Sources du Moyen-Age occidental 77–78 (Turnhout, 2000), 863–961.

Churches, Parishes,
and Daily Life: Confession
and Penance

—18—

Doing Penance

Sarah Hamilton

The ordo for giving penance begins:

Priests, when they receive the confessions of the faithful, ought to humble themselves, and in sorrow, with sighs and tears, they ought to pray not only for their own faults but also for the cause of their brethren. For the Apostle says: Who is weak and I am not weak? When therefore anyone should come to the priest to confess his sins, it is enjoined on him that he wait a little while as the priest enters the church or his closet to pray. And if there is no appropriate place to do this, (he) the priest should say in his heart this prayer:

Prayer:
   Lord God almighty be gracious to me, a sinner, that I may be able to give thanks to you worthily, who through your mercy has made me, although unworthy, a minister by the sacerdotal office, and set me up, although slight and lowly, as a mediator to pray to and intercede with our Lord Jesus Christ, your son, for sinners and those returning to penance. Therefore Governor, Lord, who wills all men to be saved and to come to the knowledge of the truth, who wishes not the death of sinners, but that they should be converted and live, receive my prayer, which I pour out in the presence of your mercy on behalf of your manservants and maidservants who come to penance, that you give them the spirit of compunction, that they be mindful of the snares of the devil, by which they are held tightly, and return to you through a worthy satisfaction. Through our Lord Jesus Christ . . .

Then those who ought to confess their sins are introduced; when they are prostrate on their faces let the priest say:

   Make haste, O God, to deliver me; make haste to help me O Lord.
   Glory be to the Father . . . (and to the Son and to the Holy Spirit)
   As it was in the beginning is now and ever shall be.

Praise to you Lord.

(Lord) rebuke me not in thine anger, neither chasten me in thy hot displeasure . . . (Psalm 6)

Blessed is he whose transgression is forgiven, whose sin is covered . . . (Psalm 31)

Lord rebuke me not in thy wrath, neither chasten me in thy hot displeasure . . . (Psalm 37)

Have mercy upon me, O God, according to thy loving kindness: according unto the multitude of thy tender mercies blot out my transgressions . . . (Psalm 50)

Hear my prayer, O Lord, and let my cry come unto thee . . . (Psalm 101)

Out of the depths have I cried to thee O Lord . . . (Psalm 139)

Hear my prayer O Lord, give ear to my supplications: in thy faithfulness answer me and in thy righteousness . . . (Psalm 142)

Christ hear us. Christ give ear to us. Christ have mercy on us. God hear our voices. God hear our prayers.

Holy of holies, Lord have mercy on us.

God of the angels have mercy on us.

Lord of all the blessed heavenly spirits have mercy on us.

God of the patriarchs have mercy on us.

God of the prophets have mercy on us.

God of all the holy innocents have mercy on us.

God of the apostles have mercy on us.

God of the martyrs have mercy on us.

God of the Levites have mercy on us.

God of the just have mercy on us.

God of the doctors have mercy on us.

God of the confessors have mercy on us.

God of priests have mercy on us.

God of monks have mercy on us.

God of virgins have mercy on us.

God of widows have mercy on us.

God of all the saints have mercy on us.

God whole and beyond all understanding have mercy on us.

God invisible and ineffable have mercy on us.

God laudable and merciful have mercy on us.

God patient and powerful have mercy on us.

God great and powerful have mercy on us.

God blessed and just have mercy on us.

God most high and magnificent have mercy on us.

God living and true have mercy on us.

God holy and splendid have mercy on us.

God great and terrible have mercy on us.

God powerful and faithful have mercy on us.

God pious and glorious have mercy on us.

God patient and clement have mercy on us.

God just and upright have mercy on us.
God the Father and Son and Holy Spirit have mercy on us.
God of all who adore you have mercy on us.
God of all who put their trust in you have mercy on us.
God of all who cry out to you have mercy on us.
God of all who hope in you have mercy on us.
God of all who praise you have mercy on us.
God of all who fear you have mercy on us.
God of all who bless you have mercy on us.
God who listens to all men have mercy on us.
St. Mary pray for these penitents.
Holy Mother of God pray for them.
Holy Virgin of Virgins pray for them.
St. Michael pray for them.
St. Gabriel pray for them.
St. Raphael pray for them.
All the holy orders of heavenly spirits pray for them.
All the sainted patriarchs pray for them.
All the sainted prophets pray for them.
All the sainted apostles and evangelists pray for them.
All the sainted martyrs pray for them.
All the sainted confessors pray for them.
All the sainted virgins pray for them.
All the sainted widows pray for them.
All God's saints, male and female, pray for them.
Kyrie eleison.
Our Father (who art in heaven) . . . And lead us not into temptation but deliver us from evil. (three times)
Save your servants and maidservants, God of mercy.
Send them help from the sanctuary. *Response.* And strengthen them out of Zion.
Be unto them O Lord a tower of strength. *Response.* In the face of the enemy.
May the enemy gain nothing over them. *Response.* And the son of iniquity have no success with them.
Lord protect them from all evil. *Response.* May the Lord protect thy soul.
Lord protect them coming in and going out. *Response.* Now and for ever more.
Lord hear my prayer. *Response.* And let my cry come unto thee.

Prayer.
God whose indulgence we all ask, be mindful of our servants who are exposed because of the slippery frailty of earthly bodies, give pardon, we beg, to those who confess, spare the suppliants so that those who are condemned by what they deserve may be saved by your mercy through our Lord Jesus Christ.

Another.

God of mercy, God of piety, God of kindness, be kind, I beg, and merciful to these your servants, that of the faults which have been orally confessed to us in the presence of this your holy altar, no stain remain in them on earth and through the sorrow of penitence may they merit to come to your promise. Through the Lord . . .

Another.

Lord under whose eyes the hearts of all tremble, and the consciences of all shake, look with favor on the laments of all and heal the melancholy of them, so just as none of us is free from guilt, so no one should be alienated from forgiveness. Through the Lord . . .

Another.

God of infinite mercy and immense truth, look with favor on our iniquities and heal the melancholy of all our souls, so that having obtained forgiveness of our iniquity, we may always rejoice in your blessing. Through the Lord . . .

These prayers having been completed the priest and those who are with him should rise. Then as the priest sits in front of his altar, with the cross of the Lord next to him on the right hand side, each of them should come to him individually, and humbly kneel before him, while the priest, looking closely, soothingly and pleasantly questions him saying:

Son, do you believe in God the Almighty Father? *Response*: I believe.

Do you believe in Jesus Christ, his only Son, our Lord, who was born and suffered? *Response*: I believe.

Do you believe in the Holy Spirit, the Paraclete? *Response*: I believe.

Do you believe that these three persons, Father, Son, and Holy Spirit, are three persons and one God? *Response*: I believe.

Do you believe that you will be resurrected in the flesh, in which you now are, on the day of Judgment and be received as either good or bad, depending on what you did? *Response*: I believe.

Do you have a complete desire to amend what you have done against God? *Response*: I have.

Have you faith that God can forgive all your sins, if you will confess and convert to him with your whole heart? *Response*: I have.

Do you wish to forgive all those men who have sinned against you for says the Lord: if ye forgive men their trespasses, your heavenly Father will also forgive you.

Then the priest prostrates himself with the penitent on the ground calling to the Lord and saying:

Kyrie eleison.

Our Father (who art in heaven) . . . (3 times)

The Lord hear thee in the day of trouble. *Response*: The name of the God of Jacob defend thee.

Send thee help from the sanctuary. *Response*: And strengthen thee out of Zion.

Grant thee according to thine own heart. *Response*: And fulfill all thy counsel.

The Lord fulfill all thy petitions. *Response*: Now know I that the Lord saveth his anointed.

The Lord shall preserve thee from all evil. *Response*: He shall preserve thy soul.

The Lord bless thee out of Zion. *Response*. (The Lord) who made heaven and earth.

Hear my prayer O Lord. *Response*. And let my cry come unto thee.

Prayer.

To you God the Father, almighty and merciful, who wishes not the death of sinners, but that they be converted and live, who through your servant, Lord, you return our lost sheep on your own shoulders, we plead to you, humbly, that you confer your forgiveness on this your servant, who has converted to the right life, and wipe him clean so that the wounds of all his faults which he incurred after baptism may be obliterated and healed through pure and true confession so that no sign remains in him. Through . . .

Another.

Lord hear the prayers of your supplicants and spare the sins of those who confess to you, so that those whom the guilt of conscience condemns, the indulgence of your piety may absolve. Through . . .

Another.

Lord holy father, almighty and eternal God, who through Jesus Christ your Son, our Lord, deigned to cure the wounds of sin, I humbly ask and petition you, that you deign to turn the ear of your piety to our pleas, so that moved by the confession of this penitent, you forgive him all his sins and pardon all his faults. Grant, we beseech you Lord, forgiveness to this your servant, rejoicing instead of grief, life instead of death, so that he who fell into ruin at the devil's persuasion may at your summons be drawn to mercy and because he has been recalled to the hope of the heavenly summit, trusting to your mercy he may merit to come through your help to salvation and eternal life. Through . . .

These prayers having been said, they should both rise up. Then the priest immediately indicates to the penitents kneeling before him the fast and alms and sacerdotal prayers as shall seem best to him. Then he should get up and say to them humbly and kindly:

Christ the Almighty God who said those who confess to me before men I will confess in the presence of my father, may he bless you and give you remission of all your sins. Amen.

May the Lord Jesus be mindful of your salvation, to pour upon you the supernatural gifts of virtues so that the enemy of mankind may not lead you astray by any deceit. Amen.

May Lord Jesus Christ protect you, who slew the devil and freed us all from the torments of sin so protect you that you may be worthy to attain the kingdom of heaven in the blessing of Christ. Amen.

May the Most High Lord protect you all the time of life. Amen. And may he defend you from all danger in this world and the world to come. Amen.

May the Lord turn his face to you and give to you peace through all the days of your life. Amen.

May the Lord himself sanctify you wholly so that you may be humble and perfect and and your whole body and spirit be preserved blameless unto the coming of Our Lord Jesus Christ who lives. . . . Amen.

May the Lord give you an angel, the guardian of peace, may he govern your heart in this world and in the next and purify your soul from all faults. Amen.

May almighty God bless you and through the abundance of the Holy Spirit strengthen your mind, sanctify your life, make you as chaste as angels, embellish your senses and mold your good works, bestow favor and grant peace, confer salvation, nourish repose, fortify charity and by his everlasting protection and virtue protect you from all attacks, diabolic or human, and mercifully grant whatever you have demanded from him, remove all the evil you have done and give you the grace which you have always desired with the support of the Lord.

Then the priest should tell him to stand up and immediately, having put on the stole, and taking the right hand of the penitent give him remission saying:

Through this true and pure confession which you made just now to me, a sinner but a priest of Christ, may almighty God absolve you from all judgments, which ought to be given you for your sins, according to the multitude of his mercies which are of old, and spare and delete all your sins and lead you to eternal life. Amen.

And I a priest of Christ through the intercession of the Blessed Peter, prince of the Apostles, to whom God gave the power to bind and to loose and through him this same power was given the bishops and priests of Christ, and according to my ministry I absolve you from all judgments, through which I have bound you for your sins, except, however, the fasts, alms or priestly prayers which I indicated and imposed on you a short while before, and if during this confession and penance you die, before you come to another confession, may Christ the divine son be merciful to you and may you remain absolved permanently in eternity. Amen.

Then the priest says:

Son I commend you to God and to this sign of the cross of our Lord Jesus Christ.

Then, with joined hands, the penitent offers himself at the feet of the cross. Meanwhile let the priest say over him:

Through this sign of salvation, the cross of our Lord Jesus Christ, and through his intercession and that of these and of all the saints, may the Almighty God be merciful to you and give you true humility, perseverance in (doing) good, and forgive you all your sins, past, present and future and free you from the snares of the devil and lead you into eternal life. Through . . .

At the end say to him:

Make the sign of the Cross against the devil and against all his temptations.

Meanwhile while he makes the sign, say to him:

May the peace and blessing of the Almighty Father and Son and the Holy Spirit descend and stay over you now and in perpetuity and may the good angel of the Lord guard you and defend you from all evil here and everywhere, now and always, with the help of our Lord Jesus Christ who lives and reigns. Amen.
    Son, go in peace.

> Source: *Die Bussbücher und das kanonische Bussverfahren nachhandschriftlichen*
> *Quellen*, ed. Hermann J. Schmitz (Düsseldorf, 1898), 402–7, corrected
> against manuscript Vatican City, Biblioteca Apostolica Vaticana ms. Vat.
> lat. 4772, fols. 190v–194v.

This rite for the giving of penance comes toward the end of a pontifical, now in the Vatican library, which was copied for the cathedral of Arezzo in Tuscany circa 1100. It has acquired considerable significance in the history of penitential practice since Josef Jungmann cited it in his 1932 study of the penitential liturgy (where he erroneously dated it to circa 1000) as one of the earliest pieces of evidence for the switch from the two-stage procedure for the awarding of penance prevalent in the early Middle Ages to one rite. Instead of the penitent visiting the priest once to confess his sins and be awarded penance, and returning after a period of time, usually Lent, to be reconciled with the Church, all three procedures—confession, penance, absolution—were combined on one occasion. Recent research suggests that early medieval practice was considerably more varied than previously thought, and it is therefore no longer possible to sustain the case for a switch to one-stage penance circa 1000. This *ordo* is therefore more typical of its time than previously thought.

The Arezzo rite was intended for use by a local priest to administer penance to his flock. Medieval Church law distinguished between public penance and secret or hidden penance. Public penance was reserved for the bishop, and was to be administered in the case of those who had committed truly scandalous sins, sins that had offended the public interest, while secret penance was for secret or private sins. This penitential dichotomy, first recorded under the Carolingians, seems always to have remained an ideal and was probably never reflected in reality. Nevertheless a distinction continued to be made in liturgical books between, on the one hand, the episcopal rites for entry into public penance on Ash Wednesday and the reconciliation of penitents on Maundy Thursday, and, on the other, the administration of what one popular tenth-century Germany rite called "penance in the usual way" by a priest. Such rites for "penance in the usual way" survive in a variety of contexts from the ninth century onward; they may be found in pontificals, in sacramentaries, collections of purely pastoral rites for baptism, penance and the last rites, and penitentials.

One of the more unusual aspects of this service, however, is that it assumes that the rite will be administered to several penitents at once. The first part of the service is very public. The penitents are introduced into the Church. This is followed by a versicle from psalm 69 followed by the seven penitential psalms, and then a litany, concluding with some responsorial versicles. Some or all of the seven penitential psalms are a standard element in these rites. But the responses to the versicles suggest a fairly elaborate ceremony with the presence of other clergy, even perhaps a choir. After the priest prays to God to show mercy to the penitents, the penitents go up one by one to the priest, who is sitting before the altar, in order to make their confession. This part of the service is one to one, but it takes place in a very public environment. Although penance administered by a priest was sometimes described as "secret" or "private" this referred to the degree of publicity attached to the sin and not to the conduct of the rite itself.

The confession rite proper begins with an interrogation about the orthodoxy of the penitent's belief, which is another standard component of such rites. The priest then petitions God formally to grant the penitent forgiveness, followed by the imposition of penance. What seems to be missing from this rite is any provision for the specific interrogation of the penitent about his sins, his confession.

According to the rite, penance may be awarded as fasting, alms, or sacerdotal prayers, the last presumably a reference to vicarious prayer. There is a good deal of other evidence for the practice of commuted penance in the eleventh century; penance could be performed through direct prayer, or commuted through payment to the poor or for prayer. After the awarding of penance the priest takes the penitent by the right hand and absolves him. The use of the indicative, "I absolve you," in one prayer is another indicator that this rite represents changes taking place in penitential practice in the tenth and eleventh centuries. In the early Middle Ages the priest merely petitioned God to absolve the penitent, but the use of the indicative began to appear in the late tenth and early eleventh centuries. The liturgy therefore echoes and perhaps even anticipates what have been perceived as changes in penitential theology, generally associated with the twelfth-century scholastic theologians' view that absolution could occur before the penitent had made full satisfaction through completing the performance of his penance. But this rite is from a time of transition; in practice the majority of references to absolution in the prayers are in the subjunctive: "may God absolve you"; there is no guarantee that He will. And this is indeed made clear in the final absolution prayer; the sacerdotal absolution will only come into effect after the performance of the penance, and this penance will not be formally lifted through absolution until the penitent's next confession, but if he dies before that provisional absolution is awarded him.

Although this rite mentions only male penitents when referring to them in the singular, it was obviously intended for administration to both men and women, as is clear from the priest's preparation prayer and the litany. It was normative for such rites to be written exclusively in the masculine, but we need not assume that they were only intended to be administered to men. There are examples of rites where alternative feminine endings for gendered nouns and pronouns were added above the line. The absence of such indicators in the Arezzo rite suggests, perhaps, that is was intended for an educated minister, one whose knowledge of grammar, it was assumed, was sufficient to make the transition from male to female penitent where necessary and without guidance.

The service concludes with a section centered on the cross: the penitent prostrates himself before the cross, the priest makes the sign of the cross over him, then the penitent makes the sign of the cross. It is a novel element, not found in other penitential rites of the period. The biblical significance of this act is clear: it is a reminder of Christ's sacrifice, which ensured the salvation of all sinners. But it is also reminiscent of early medieval private devotions to the cross. This rite clearly shows how devotion and prayer, rather than confession and penance, dominated the penitential services of the central Middle Ages. This was first and foremost a religious service in which the hearing of confessions, the awarding of penance, and pronouncing of absolution played relatively minor, albeit essential, parts. Rites such as these therefore provide a necessary balance to the more prosaic picture of question and answer, of an emphasis on the act of confession itself, suggested by the texts of early medieval penitentials.

Further Reading

Sarah Hamilton, *The Practice of Penance, 900–1050*, Royal Historical Society Studies in History, n.s. (Woodbridge, 2001).

Josef A. Jungmann, *Die lateinische Bussriten in ihrer geschichtlichen Entwicklung* (Innsbruck, 1932).

Mary C. Mansfield, *The Humiliation of Sinners: Public Penance in Thirteenth-Century France* (Ithaca, NY, 1995).

Rob Meens, "The Frequency and Nature of Early Medieval Penance," in *Handling Sin: Confession in the Middle Ages*, ed. Peter Biller and A. J. Minnis (Woodbridge, 1998), 35–61.

—19—

A Penitential Diet

Rob Meens

Chapter 16: On the distinctions of food, clean and unclean.

1. Whoever eats unclean meat, or carrion, or meat from an animal on which beasts have fed, should do penance for forty days; if he does so in the case of necessity forced by hunger, he should fulfill a much lighter penance.

2. When a mouse falls into a liquid, it should be taken out and the liquid should be sprinkled with blessed water. If the animal is alive, the liquid may be used; if it is found dead, all the liquid should be thrown away and it may no longer be consumed by men, whether it was milk or beer or anything else, and the vessel should be cleansed.

3. But if it is a lot of liquid in which a mouse or a weasel falling into it has died, it should be cleansed, a tenth part should be offered to the church, it should be sprinkled with holy water and then it may be consumed, if necessary.

4. When a dead mouse or a weasel is found in flour or in another dry form of food, or in honey or in curdled milk, that part which is near to the dead body should be thrown away; the rest should be sprinkled with holy water and be used.

5. When birds leave excrements in some liquid, the excrements should be taken away and the food may be eaten after it has been cleansed with holy water.

6. When a hen or any other animal falls into a well and is found dead there, the well should be emptied and it should be cleansed with holy water and prayer. If someone drinks from the well knowing [it to be unclean], he should fast for one week; if he did so, however, without knowing, he should fast on Wednesday and Friday until the ninth hour.

7. When a man falls into a well or a cistern and is found dead therein, the well or cistern should be emptied and it should be cleansed with holy water and prayer. If someone drinks from this well knowing [it to be unclean] he should abstain from drinks for forty days and fast on Wednesday and Friday until the ninth hour; if he did so, however, without knowing that the water was unclean, he should fast for twenty days.

8. When a mouse or a hen or another animal has fallen into oil or honey and is found dead therein, the oil should be used for a lantern; honey, however, for medicine or for some other vital purpose.

9. Animals that have been caught by wolves or dogs should not be eaten unless they were killed by men while still alive, but they should be given to the pigs and the dogs; similar a deer or a goat when they are found dead.

10. Birds and other animals that are strangled in nets should not be consumed by men, since they are suffocated.

11. An animal or a bird that has been killed by a dog, a fox, or a bird of prey such as a falcon, or by a stick, a stone, or an arrow that has no iron [arrowhead], these are all suffocated and should not be eaten, for the chapter of the Acts of the Apostles demands that we will abstain from four things: fornication, things suffocated, blood, and idolatry (Acts 15:29). And who eats from such things should abstain from meat for five weeks. When someone did so forced by hunger, he should fast for one week.

12. When an arrow hits a deer or another animal or bird and it is found after three days and [in the meantime] a wolf, bear, dog, or fox has feasted on it, nobody should eat from it and who does so, should fast for four weeks.

13. Who out of necessity eats an animal that seems to be unclean, bird or beast, this doesn't matter.

14. When we say this, we have not forgotten the Lord's word when he said: Hear, and understand: Not that which goeth into the mouth defileth a man; but that which cometh out of the mouth, this defileth a man (Matt. 15:10–12). And the Apostle said: Let no man therefore judge you in meat, or in drink, or in respect of the new moon: Which are a shadow of things to come (Col. 2, 16–17), and again: For every creature of God is good, and nothing to be refused, if it be received with thanksgiving (I Tim. 4:4). But in this case we should not neglect the old religious custom that was handed down to us and is well preserved by the holy fathers since it is clear that it does not deviate from the true faith. Truly, as the apostle says: Each one should be fully convinced in his own mind (Rom. 14:5). With a sound faith and religion we do not condemn anything from this.

15. It is allowed to eat fish even when they are found dead. But some are of the opinion that these should not be eaten; and who eats them, should fast for three weeks.

16. One may eat hare.

17. A horse we do not forbid, but it is no custom to eat it.

18. When bees kill a man, they themselves should be killed quickly, but their honey may still be used for medicine or for other vital purposes.

19. When by chance pigs or hens eat blood, we believe that these should not be rejected, but may be consumed, yet their flesh should be salted with blessed salt and sprinkled with holy water.

20. When a pig or hen eats from a human corpse, these should not be eaten, nor be used for breeding, but they should be killed and fed to the dogs. Others

say that it is allowed to eat these animals after they have slimmed down and after the lapse of a year.

21. Animals polluted by sexual intercourse with men should be killed and nobody should use their milk and their flesh should be thrown to the dogs; their fat, however, one can use for vital purposes and their hides may be used. Where there is doubt, they should not be killed.

22. Who eats or drinks from a substance tainted by a familiar animal, that is a dog or a cat, and knows this, should sing hundred psalms or fast for two days. If he did not know, fifty psalms or a one-day fast.

23. When someone offers someone else a drink in which a dead mouse or weasel has been found, laypeople should fast for seven days, members of the clergy should sing three hundred psalms. Who afterward finds out that he has drunk such a beverage should sing a psalter if he is a cleric, and fast for three days until the ninth hour and abstain from drink if he is a layman.

24. When someone eats something half-cooked without knowing this, he should fast for three days or sing a psalter; if he knew, he should fast for seven days.

25. When someone touches food with an unclean hand, this does not matter.

26. If someone on purpose taints some liquid foodstuff with his hand, it shall be cleansed with a hundred psalms.

27. Who consumes his own blood together with his saliva, without being aware of it, is not to be blamed; if he knew, he has to do penance according to the measure of defilement.

28. When a certain animal kills another animal, it should not be eaten and he who eats from it should fast for two weeks until the ninth hour.

29. When someone eats something which is polluted by blood or by something unclean, it is nothing; but if he knew, he should do penance according to the measure of defilement.

30. Who drinks blood or semen, should do penance for three years.

31. When someone, acting contrary to the precept of the Lord [e.g., Lev. 17:11], of his free will eats animal blood, he should do penance for three years, one of which on bread and water.

32. When someone eats skin or a scab from his own body or vermin which are called lice, or excrements or dirt or drinks his urine, he should do penance for a whole year on bread and water; but if it concerns a child or a boy, he should be corrected with the discipline of the whip.

Source: *Paenitentiale Pseudo-Theodori,* in *Die Bussordnungen der abendländischen Kirche,* ed. H. Wasserschleben (Halle, 1851), 566–622.

The Bible, particularly the books of Leviticus and Deuteronomy, contains several rules regulating the kinds of food one is allowed to eat. In the New Testament some passages suggest an abandonment of these rules, and early Christian exegetes sought to interpret the Old Testament precepts in a spiritual sense. Although one could conclude from this that Christians were allowed to eat everything God had created (I Tim. 4:4), in the Middle Ages in the West this was not the case. Some rules tried to regulate the kinds of food one was allowed to consume, while a certain penance was prescribed for breaking these dietary rules. In the text discussed here some of the New Testament passages are quoted, in which compliance to the rules of the Old Testament is abandoned, yet, so our text stresses, Christians should adhere to the dietary rules which are proposed here because of their old and venerable age (no. 14). Dietary rules are not to be found in the more formal and official ecclesiastical legislation, such as canons decreed by councils or in papal letters, but mainly show up in a more practical sort of text, intended to be used by a priest when hearing confession: a penitential handbook. Since the existing sermon literature also refrains from discussing such a topic, we have to infer that ordinary Christians were mainly confronted with such rules in the process of confessing their sins, a process that apparently entailed some instruction. The chapter titled "On the distinctions of food: clean and unclean taken from the penitential known as Pseudo-Theodore," because of its false attribution to the seventh-century Archbishop Theodore of Canterbury, is of particular interest for its inclusive character, thus proffering a nice picture of the general character of such rules.

Penitential handbooks, of which Pseudo-Theodore is an example, were composed since the sixth century to help priests in their task of hearing confession. Although we do not know how often ordinary believers confessed their sins (probably no more than three times a year) and whether they lived according to the rules that can be deducted from penitential handbooks, the practical nature of these texts and several indications concerning the books a priest had to own and know, strongly suggest that there was a close relationship between these texts and Christian practice. A penitential was intended to offer guidelines to the priest hearing confession, suggesting the kind of penance he should impose on a sinner. Since the system of secret, tariffed penance provided the possibility to receive absolution for minor sins as well, penitentials treated not only major sins, like murder, adultery, or sacrilege, but also smaller offenses like breaking a dietary rule. That such offenses were regarded as minor is clear from the fairly light penances required for such an offense, ranging in general from a few days of fasting (nos. 22, 23) to forty days (no. 1). The fact that some penitential handbooks refrain from discussing this topic further suggests that some compilers of these texts did not regard it important enough to warrant inclusion in their work. Obviously, this was not the case with the author—or better compiler, since he took almost all of his material from existing texts—of the Pseudo-Theodore penitential. He evidently culled several earlier texts looking for dietary rules, which he also defended against criticism based on the New Testament.

The Pseudo-Theodore penitential is considered to be a Carolingian text, written in the first half of the ninth century, influenced by efforts of the Carolingian bishops trying to reform penitential practice. The manuscripts of this text suggest that it was well known in England in the tenth century. This penitential, like many others, deals mainly with themes like murder, theft, fornication, perjury, drunkenness, various forms of religious behavior that the Church regarded as superstitious or magical, and dietary rules. The latter are sometimes regarded as rules of hygiene, by which the Church tried to correct "barbarian" modes of behavior in this field. Another interpretation regards them as a reaction to pre-Christian uses of food in sacrificial rites. It seems better, however, to regard them as an expression of a cultural system in which purity was of particular importance. Sexual purity was a central theme for clerics in their contact with the holy, while there was discussion about the question whether menstruating women were allowed to enter church in such an impure condition. Similar anxieties about pollution and taboo also motivated the dietary prescriptions we find in penitential texts, texts that are generally pervaded by a concern for purity.

Despite the fact that the rules presented by Pseudo-Theodore are not presented in a particularly orderly pattern, some categories of forbidden food can nevertheless be discerned. Liquor was prone to become unclean when something had fallen into it: a mouse, excrements of passing birds, a dead hen, or a dead man (nos. 2, 3, 5–8). Solid substances posed less of a problem since in them the polluted parts could more easily be removed (no. 4). Apparently contact with a dead animal was considered to be much more polluting than contact with a living creature (no. 2). Animals also had to be killed in the right way: when captured by wolves or dogs they were not to be eaten. Animals caught in the hunt by other animals, by dogs or birds of prey, were taboo as well. The same was true for animals caught in nets or killed with a stick, a stone, or an arrow without a metal arrowhead. That such rules had some bearing on aristocratic life is shown by an anonymous ninth-century letter written in response to a royal query, in which hunting animals are defined as passive instruments for human agency, thus opening up the possibility to enjoy meat from creatures captured in the noble hunt. Two motives seem to play a role in these rules. First, there was a fear of predators, which were held capable of contaminating human food, probably because of their frequent contact with blood. The other motive is suggested by the rules where the food in question is characterized as being strangled (*suffocata*) (nos. 10–11). Animals had to be hunted in such a way that they would bleed; if not, they were regarded as suffocated and therefore taboo. This rule, again motivated by a fear of the polluting power of blood, was supported by a reference to the Acts of the Apostles (15:29), forbidding Christians to partake from blood and from suffocated animals.

Then there is the category of misbehaving animals. Meat from animals that had in some way been in contact with death, for example, by killing an animal or a man or by tasting from human corpses, should be avoided (nos. 18, 20, 28). Animals that had had sexual contact with man, and thereby overstepped the boundary between man and animal, should be killed, and their meat and milk were no longer to be used for human consumption. These animals were regarded as unclean only when they misbehaved, but other animals apparently were inherently unclean. Pseudo-Theodore mentions animals that seem to be unclean and that one may eat only in the case of necessity. Unfortunately he did not specify which animals fitted into this category (no. 13). Interestingly, it is in this context that he feels obliged to justify obedience to these rules with regard to biblical arguments (no. 14). He does not inform us about the animals which are unclean of themselves—although one can assume that animals of prey and scavenging animals belong to this group—but merely takes away doubts which seem to have existed in the cases of the hare, the horse and fish (nos. 16, 17, 15).

Not only animal blood seems to have been problematic. Pseudo-Theodore felt obliged to declare that absorbing one's own blood together with one's saliva did no harm, if done unknowingly; but in the case of doing such a thing consciously a certain penance had to be fulfilled (no. 28). The same rule applied to food that had been tainted by blood or by something unclean (no. 29). The rule forbidding the consumption of half-cooked food probably also aimed at the problem of eating blood (no. 24). Drinking blood on purpose entailed a far heavier penance: three years of fasting. Such a heavy penance indicates that such an act played a part in some ritual act that the Church had prohibited. The drinking of semen, for example, which is forbidden in the same rule, is known to have figured in love magic (nos. 30–31). Semen and blood are bodily products. Other bodily substances such as parts of one's skin, excrements, and urine as well as lice and earth (dirt, mud?) are also taboo (nos. 5, 32).

It has to be said that these rules form no coherent set of prescriptions, but behind their patchy nature some general patterns can be discerned. First of all they show that it was mainly meat that was problematic. Other food could be contaminated by animals or by men, but in almost all of the prohibitions of Pseudo-Theodore, animals figure in a prominent way. Second, three major concerns emerge: death, sex, and blood. The impurity pertaining to death is clear from the rules prohibiting contact with dead animals, while blood and semen seem to be the most problematic bodily fluids (as they were for clerics and monks in general). Since blood was regarded as the essential life-giving substance (e.g., Lev. 17:11) it can also be related to death, once it leaves the body. Pollution was normally transmitted through contact, which explains, for example, the clause that the honey of bees having killed a human being can only be used for medicinal or other vital purposes, thereby excluding normal uses of honey as food (no. 18). Impurity was unrelated to man's intentions. Even if he ate or drank from a forbidden substance without being aware of the fact that he was partaking of something unclean, he had to fulfill a certain penance upon finding out (nos. 6, 7, 24). Finally, it is clear that, in general, breaking a dietary rule was regarded as a minor offense. Penances prescribed for transgressions are fairly light, while it is repeatedly explained that one may safely break these rules in times of need (nos. 1, 11, 13). Moreover, one can purify food polluted by animals by sprinkling it with holy water and saying a prayer. Prayers specifically aiming at this have been preserved in some liturgical manuscripts. This possibility to purify polluted food clearly shows that we are not dealing with matters of hygiene and health, although ultimately such considerations may have been of some influence, but that the pollution was conceived first and foremost as a ritual feature needing a liturgical cure.

Pseudo-Theodore's text is known from seven medieval manuscripts dating from the tenth to the beginning of the thirteenth century. This manuscript tradition demonstrates that it had considerable influence, mainly in England, although it was not an extremely popular penitential handbook. Pseudo-Theodore is remarkable for his comprehensive treatment of dietary rules, but it is not the only text propagating such rules. To go by penitential handbooks, many medieval Christians were confronted with these rules in the process of confessing their sins. Priests were not content with knowledge of sins of a sexual or violent nature, but were also eager to bring their flock to an appropriate conduct regarding food. The purity of Christian believers, an essential prerequisite for contact with the Holy, was also to be maintained through the food they consumed. If this purity was threatened by some kind of intake of polluted substances, a proper penance should restore the former relation with God. Pseudo-Theodore helped priest-confessors in showing how this could be done.

Further Reading

The best introduction in English to penitential handbooks is A. Frantzen, *The Literature of Penance in Anglo-Saxon England* (New Brunswick, 1983). For a selection of translations from these texts, see J. McNeill and H. Gamer, *Medieval Handbooks of Penance* (New York, 1938, 2nd ed., 1990).

P. Bonnassie, "Consommation d'aliments immondes et cannibalisme de survie dans l'occident du Haut Moyen Age," *Annales ESC* 44 (1989): 1035–56.

H. Lutterbach, "Die Speisegesetzgebung in den frühmittelalterlichen Bußbüchern (600–1200). Religionsgeschichtliche Perspektive," *Archiv für Kulturgeschichte* 80 (1998): 1–38.

R. Meens, "Pollution in the Early Middle Ages: The Case of the Food Regulations in Penitentials," *Early Medieval Europe* 4 (1995): 3–19.

— 20 —

A Layman's Penance

Joseph Goering

Perceval, so the story says, had so lost his memory that he no longer remembered God. Five times April and May had passed; five whole years had gone by since he worshiped God or His cross in church or minster. . . .

At the end of five years, he happened to be making his way through a wilderness, fully armed as he usually was, when he met five knights escorting as many as ten ladies, their heads hidden within their hoods. All were walking barefoot in woolen rags. The ladies were astonished to see a man coming in armor bearing shield and lance. As penance for their sins, they were themselves proceeding on foot for the salvation of their souls. One of the five knights stopped him and said: "Stay back! Do you not believe in Jesus Christ, Who wrote down the new law and gave it to Christians? It is surely neither right nor good, but most wrong to bear arms on the day Jesus Christ died."

And he who had given no thought to day, hour, or time, so distressed was his heart, answered: "What is today then?"

"What, sir? Do you not know? It is Good Friday, when one should venerate the cross with a pure heart and weep for His sins. This is the day when He Who was sold for thirty pieces of silver was hung on the cross. Free of all sin Himself, He became man for the sins of the entire world, be certain of this, for the entire world was corrupt. It is true that He was God and man, and was conceived by the Holy Spirit and born of the Virgin. In Him God took on flesh and blood, and His divinity was clothed in the flesh of man. This is a certainty, and he who will not believe it will never look on His face. He was born of the Virgin Lady and took on both the form and the soul of man with His holy divinity. Truly on such a day as this was he put on the cross. He then delivered all His friends from hell. Most holy was that death that saved the living and the dead, bringing them from death to life. In their hatred the wicked Jews, who should be killed like dogs, forged their own evil and our great good when they raised Him on the cross. They damned themselves and saved us. This is the day all those with faith in Him should do penance."

"And where do you now come from like this?" Perceval asked.

"Sir, from over there, from a good man, a holy hermit, who lives in the forest. Yet such is his holiness that he lives only by the glory of God."

"In God's name, sir, what did you seek there? What did you ask? What did you do?"

"What, sir?" replied one of the ladies. "We asked counsel there for our sins and made our confession. We did the greatest work that may be done by a Christian who would please God."

What Perceval heard made him weep, and thus he wanted to go and speak with the good man. "I would go there to the hermit if I knew the pathway," he said.

"Whoever would go there, sir, let him follow the straight path and look for the branches we knotted with our hands when we came. We made such signs there that no one going to this holy hermit would lose his way."

Then, without further questioning, they commended one another to God.

As Perceval started on the path, he felt his very heart sighing because he knew he had sinned against God and was sorry. In tears he made his way toward the forest, and when he reached the hermitage, he dismounted and disarmed. He tied his horse to a tree, then entered the hermit's dwelling. In a small chapel he found the hermit and a priest and a young cleric, this is the truth, who were beginning the service, the sweetest and most beautiful that may be said in the holy church.

Perceval fell to his knees the moment he entered the chapel. The good man called to him, seeing his innocence and his weeping, and noticing the tears running down from his eyes to his chin. Deeply afraid of having offended the Lord God, Perceval first grasped the hermit's foot, then bowed to him and, with hands joined, begged him to counsel him in his great need. The good man bade him make his confession, for unless he confessed and was repentant, he would never receive communion.

"Sir," he said, "it has been five years since I have known where I was. I have not loved God or believed in Him. Since that time I have done nothing but ill."

"Oh, dear friend," said the worthy man, "tell me why you have done so. I beg God to have mercy on the soul of His sinner." . . . [Perceval confesses his sins, and then the hermit, after counseling him, adds:]

"Since Pity takes hold of your soul, have in you Repentance as well. Every day go as a penitent to the minster before you go anywhere else. This will be for your benefit. Let nothing deter you. If you are in a place with minster, chapel, or church, proceed there when the bell rings, or earlier if you have risen. This will never be to your disadvantage, but for your soul's benefit. And if mass has begun, to be there will be all the better. Stay there until the priest has finished all his prayers and his chants. If you do this willingly, you may yet rise in merit and have a place in paradise. Believe in God, love God, worship God. Honor worthy men and good women. Stand in the presence of priests. These are observances that cost little, and God truly loves them because they spring from humility. And if a widow, maiden, or orphaned girl seeks your assistance, help her, and you will be the better for it. This is the highest act of charity. So assist them, and you will do well. Take care never to fail in this on any account. It is my wish that you do this for your sins if you would have the graces of God that once were yours. Now tell me if this be your will."

"Yes, most gladly," he answered.

"Now I pray you stay here with me two full days, and in penance eat such food as is mine."

Perceval agreed to this, and the hermit whispered a prayer in his ear, repeating it until Perceval learned it. This prayer contained many of the names of Our Lord, all the highest and the greatest the tongue of man never dared pronounce except when in peril of death. When he taught him the prayer, he forbade him to utter it except in times of grave danger.

"I will not, sir," he said.

So with delight Perceval stayed and heard the service. After the service he venerated the cross and weeping for his sins humbly repented. . . .

Thus Perceval came to know that God was crucified and died on the Friday. On Easter, Perceval received communion with a pure heart.

> Source: David Staines, ed., *The Complete Romances of Chrétien de Troyes* (Bloomington, IN, 1990), 414–18.

Medieval religious instruction may usefully be divided into two broad categories: that which relied heavily on written materials and that which did not. The first type came to be associated primarily with monks and then with clerics; the second was the common expectation of the laity throughout most of the Middle Ages. While the first type of education has left for us a great many written documents revealing its content and practice, the unwritten instruction of the vast majority of medieval people is more difficult to reconstruct and to appreciate.[1]

As is often the case in history, the things about which we know the least are easily overlooked and relegated to a position of unimportance in scholarship that they would never have held in life. The unwritten educational regime of the laity has suffered such a fate. Even popes, bishops, priests, and monks began life as laypeople and received their earliest and most formative education from other laypeople. For those who did not go on to become monks and priests, the religious education provided in the local community by other laypeople was far and away the most important element in their religious formation. They may have been influenced occasionally by the preaching of a priest or friar, or by the example of a particularly holy (or unholy) monk or cleric, but for the most part, religious education of the laity in medieval Europe was a family and a community responsibility.

Because it leaves so few traces in the historical record, the unwritten religious education of the laity is difficult to study. We can perhaps discern some of its general contours and specific interests through a careful study of modern folklore and folk practices that have survived from the Middle Ages into the nineteenth and early twentieth century.[2] There are also some manuscript sources from the Middle Ages, written for clerical audiences, that bring us closer to the lay religious experience than do the more formal treatises of the schoolmen. I think here of the thousands of small *summae* and handbooks, composed in Latin and eventually in the various vernacular languages, for the education of parish priests.[3] Finally, I should mention the school-books or primers, written by clerics, for the education of young children in the local song and grammar schools throughout Europe.

But if we are to learn to appreciate and understand the religious education received by laypeople from other laypeople in the Middle Ages, we will need to make full use of all the available evidence. The example I have chosen here is an excerpt from a popular story, a "romance," written in the French vernacular around 1180 by Chrétien de Troyes, called the *Contes du graal* (*Story of the Grail*). As an example of lay religious instruction, this source is not without its problems. First, of course, it is a written source and, even if it was recited orally, it is not itself an example of the kind of unwritten instruction and education that interests us most directly. Nor is it unambiguously an example of lay instruction. Scholars continue to debate whether Chrétien was himself a cleric (that he had a "clerical" education and considerable book learning is beyond doubt). For my part, I am convinced that Chrétien is a lay-man, and that his particular form of lay piety is quite characteristic of laypeople throughout Europe. He might be called "anticlerical"—bishops, priests, and monks rarely appear in his stories—but he is deeply and thoughtfully religious, and his pieties are intended to be in keeping with those of what he would call "holy mother church." I take him, rightly or wrongly, for a typical catholic layperson of his time.

However that may be, I have chosen him not because he is a layman instruct-ing other laypeople, but because his *Story of the Grail* is, in large part, a depiction of the religious education of a layperson. For all the fame of the fictional hero, Perceval, and all the fancy of the storyteller, one can still find here a plausible and circumstantial presentation of lay religious instruction. In Chrétien's tale we are allowed to see how the hero, Perceval, came to learn all he needed to know about the Christian story without ever opening a book.

When we first meet Perceval in the extract above, he has already learned a great deal from various mentors, friends, and lovers about knightly comportment, but his religious education has not yet been similarly advanced. That is not to say that he was uninstructed. His mother (may she rest in peace) had taught him well as a child. Perceval refers now and again to the crucial things he had learned from her. Early in the tale he hears a fearsome noise in the forest: "On my soul, my lady my mother spoke truth to me when she told me that there is nothing in the world more terrifying than devils. She instructed me that one should make the sign of the cross to guard against them. But I shall scorn this teaching. Never will I cross myself. In-stead, with one of these javelins I carry, I shall strike the strongest. . . . [4] But when he discovers that the terrible clamor was made by a small troop of beautifully ap-pointed knights, he exclaims: "Ah, Lord God, have mercy! These are angels I behold here. Alas, truly, now I have sinned grievously, now I have behaved most badly to call them devils. My mother told me no idle tale when she said to me that angels were the most beautiful creatures there are except for God, Who is the most beautiful of all.

Here I behold the Lord God, I believe. . . . My mother said herself that one must believe in God and worship Him, honor and pray to Him. And I shall worship that one and all the others with him."[5] Despite Perceval's youthful exuberance and his boyish transgressivity, one can see that his mother has taught him well and truly, from an early age, some of the most important elements of the Christian faith.

When Perceval decides to leave home and seek out King Arthur's court to become a knight, his mother offers him more detailed instruction in parting: "Darling son . . . above all I would beg you to enter church and minster to pray to our Lord that He grant you joy and honor, and help you conduct your life so that you may come to a good end." Perceval replies: "Mother, what is church?" And she answers: "A place where the service is celebrated to the One Who created heaven and earth, and there placed men and beasts." Perceval persists: "And what is minster?" And his mother replies: "The same. A beautiful and sacred house filled with holy relics and treasures, where the sacrifice of the body of Jesus Christ occurs, the holy prophet whom the Jews treated so shamefully. He was betrayed and wrongly condemned and, for the sake of men and women, suffered the agony of death. Until that time, souls went to hell when they left their bodies. He was the one who delivered them from there. He was bound to the pillar, scourged, and then crucified, wearing a crown of thorns. I counsel you to go to the minster to hear masses and matins and to worship the Lord." Perceval responds: "Then I shall be delighted to go to churches and minsters from now on, I promise you."[6]

My text relates events that occur five years later.

Notes

1. Joseph Goering, "The Thirteenth-Century English Parish," in *Educating People of Faith: Exploring the History of Jewish and Christian Communities*, ed. John van Engen (Grand Rapids, MI, 2004), 208–22.

2. Joseph Goering, "The Changing Face of the Village Parish: The Thirteenth Century," in *Pathways to Medieval Peasants*, ed. J. A. Raftis (Toronto, 1981), 323–33, at 323–27.

3. Ibid., 328–31; W. A. Pantin, *The English Church in the Fourteenth Century* (Notre Dame, IN, 1963), 189–243.

4. David Staines, ed., *The Complete Romances of Chrétien de Troyes* (Bloomington, IN, 1990), 340.

5. Ibid., 341.

6. Ibid., 346–47.

Further Reading

Peter Biller and A. J. Minnis, eds., *Handling Sin: Confession in the Middle Ages* (Woodbridge, 1998).

Eamon Duffy, *The Stripping of the Altars: Traditional Religion in England, 1400-1580* (New Haven, CT, 1992).

Churches, Parishes, and
Daily Life: Prayer

—— 21 ——

Prayers

Virginia Reinburg

PRAYER TO OUR LADY OF LE PUY

This is the prayer that Our Lady of Le Puy holds in her hand.

Very holy and glorious mother of God, my dearest Lady, Lady of all, you who are the true cause of our redemption, and our hope for heavenly life. I offer myself to you in soul and in body, and all that I have, together with our people. And I pray you to guard me from mortal sin, and that you pray your dear Son, Our Lord Jesus Christ, that He grant me to work in such a way in this mortal life, and to die as a good Christian, that neither I nor your people may be damned. And give us all true comfort in soul and body. Amen.

> Source: Paris, Bibliothèque nationale de France, MS. nouv. acq. lat. 592, fol. 120v.
> Manuscript book of hours, use of Paris, fourteenth century. Original in French.

PRAYER TO OUR LADY OF LORETO

O Mary.

Pope Sixtus IV composed this prayer and granted to everyone who says it devoutly before the image of Our Lady of Loreto 11,000 years of true pardon [indulgence].

Prayer. Hail most holy Mary, Mother of God, Queen of Heaven, Gate of Paradise, Mistress of the Earth, you are virgin most pure, you conceived Jesus without sin, free me from every evil and pray for my sins. Amen. Pater Noster. Ave Maria.

> Source: *Ces presentes Heures, à l'usaige de Paris . . . aveq les heures saincte*
> *Geneviefve et la commemoration saint Marcel et de saint Germain* (Paris: Thielman
> Kerver for Guillaume Eustache, 1500), last page. Paris, Bibliothèque nationale

de France, Réserve des Imprimés, Vélins 1634. Printed Book of Hours,
use of Paris. Text is handwritten, added by the book's owner. Original in
French (rubric) and Latin (prayer).

THE SEVEN PRAYERS OF SAINT GREGORY

O Lord Jesus Christ, I adore you hanging upon the cross, wearing a crown of
thorns upon your head. I pray you that your cross might free me from the de-
ceiving angel. Amen. Pater Noster. Ave Maria. O Lord Jesus Christ, I adore you
hanging on the cross, wounded, drinking vinegar and gall. I pray you that your
wounds might be the remedy for my soul. Amen. Pater Noster. Ave Maria. . . .

Saint Gregory, who while living was pope of Rome, and several other subsequent popes, [plus] cardinals, archbishops, and bishops granted to everyone in
a state of grace who says the prayers written above, while kneeling before the
figure, 56,000 years of true pardon [indulgence] for each time [it is] recited.

> Source: Paris, Bibliothèque nationale de France, MS. nouv. acq. lat. 3208, fol.
> 76r–77v. Manuscript book of hours, use of Coutances, mid-fifteenth century.
> Original in Latin (prayer) and French (rubric). Illustrated with a miniature
> of *The Mass of Saint Gregory.*

THE SEVEN PRAYERS OF SAINT GREGORY

Whoever in a state of grace says the following prayers before [an image of] the
vision of Saint Gregory will earn 46,000 years of true pardon [indulgence],
granted by this and other popes. And at the end of each prayer say Pater Noster [and] Ave Maria. And each and every time that one says these prayers, or
says the Pater Noster and Ave Maria fifteen times, he will earn the above-
mentioned indulgence, but he must pause between each time.

O Lord Jesus Christ, I adore you hanging upon the cross. . . .

> Source: Baltimore, Walters Art Gallery, MS. W. 245, fols. 62r–62v.
> Manuscript book of hours, use of Angers, last quarter of the fifteenth
> century. Original in French (rubric) and Latin (prayer). Illustrated
> with a miniature of *The Mass of Saint Gregory.*

PRAYER TO BE SAID DURING MASS, AT THE ELEVATION
OF THE HOST

To all those who say the prayer that follows are granted 2,000 years of true pardon [indulgence]. And it should be said between the elevation of the body of
Our Lord and the third Agnus Dei. Pope Boniface VI gave this pardon, at the

request of Monsieur Philippe, King of France. Pope Clement gave it to Monsieur Pierre de Gyac, [who was] then chancellor of France. And it is written in Jerusalem, near the altar of the holy sepulcher. And so that this might be more firmly believed, the bulls are in Paris, in the King's treasury.

Lord Jesus Christ, who assumed this your most holy flesh in the womb of the glorious Virgin Mary, and shed your most precious blood from your most holy side on the tree of the cross for our salvation, and in this glorious flesh rose from the dead and ascended into heaven and is to come in this flesh to judge both the living and the dead. Deliver us by this your most holy body [which now] is held on your altar, and from all evil and danger, now and forever. Amen.

> Source: Paris, Bibliothèque nationale de France, MS. lat. 13277, fols. 158r–159r.
> Manuscript book of hours, use of Rouen, mid- or second half of the fifteenth
> century. Original in French (rubric) and Latin (prayer).

PRAYER OF CHRIST'S WOUND OR "CHARLEMAGNE'S PRAYER"

Here above is the portrait of the measure of the body of Our Lord, reduced twenty-one times, brought long ago from Constantinople in a golden cross. Whoever sees this measure will not die a sudden death on that day; nor can he be harmed by fire, water, nor the devil, nor by lightning nor storm; nor can an evil judge sit in judgment of him if his case is just. And if a woman wears it she will not perish in childbirth, if she looks at this measure having firm faith in Our Lord Jesus Christ. This sign was brought by the angel to King Charlemagne while he was in battle, so that he could not be harmed. And below is the measure of the wound of Our Lord which Longis [Longinus] made with the lance in Jesus Christ's side, and [it] has as much honor and power as the measure of the cross.

The blessing of God the Father with his angels be upon me. Amen. The blessing of Jesus Christ with his apostles be upon me. Amen. The blessing of holy Mary with her Son be upon me. Amen. The blessing of the holy, eternal Church be upon me. Amen.

> Source: Paris, Bibliothèque nationale de France, MS. lat. 1359, fol. 213v. Manuscript
> book of hours, use of Paris, fifteenth century, with sixteenth-century
> additions. Original in French (rubric) and Latin (prayer). The text is
> accompanied by a pen-and-ink drawing of a chalice, the rim of which
> is represented as an open wound.

In 1401 the theologian and author Jean Gerson explained to his French readers why and how to pray. "Know therefore that God has ordered that we make our salvation, and we acquire it by devout prayers and by fulfilling His commandments." Moreover, "He often wishes to give us through the prayers of the saints" what He had already decided to grant before the saints existed. Certainly God can do all things Himself. Nevertheless, He wishes that "secondary causes have their role." Thus one human being can be saved through the spiritual aid of another. "This is why we pray to God and His saints." As in the other vernacular religious tracts he addressed to lay and female readers, Gerson here offered practical advice rather than learned theology. Thus he described prayer as "spiritual begging," and compared the devout Christian to a beggar forced to appeal to the more fortunate—especially the spiritually "rich"—for sustenance. Prayer, Gerson explained to his readers, was nothing other than the habit of asking God for His grace, and appealing to the Virgin Mary, saints, and angels for their prayers ("alms") and intercession before God. This is the human path to eternal salvation.

Gerson beautifully captured the internal logic of late medieval practices of prayer. Key to the logic of prayer were pious acts and human effort, linked to the intercession and spiritual patronage of the Virgin and saints. The same logic also lies behind the many texts and images late medieval Christians used to guide their prayer, especially those included in books of hours. Books of hours were prayer books used by literate laypeople and nuns, as well as some monks and priests, from the thirteenth through the sixteenth century. A compilation of offices, hymns, and prayers originally adapted from the clergy's breviary, the book of hours eventually surpassed the psalter as the laity's prayer book. It was also the single book most commonly owned by laypeople, and often served as a primer for literacy as well as a guide to devotion. Most owners acquired their books of hours through inheritance or gifts, commissioned them from scribes, or purchased them nearly ready-made or secondhand. To the book's standard array of texts and images, owners often added personal material: their names, monogrammed bindings, family chronicles, handwritten prayers, pious legends, images of patron saints, or small objects like pilgrim souvenirs. Hence books of hours are a rich mine of information about popular cults and saints, late medieval practices of prayer, and the religious choices book owners made.

The texts are translated from the original French and Latin prayers in French books of hours. They exemplify common features of late medieval practices of prayer: devotion to the Virgin Mary and crucified Christ; relics and images; prayers to be said in church, before an image of Mary or Christ, during mass. Instructions appended to several of the prayers note that a pope, cardinal, or bishop promised to pious Christians who say the prayer an indulgence, or stated term of time off from purgatory. Rubrics to several prayers include a legend explaining the prayer's origin, or linking it to revered secular and ecclesiastical authorities. One prayer of protection, commonly called "Charlemagne's prayer," has sometimes been labeled "magical" by modern historians because in the text the devotee directly calls upon himself or herself the blessings of God, the Virgin, apostles, angels, and the Church. Two texts refer to famous shrines of the Virgin Mary—the "black virgin," a sculpted image of Our Lady in Le Puy (France), and the relic of the "holy house" of Our Lady in Loreto (Italy)—popular with late medieval pilgrims. Both texts may have been confraternity prayers, learned by owners of books of hours through membership in pious associations organized locally to pay homage to images of the Virgin honored at distant shrines.

Beyond exemplifying common modes of late medieval devotion, all the texts show that for the Christian faithful prayer was not just a pious act but also a dialogue or communication with God or a saint. The devotee's goal was to cultivate with God, the Virgin, or a saint a relationship of reciprocity, in which the devotee petitioned for assistance or intercession and offered in exchange praise, tribute, or donations. In return for such faithful service devotees hoped to be rewarded with help "in this life" and salvation in the life to come. Readers of books of hours used the texts and images in their books to help them cultivate bonds with God and the saints. Praying with the aid of a book brought a reader's experience into dialogue with a text, and through the text with a religious tradition. Although not every medieval Christian could read or own a book, the practices of prayer preserved in books of hours were not restricted to the literate or wealthy. As Gerson noted, it is no more necessary to have books than to go overseas on pilgrimage. "We have the books in ourselves," he explained. But fortunately for the twenty-first-century history student, some of the books have survived to preserve for distant generations the practices of prayer cherished by medieval Christians.

Further Reading

Jean Gerson, *Oeuvres complètes*, VII, ed. Palémon Glorieux (Paris, 1966), no. 317, pp. 220–81.

Kathryn A. Smith, *Art, Identity, and Devotion in Fourteenth-Century England: Three Women and Their Books of Hours* (London and Toronto, 2003).

R. W. Swanson, *Religion and Devotion in Europe, c. 1215–c. 1515* (Cambridge, 1995).

Roger S. Wieck et al., *Time Sanctified: The Book of Hours in Medieval Art and Life* (New York, 2001).

—— 22 ——

Two Healing Prayers

Eamon Duffy

A good prayer for fevers through the thousand names of the Lord.
Theobal quith et quth Kanai[1]

Through the truth of our Lord Jesus Christ may all malignant spirits flee from me. In the name of my Lord Jesus Christ sign me + with this sign + A Ω. In the name of the Father and of the Son + and of the Holy Spirit + Amen + whatsoever the Father is, Alpha and Omega, that also + is the Son + and that also is the Holy Spirit.

Remedium, Tetragramaton, Hosyon.[2]

The truth of Christ, the peace of Christ, the labors of Christ:[3] Christ have mercy through the thousand names of the Lord.

In the name of the Father and of the Son and of the Holy Ghost Amen.

Before the gates of Jerusalem Saint Peter lay oppressed with fever. Jesus came and said to him, "Peter, what is the matter that you lie here?" Peter said, "Lord I lie here pierced and oppressed with fever" and the Lord said to him, "Peter, arise, nevertheless, [autem] and receive your health." And he was freed [from the fever]. And Peter said to him, "I beseech you, Lord, that whoever carries this written upon them may be strengthened against troublesome or harmful fevers." And he said to Peter, "Let it be done according to your word."

Amen Tetragramaton.

In the name of the Father and of the Son and of the Holy Spirit.

Saint Peter lay upon a marble stone and, coming upon him, Jesus said, "Peter, why are you lying here?" and Peter said "Lord I lie here because of evil fevers." Then Jesus said "Arise and scatter them" and at once he rose and scattered them. Then said Peter "Lord I wish that if anyone carries about them this written in your name, no fevers may harm them, whether cold or hot, whether double, or tertian, or quartain, or quintain, or sextain, or septain [fevers]. Then Jesus said "Let it be [so] to you. Amen."

+ Christ conquers, + Christ reigns + Christ rules + Christ rule me, Amen.

In the name of the Father and of the Son and of the Holy Spirit Amen.

You shall say a thousand times Ave Maria [Hail Mary] and you shall say them in ten days, that is every day a hundred, and you shall say them standing, going, kneeling, or sitting, and you shall have a certain alms in your hand while you make your prayer, and after, say this orison or prayer that followeth.

O Adonai, Lord, great and wonderful God, who gave the salvation of human kind into the hands of the most glorious Virgin, your Mother Mary: through her womb and merits, and through that most holy body which you took from her, in your goodness hear my prayers and fulfill my desires for [my] good, to the praise and glory of your name. Liberate me from every tribulation and assailant, and from all the snares of my enemies who seek to harm me, and from lying lips and sharpened tongues, and change all my tribulation into rejoicing and gladness. Amen.

And when you have said this orison kiss your alms, and after, give it to a poor man or woman in honor of that blessed joy that Saint Gabriel greeted our Lady [with], and for what[ever] thing you do this ten days together, without doubt you shall have that thing you pray for lawfully, with God's grace.

[Added in English in a later hand, "I used this prayer well ten days, Edmund Roberts *inquit* (says)."

In the same hand over the last two pages of the prayer Edmund Roberts has written in Latin *The year of our Lord 1553, the first year of Mary our Queen.* Anno 1553].

> Source: Cambridge University Library, Ms. Ii.6.2. fols. 108–9. Passages in italic have been translated from Latin; passages in Roman script are in English in the original. Words in square brackets have been supplied to clarify the sense.

Cambridge University Library, Ms. Ii.6.2 is a manuscript book of hours, a Latin prayer book organized around the eight monastic "hours" or prayer times of the day, produced for the English market in Bruges around 1400, making it one of the earliest surviving examples of a wave of mass-produced and relatively inexpensive prayer books produced in the Low Countries for an export market catering to an expanding literate class in England. It is brightly illustrated, but the pictures are of poor quality, of a type produced in sets by artists of modest talent, and bought by stationers, for inclusion in production-line manuscript prayer books. These bound-in pictures, painted on vellum, were blank on the back, and as was common practice, in the course of the fifteenth and early sixteenth centuries successive owners of this particular book customized their inherited prayer book by adding their own material, in Latin, English, and a sometimes-odd mixture of the two, using the backs of the pictures to write on. The added material in this book is typical of such late medieval devotional additions, though few books have quite so many. It includes obits (commemorations added to the liturgical calendar of the anniversaries of the deaths of friends and relatives, as a reminder to pray for them). These obits relate to gentry from the parishes of Badingham and Hevingham in Suffolk, and so locate the early use of the book in East Anglia. One of them commemorates Margaret Redisham, the mother of the Sir John Hevingham, whose sudden death while reading a "little devotion" in his orchard, is recorded in the Paston letters; he may even have been using this very book to say his "little devotion."[4]

In the course of the fifteenth century the book passed into the hands of the Roberts family of Middlesex, prosperous landowning gentry in Willesden and Neasden since the thirteenth century, latterly acting as bailiffs for the Dean and Chapter of St. Paul's Cathedral in London.

The manuscript additions to the book offer a good cross-section of the sort of material which later medieval people liked to have in their books of hours. In addition to the obits in the calendar they included practical material such as medical recipes; a table to calculate the conjunctions of the moon; a coat of arms, with quarterings of Norfolk families from Tilney and Thorp; and notes on the precise time and date of the births of some of Thomas Roberts' twenty-four children, to enable their horoscopes to be cast with precision. There are Latin or English devotional verses—on the life of the Virgin, the passion of Christ, and the merits of the Mass—and a long devotional instruction (in the same hand as that in which our prayer is copied) on the need to prepare against sudden death by constantly renewed acts of contrition and resolutions of amendment, and by undertaking to make a sacramental confession at the first opportunity "accordyng to the commandments of all holy church," and to give alms and do good deeds. There are several Latin prayers to St. Dorothy, one of which includes a Latin rubric on the benefits of keeping an image of the saint in your house (Dorothy is a Roberts family name), as well as prayers to St. Cornelius, St. John the Baptist, St. George, St. Erasmus, St. Frideswide, and St. Michael the Archangel (some of these are, like the prayers to St. George and St. Michael, appropriately copied onto the back of pictures of the relevant saint). There is also a Latin prayer to the Virgin as Empress of hell for help at the hour of death; a prayer in English to Jesus and Mary for forgiveness of sins, beginning "O my sovereign Lord Jesus"; a short scheme of meditation, in Latin, on sin and the brevity of life; and a number of prayers, in Latin and English, for use at Mass: these include two versions of the very common elevation prayer,

> O Jesu Lord, welcome thou be,
> in form of bread as I thee see,
> Jesu for thy holy Name,
> shield me this day from sorrow and shame . . .

A characteristic Tudor moralizing rhyme urges the reader:

> Joy in God whose grace is beste
> Obey thy prince and live in awe
> Help the poor to live in rest
> And never sin against the law.

Some of the added devotional material has a broadly "magical" character; these items include a prayer "for women to conceyve a childe," a version of the so-called

Charlemagne prayer, which was a charm based on the names of Jesus, attributed in this instance not to Charlemagne but to Joseph of Arimathea, a short spell in English to quench flames if your house should happen to catch light, a Latin prayer against the pestilence, and an injunction to say the well-known plague hymn *Stella Coeli extirpavit* at the elevation in the Mass.

Our text clearly belongs with this assortment of quasi-magical material, with its Latin charm using the sign of the cross, the titles of Jesus, and an anecdote about two episodes of the healing from fever by Christ of the Apostle Peter, designed to protect the user against various types of fever.

Prayers of this kind, straddling somewhat uneasily the dividing line between magic spell and petitionary prayer, were very common and very popular in the later Middle Ages. Examples are found in most manuscript devotional collections, and were often copied into books of hours and other lay prayer books. Theologians and bishops condemned them as superstitious, and they were regularly denounced in sermons and pastoral textbooks, in itself a sign of their widespread popularity. It was routine for the authorities to represent such prayers as products of lay ignorance, and to associate them in particular with the lower orders, "ignorant of simplesse." In fact, they were popular with laypeople of all classes, as is indicated by the presence of a number of such prayer-charms in the prayer book of Henry VII's mother, Lady Margaret Beaufort. The Roberts family, a member of which copied our text into the book, were powerful and well-educated people.

The prayers employ a number of religious strategies found also in the official liturgy of the Church—invocation of holy names, the use of the sign of the cross (traced with the fingers of the right hand downward from forehead to breast and across from the left to the right shoulder), a divine guarantee attached to specific devotions. It was believed that the use of good names and holy gestures and objects (holy water, blessed candles) drove away evil and protected the user from harm; such sacred signs and words were known as "sacramentals," and were used in the Church's official liturgy of exorcism, in baptism, and many other official ceremonies. Made by the clergy in the course of the liturgy, such sacramentals could then be deployed by laypeople, without clerical supervision. This lack of clerical supervision and the opportunities it offered for unauthorized and unorthodox elaboration, worried some churchmen, and theologians anyway disagreed over whether sacramentals were simply elaborate prayers of petition, or whether they in fact carried some divine guarantee of benefit. Essentially, however, the Church authorities approved and encouraged such material symbolism so long as it was not imagined to be essential for the success of a prayer, or was not believed to work automatically, or to coerce God or his angels or saints into granting the user's requests.

The text falls into two halves. The first part opens with the solemn invocation of the holy names of God, and the repeated use of the sign of the cross; it then continues with a form of sympathetic magic, in which two apocryphal stories (i.e., not found in the New Testament), about the healing of St. Peter from fever are said to protect from fever anyone who carries around with them a paper on which the stories have been written out.

In the premodern world the possession of a person's true name was held to give one power over them; hence, the devils in the Gospels call out Jesus's name. The invocation of the various names of God, above all the Tetragramaton, the Hebrew name of God Jews were forbidden to utter or to write down, was specially powerful, and was sometimes employed in the liturgy itself, while aspects of these sorts of beliefs were embodied in the growing devotion in the late Middle Ages to the holy name "Jesus." In magical practice also, the names of God, of angels, and of other supernatural figures were a regular part of the making of spells, and magical and protective names and words, like "Anazapta," are found both in magical and in more conventionally religious texts, jewelry, and amulets. The prayer here mixes real biblical names for God (Adonai, the Tetragramaton) and names adapted from pagan sources (the Greek word "Hosion"). Linked to this use of the holy names of God, is the repeated use of the sign of the cross, which the medieval Church believed banished evil.

Although often condemned, the use of appropriate texts or stories in sympathetic magic also had some official connivance. It was commonly believed, for example, that the recitation of or allusion to the account of Christ's miraculous escape at Capernaum from an attempt by a hostile crowd to lynch him, would protect the believer from dangers and hostile forces. The concluding verse of the story in St. Luke's gospel "Jesus however, passing through the midst of them, went on his way" (Luke 4:30) was often inscribed on amulets designed to protect against danger, and is found in some very elite contexts. It is embroidered, for example, on the halo and cuff of the robe of God in the judgment scene in the magnificent (and theologically sophisticated) Rohan Hours.[5] Similar anecdotes, with a promise of blessing for anyone who wrote the story down and carried the writing around or kept it in their house, were attached to the legends of a number of late medieval saints, including the Roberts' family favorite, St. Dorothy.[6]

The second prayer then moves away from protection from fever to more universal benefits. In this case the second half of the prayer focuses on the story of the Annunciation, when the Angel Gabriel greeted Mary, the precise moment when Christ took human flesh and became an embryo in the Virgin's womb. The most popular prayer to Mary, the Hail Mary, begins with the Angel's greeting "Hail Mary, full of grace, the Lord is with thee"; the devotee is to recite a hundred Hail Marys (the equivalent of two rosaries) every day—the prayers can be recited while the devotee goes about their ordinary business, "standing, going, kneeling, or sitting," but is linked to the late medieval preoccupation with the works of mercy listed by Christ in the parable of the sheep and the goats (Matthew 25) as a means of salvation. It was believed that everyone would be judged at the Last Judgment not by words of faith or homage, but by whether or not they had concretely helped the poor and suffering. The success of the prayer is said to depend on its being accompanied by the relief of the poor, in honor of the Annunciation. But the link is made in a quasi-magical way, which the Church authorities would certainly have condemned—holding money in the hand while the thousand Aves are recited, then kissing it before giving it to the poor recipient. To the thousand Aves is added a Latin prayer that emphasizes the centrality in the salvation of mankind of the physical reality of the Incarnation at the Annunciation—Christ is invoked by his Mother's womb and by the flesh he took on in that womb. That flesh is declared to protect the user of the prayer especially from their enemies. Prayers against enemies—corporeal and incorporeal—were a prominent feature of late medieval piety. Characteristically, the English instructions attached to the prayer display some awareness of the precarious line being trod between "legitimate" prayer and forbidden "magic"; success is guaranteed if the prayer is rightly used (a guarantee theologians rejected as magical), but that guarantee is softened by the reference to praying "lawfully, with God's grace."

Historians often claim that the growing popularity in the late Middle Ages of the books of hours, essentially a boiled-down version of the monastic round of liturgical prayer, mostly in Latin and increasingly fashionable among wealthy, well-educated laypeople from the fourteenth century, marks the parting of the ways between elite and popular religion. The users of these books were, it is claimed, imitating the clergy and monastic orders, looking for a text-based, privatized, and subjective religiosity, unlike the rote prayer and externalized religion of the mass of laypeople, who could not read and whose prayer life consisted of the repetition of a few half-understood formulas like the Hail Mary. They were, so to speak, getting their heads down, turning their eyes from the distractions posed by their fellow worshippers, [and] at the same time taking them off the priest and his movements and gestures. Such folk, in becoming isolated from their neighbors, were also insulating themselves against communal religion, possibly even religion per se, for how can you be religious on your own?

There are many reasons for questioning these assumptions, and at any rate it will be evident from these additions to the Roberts family book of hours that we are not dealing here with amateur mystics, or men and women in flight from the world and their neighbors, which historians have tended to think books of hours encouraged. As we have seen, the added prayers range from devotions to named saints or prayer to be said at the elevation of the host at Mass, to elaborate penitential prayers to be used as a temporary substitute for the sacrament of penance, and several prayers that are in fact apotropaic charms, designed to fend off evil or procure material good. The prayers are churchly, sacramental, attentive to the saints, concerned with meritorious acts of charity; they are highly supernatural, but in no sense otherworldly. There are prayers here to stop your house burning down or to help a woman to conceive a baby, and it is quite clear that some of these prayers are thought of as instrumental rather than merely supplicatory; done properly, they are guaranteed to work. Yet these are not the prayers of ignorance. They were written into the book made for and used by wealthy and influential men and women. Thomas Roberts, who died in 1543, and whose surviving children are all listed in the book, was Clerk of the Peace and coroner of Middlesex, a man of weight and education. The prayers come from a repertoire of such things as were appreciated at the very top of the social ladder; one of the additional prayers in the book, the English invocation to Jesus and Mary, "O my sovereign Lord Jesus," is a favorite item in many books of hours, and is also found, for example, in the sumptuous Talbot Hours, now in the Fitzwilliam Museum, and made for the 1st Earl of Shrewsbury. And the same devotees who dutifully recited their thousand Aves in the hope of guaranteed protection from enemies or disease, might also practice an impeccably orthodox penitential regime, involving fervent acts of contrition, self-examination and introspection, and sacramental confession. Our magical prayers are in the same handwriting as a long devotional instruction on the importance of obedience to the Church, sincere sorrow for sin, and the devout use of the sacrament of penance. Official religion and popular religion are inextricably intertwined.

Notes

1. The handwriting here is clear and there seems little doubt that the words are written as transcribed here, but I can make no sense of them. Theobal appears to be a proper name; the copyist of the prayer was perhaps rendering an imperfectly understood formula of invocation.

2. *Remedium* = Latin for healing or remedy; *Tetragrammaton* = the Hebrew Name of God, YHWH, often invoked in magical prayers of this kind; *hosion* = Greek for Holiness. All three words are being invoked as *names* of power.

3. The Latin at this point (as in others) is obscure and perhaps in part nonsensical; it runs: "Christi veritas, pax Christi, molitus Christi eleyson." Molitus is a participle of the verb Molior, to strive or labor, and is just possibly being used as if it were a fourth declension noun, and I have translated it accordingly. This ingenious conjectural solution to a puzzle, which admittedly may simply be due to the undoubtedly shaky latinity of whoever copied the prayer, was suggested to me by my colleague Dr. Richard Rex, to whom I am grateful.

4. N. Davis, ed., *Paston Letters and Papers of the Fifteenth Century* I (Oxford, 1971), 39, 250.

5. Marcel Thomas, *The Rohan Master: A Book of Hours* (New York, 1973), plate 63. The commentary on this plate should be disregarded, being seriously misleading.

6. Ibid., 175.

Further Reading

Eamon Duffy, *Marking the Hours: English People and Their Prayers, 1240–1570* (New Haven, CT, 2006).

———, *The Stripping of the Altars* (New Haven, CT and London, 1992).

Nicholas Rogers, "Patrons and Purchasers: Evidence for the Original Owners of Books of Hours Produced in the Low Countries for the English Market," in *Corpus of Illuminated Manuscripts*, ed. Bert Cardon, vols. 11–12, Low Country series 8 (Leuven, 2002), 1165–81.

Churches, Parishes, and Daily Life: Devotional Behavior

— 23 —

Images in the World: Reading the Crucifixion

Sara Lipton

I.

Jesus is sweet in the inclination of his head and in death, sweet in the extension of his arms, sweet in the opening of his side, sweet in the piercing of his
feet with one nail.

Sweet in the inclination of the head; for, inclining his head on the cross, he
seems as if to say to his beloved: O my beloved! How often have you desired
to enjoy the kiss of my mouth, announcing to me through my companions, *"he
kissed me with the kiss of his mouth?"* (Cant. 1:1). I am ready, I incline my head, I
offer my mouth, to kiss [you] however many times; and do not say in your heart,
"I do not seek that kiss, which is without beauty and loveliness; but rather that
glory, which the angelic citizens always desire to enjoy." Do not err thus, because, unless first you will have kissed that mouth, you will not be able to arrive
at that [glory] at all. Therefore, kiss that mouth, which now I offer to you, since,
though it may be without beauty and loveliness, nevertheless, it is not without
grace.

Sweet in the extension of his arms: for, extending his arms to us, he shows that
he himself desires our embraces, and seems as if to say: *O you who are worked and
burdened, come and be refreshed* (Matt. 11:28) within my arms, within my embraces; see that I am prepared to gather you within my arms, come, therefore, all:
let no one fear to be repulsed, *because I do not want the death of the sinner, but that
he be converted, and live* (Ezek. 33:11). *Indeed, my delights are to be with the sons
of men* (Prov. 8:31).

Sweet in the opening of his side; inasmuch as that opening revealed to us the
riches of his goodness, you may know the love for us in his heart.

Sweet in the piercing of his feet with one nail; because through this, it is as if
he speaks to us thus: Behold! If you judge that I might flee, and you therefore
tarry in coming to me, knowing that I am exceedingly swift, and like a young
stag; see that my feet have been fixed by one nail thus, so that I can in no way

flee you, because pity holds me entirely bound; nor will I flee you as your sins merit, because my hands are pierced with nails.

Kind Jesus, humble Lord, pious Lord, sweet in mouth, sweet in heart, sweet in ear, inscrutably and undescribably delightful, pious and merciful, powerful, wise, kind, generous but not profligate, exceedingly sweet and pleasant! Only you are the highest good, *beautiful in form beyond the sons of man* (Ps. 44:3), pretty and lovely, and *chosen among the thousands* (Cant. 5:10), and *wholly desirable* (Cant. 5:16). Beautiful things are becoming to the beautiful. O my Lord, now all my soul desires [to be] in your embraces, and kisses. I seek nothing except you, though no reward be promised in return; even if there is no hell and no paradise, nevertheless thanks to your sweet goodness, for the sake of you yourself I wish to cleave to you. You are my ceaseless meditation, my word, my work. Amen.

Source: Cistercian, although attributed to Saint Anselm, Meditation X:
On the Passion of Christ, ca. 1175, PL 158, cols. 761–62.

II.

For then [in 1096] rose up the arrogant, the barbaric, a fierce and impetuous people, both French and German. They set their hearts to journey to the Holy City, which had been defiled by a ruffian people, in order to seek there the Sepulcher of the crucified bastard and to drive out the Muslims who dwell in the land, and to conquer the land. They put on their insignia and placed an idolatrous sign on a horizontal line over a vertical one on the clothing of all the men and women whose hearts impelled them to go on the stray path to the grave of their Messiah. . . .

It came to pass that, when they traversed towns where there were Jews, they said to one another, Behold we journey a long way to seek the idolatrous shrine and to take vengeance on the Muslims. But here are the Jews dwelling among us, whose ancestors killed him and crucified him groundlessly. Let us take vengeance first upon them. . . . [The crusaders began attacking and attempting to kill or forcibly convert Jews in cities along the way.] . . . The young [Jewish] women and the brides and the bridegrooms gazed through the windows [at crusaders who were attacking them] and cried out loudly, Behold and see, O our God, what we do for the sanctification of your holy Name, rather then deny you for a Crucified One, a trampled and wretched and abominable offshoot, a bastard and a child of menstruation and lust.

. . . The enemy [crusaders] did battle against [the Jews] till evening. When the saintly ones saw that the enemy was stronger than they and that they would be unable to withstand them any longer, they bestirred themselves and rose up men and women and slaughtered the children first. Subsequently the saintly women threw stones through the windows against the enemy. The enemy threw stones against them. The [Jews] took the stones, until their flesh and faces became shredded. They cursed and blasphemed the crusaders in the name of the Crucified, the impure and foul, the son of lust: Upon whom do you trust? Upon a rotting corpse! The crusaders advanced to break down the door. . . .

Source: *The Chronicle of Solomon bar Simson,* ca. 1140, describing events of 1096, adapted from a translation by Robert Chazan, in *European Jewry and the First Crusade* (Berkeley, Los Angeles, and London, 1987), 243–44, 255, 258.

III.

Whoever sees on the cross
Jesus, his beloved,
(Sorrowful stood by him, weeping,
Saint Mary and Saint John),
His head all around him
with thorns a-pricked,
His fair hands and his fair feet
with nails a-sticked,
His back by rods scourged,
   His side with spear wounded,
All for the sins of man,
Grievously may he weep
and bitter tears shed,
The man who knows about love.

Source: *Man's Leman (Beloved) on the Rood,* thirteenth-century Middle English lyric, from Trinity College, Cambridge, MS 323, edited in *English Lyrics of the Thirteenth Century,* ed. Carleton Brown (Oxford, 1932), 34: 61–62.

IV.

Behold now, mortal wretches, he who suffered death for your life. Thus [suffered] sweet Jesus, the king of heaven, to win your love as knights used to do. He came to the tournament and for the love of his love, which is our soul, bore his shield in all parts of the battle as a valiant and a hardy knight. His shield, which covered the Godhead, was his blessed body that was spread upon the hard cross. There he appeared as a shield in his arms, with his hands strained and pierced and his feet nailed down, as some men say, the one upon the other. . . . This shield is given us for our defense against all evils and all temptation. As Saint Jerome says, Lord Jesus, thanked be you, you have given us a shield for our hearts, which is the thought of your painful travail [Lam. 3:65]. . . .

In this shield are three things: the one is the wood, the other is the skin, and the third is the color. This it what Jesus Christ has left you of the shield: the wood of the Cross, the leather of the painful passion of our Lord, and color of his red blood. Then, the skin of his precious body was all torn and broken, and colored with his red blood, and the cross also. The third reason for this shield is that after the death of a valiant knight, men should show his shield in remembrance of him. This shield is the crucifix that is set in the church where men may see and think of the chivalry that our Lord Jesus Christ did on high upon the cross on the Mount of Calvary before the eyes of his blessed mother. This shield is hung up in every church so that his love, which is our soul, may behold how dearly he has bought her. He does not refrain from bearing his shield and opening his side to show his heart, and showed all openly how entirely he loved her [the soul], and how she ought often to think of the tokens of love that are signified in this shield. Whereof said Saint Bernard, O! You blessed and happy spouse of Christ, behold on the crucifix the shield of Jesus Christ your spouse! And see the inclination of his head to kiss you; see the spreading of his arms to embrace you; behold the opening of his side and the crucifying of his fair body; and with great affection of your holy love, turn it and turn it again from side to side, from the head to the feet, and you shall find that there never was sorrow or pain like to that pain our lord Jesus Christ endured for your love.

Source: *Tretyse of Love*, late fifteenth-century English translation of a French devotional tract, *The Tretyse of Love*, ed. John H. Fisher (London, 1951), 12–16.

The Body of Christ nailed on the Cross was surely one of the most familiar of all artistic images in the high Middle Ages. In keeping with a growing trend from around 1100 on to focus religious devotion on the humanity of Christ, depictions of Christ's passion and death proliferated across Europe in both public and private settings. Crucifixes of wood, metal, enamel, ivory, cloth, or paint stood on the altars of churches (fig. 23.1), were suspended on choir screens (fig. 23.2), housed relics of the saints, were embroidered on liturgical robes, and decorated the covers of Bibles and the pages of prayer books (fig. 23.3). But how was this image read? How did images of the Crucifixion help medieval Christians understand or conceptualize their faith?

The artworks reproduced here give an idea of the range and quality of crucifixion imagery available to medieval Christian viewers. Figure 23.1 is a particularly fine but still representative example of a Romanesque crucifix that is a type of image made between about 800 and 1150. In keeping with contemporary religious texts that extol the glory of the Savior, this Christ, although suspended on the Cross that was the instrument of his death, is more Lordly Redeemer than Suffering Servant. His body is strong and whole, his posture is erect and alert, his eyes are preternaturally large and conspicuously open, his mustache and beard are smooth and well groomed, his expression is proud and composed. His head seems to bow in gracious acknowledgment of the viewer's regard. The cross on which Christ is

Fig. 23.1. Ivory Altar Cross, Spain, ca. 1063, height 54.2 cm, Archivo Fotográfico, Museo Arqueológico Nacional, Madrid.

Fig. 23.2. Wooden Crucifix, South Scandinavia (Skaane?), ca. 1250, height 231 cm,
The Statens Historiska Museum, Stockholm. Courtesy of the
National Heritage Board, Stockholm.

Fig. 23.3. Painting of the Crucifixion, Missal (Book of Prayers for the Mass), Paris, ca. 1315–30, height 300 mm, Bibliothèque nationale de France, Paris, ms. Lat. 861, fol. 147v.

suspended—or rather, against which he stands erect—is decorated with scenes of the Resurrection and the Last Judgment over which he presided. The Christian who looks upon this image finds his gaze returned by a figure who has stared down what all mortals fear most, death, and has emerged triumphant.

Figure 23.2, which dates to the middle of the thirteenth century, illustrates how this basic form took on a new aspect as the Middle Ages progressed. This Christ echoes the Romanesque ivory in projecting a sense of restraint and dignity, and the angle of his head and the position of his arms and feet are almost identical to those in the earlier piece. Yet the feel of this crucifix is completely different. Christ's brow furrows faintly over half-closed eyes, his cheeks are sunken, his head bows involuntarily with fatigue, his arms pull at his emaciated torso so that each rib shows clearly, and on his head rests a Crown of Thorns, sign of his tormentors' mockery and token of his own humility. In his calm avowal and acceptance of affliction, this crucified Christ appears a perfect balance of the human vulnerability and divine compassion central to a religion that situates salvation in the story of God-made-Man.

Figure 23.3 epitomizes the intensified emphasis on Christ's suffering and death apparent in art and devotional literature from about 1250 or 1300 (see text III). Christ's shoulders now sag far below the horizontal bar of the Cross, dragged down by his own weight and weakness. His eyes are fully closed and his head hangs heavily to one side. His slumping body is forced into an exaggerated curve by the single nail that pierces both his feet. Blood gushes conspicuously from the wounds in his side, hands, and feet. Christ's tortuous pose is echoed in the mannered stances of the figures at his side, revealing that the agony of the dying man is matched by the grief of the Virgin Mary and Saint John. The diagonal slant of the placard attached to the top of the cross (INRI, the Latin abbreviation of the derisive phrase Jesus of Nazareth, King of the Jews) and the storm clouds gathering in the sky overhead contribute to the sense of dislocation and anguish conveyed by this rendering of God's ultimate sacrifice.

And yet, although we can trace in these different depictions of Christ's body visual characteristics that seem to align with contemporary devotional trends, these iconographical features and doctrinal developments by no means fully limited or controlled how the images were read. The texts translated above provide tantalizing clues concerning what medieval people saw when they contemplated the crucified Christ, and suggest that the range of ideas and emotions evoked by such images is far broader and more fluid than we might expect.

Text I is a devotional meditation attributed to, and certainly influenced by, Saint Anselm of Canterbury (d. 1109); it has also been attributed to Saint Bernard of Clairvaux (d. 1153). An Italian nobleman who became a monk, then abbot at the Norman monastery of Bec before being named Archbishop of Canterbury, Anselm is credited with first articulating, if not with initiating, the shift to a more intimate and human-centered conception of Christ. His most famous work, the treatise *Cur Deus Homo?* (*Why Did God Become Man?*) emphasizes the love expressed by God in taking human form and sharing human mortality. Both emotional and bodily experience consequently permeate the Anselmian approach to Christianity.

In text I, *Meditation X,* one gets a sense of the extent to which these emphases heightened the author's awareness of and response to visual representations of Christ. The very first sentence makes it clear that the author has in his mind, and probably before his eyes, an image of the Crucifixion. He *sees,* in a very vivid way, the slant of Christ's head, the spread of his arms, the details of his punishment. And yet, what he sees is not necessarily what we would expect him to see. The author displays a remarkably fragmented and piecemeal visual method; rather than taking in the ensemble as a whole, he passes from individual element to individual element, just as he extracts isolated and originally unconnected biblical verses from the totality of scripture.

One notes also that although he is looking at an inanimate object, a passive figure, a crucified corpus, the author's description focuses on *motion.* Jesus speaks, embraces, displays love. But just as the author interprets his biblical texts spiritually or allegorically, disregarding their apparent surface meaning and narrative context, so his reading of these motions is by no means a literal one. If he were regarding a crucifix similar to figure 23.1 (as is most likely, given the date of the *Meditation)* he might be expected to read the slant of Christ's head as the stately and condescending nod of a victorious lord. If he were contemplating an early version of an image along the lines of figures 23.2 or 23.3 (as the reference to a single nail in Christ's feet and the opening in his side suggests), he might be expected to read the droop of the head as an expression of pain, fatigue, or despair. Instead, the author interprets Jesus's gestures in ways unconnected to the ostensible subject of the piece, the act of dying. Christ's head does not sag; he graciously inclines it. His arms are not fixed to the cross; he freely extends them. No tormenter pierces Christ's side; he himself voluntarily opens it.

This is a process of seeing infused with imagination. The author starts from a theological position that regards Christ's sacrifice as an act of love, and builds upon an exegetical tradition of construing the bridegroom in the *Song of Songs* as a figure for Christ, but he does not stop there. He goes beyond his textual sources to draw inspiration and emotional power from his own lived experience, evoking feelings associated with kin relations, social rituals, even hunting. The guiding principles that facilitate this creative rereading of the Crucifixion are *resemblance* and *relation.* The author roams freely within his visual memory, personal history, and cultural world in his search for images and gestures similar to, and therefore of significance for, the images and gestures of the artwork before his eyes. The likeness of the executed Messiah is accordingly transformed into a variegated portrait of a generous and loving kinsman, a beneficent dispenser of largesse, and a captured quarry. The reader/viewer is encouraged to feel, in turn, the warmth and joy associated with parental love, the anticipation and pleasure associated with the receiving of gifts, and the mixed sorrow and triumph felt when cornering a noble beast. The ultimate effects of this technique of image reading are to deepen and expand the emotional power of art, to turn a silent and static image into a vehicle for communication with the divine. This *Meditation on the Passion* uses visual contemplation of the Body of Christ to simultaneously arouse in the viewer/reader love and pity *for* God, while feeling himself loved and pitied *by* God.

Beauty plays a prominent role in the *Meditation* on the image of the Crucifixion. The final paragraph is a paean to Christ's loveliness, and indeed it is not surprising to find beauty so important to a man who clearly drew powerful inspiration from visual experience. Yet the *Meditation* begins with an acknowledgment that Christ's mouth and kiss are *not* visually appealing, that they are, in fact, without beauty and loveliness. Why would the author write this? This concession, and the author's very need to transform the surface meaning of the image to reinterpret a crucified body as an embracing one, suggests that not all who looked upon the crucifix naturally and inevitably felt ardently positive emotional or aesthetic responses to the sight. Indeed, Saint Anselm tells us at the beginning of *Cur Deus Homo?* that he was prompted to compose the treatise because unbelievers had scoffed at various aspects of Christian faith, most especially the idea that God would become a man and suffer death. Text II, selections from a Hebrew chronicle describing crusader attacks on Jews in Germany during the First Crusade (1096), testifies to some very different reactions to images of the Crucifixion.

Although the Jewish author is obviously and understandably hostile to the crusaders, his version of their motives for going on crusade closely matches those articulated in Christian crusading texts. The Crucifixion loomed large in crusader ideology and rhetoric—the very name the warriors assumed recalled the instrument of Christ's death, the insignia of the cross was sewn on their clothing, and the site of the Crucifixion was their ultimate destination. These crusaders, mostly lesser noblemen and knights from northern France and Germany, lived at exactly the same time as Anselm, yet when they looked upon a crucifix, they did not seem to see the same pretty thing infused with sweet goodness. Although they, too, saw a human body, and were deeply moved and inspired by it, they did not see beauty, hope, or the grace of a loving father. Inflamed by the preaching of crusade promoters, and reading the body of Christ in light of their own warlike lay culture and brutal daily experience, they saw instead a disturbing and infuriating sight—a murdered lord whose blood cried out for vengeance.

To Jews, this same image had a different meaning still. Like many contemporary Christians, these Jews valued suffering as an expression of sanctity, holding up their own shredded flesh as an emblem of their devotion to God. They, too, saw the wound in Jesus's side and the sag of his head as evidence of his humanity. But to the victims of Christian persecution, this was not the sign of a compassionate God-made-Man, but proof of the mortality of a very flawed, and ultimately failed, human being: a rotting corpse. For both the crusaders and these German Jews, caught up in swift-moving events and violent times, images seemed not to open new vistas or spur serious reflection, but to provide affirmation for decisions already made, paths already chosen.

Text III is a devotional lyric (song) in the vernacular, probably composed by and for laymen to be sung at village or town festivities and religious celebrations. It reveals explicitly what the other texts examined so far have implicitly suggested: that gazing at a crucifix provided a powerful stimulus for devotion. It also suggests a particular approach to be adopted in contemplation of the crucifix. The figure on the cross is called by his given human name, Jesus, rather than by his religious title, Christ, and throughout the lyric an intimate tone is maintained. The lyric first advises the viewer/singer/listener to note the behavior and expressions of Mary and John, located on either side of the crucified Christ (see figure 23.3), in order that he might take emotional cues from their anguish and feel kinship with them in their shared grief. The viewer/singer/listener then is to consider each detail of Jesus's torment, paying particular attention to the wounds caused by the crown of thorns, the nails, the rods, and the spear. The rhythm and rhymes of the song at this point seem to echo the very actions of Jesus's persecutors—i-prikit, y-stickit, y-vundit—heightening through sound as well as through sight the emotional impact of the experience. Having duly seen, heard, and thus imaginatively felt Jesus's affliction, the viewer/singer/listener is reminded of its source—his own and his neighbors' sins. Finally, the viewer/singer/listener is invited to display his understanding of true love by emulating Mary and John in weeping for his dead beloved. In this way, the viewer/listener has become one with the group of mourners around the Crucifixion. The barriers between object and audience, between art and life, have been lowered, and through singing or hearing this song, and gazing at a crucifix, a thirteenth-century Christian feels himself closer to Christ.

Text IV is a late fifteenth-century English version of a French devotional tract that was a translation of a thirteenth-century treatise written for three noble anchoresses (female religious recluses). It is in many ways quite similar to Text I—in fact, it concludes with a paraphrase of the Anselmian *Meditation*, which it mistakenly attributes to Saint Bernard—yet the techniques and emphases of this *Tretyse* are quite different from the earlier work, just as the techniques and emphases of figure 23.3 are different from those of figure 23.1. The audience for this *Tretyse of Love* was almost certainly elite laywomen, and throughout the text one can see how religiosity is contoured to suit their society and tastes.

   Although this excerpt opens with an exhortation to the reader to behold the crucified body of Christ, and is contemporary with images that very vividly depicted Christ's tribulations, the subsequent passage does not dwell only or even primarily upon his agonies. Instead, the treatise harnesses the language and imagery of courtly love to present sweet Jesus as a valiant and hardy knight jousting to win the viewer/reader's heart. But if the recasting of the dying Jesus as chivalrous champion arose from the reader's social milieu and literary culture, and seems unrelated to contemporary representations of the Crucifixion, the specifics of the allegory are nevertheless dictated by visual considerations. In fact, throughout the treatise sentimental emotionalism is combined with a refined aesthetics, and as much attention is paid to the shape, materials, colors, and placement of the crucified Christ as to his pain and torment. One gets the sense that the reader of this treatise is being trained to contemplate the crucifix as carefully as she might choose an expensive fabric or decorate an elegant apartment.

The comparison of Jesus to a shield, although prompted by the cited verse from Lamentations, is energized by the fact that it matches the reader's knowledge of actual shields. The treatise points out that the triangular shape created by Jesus's limbs on the cross traces the shape of a shield; Jesus appeared as a shield in a direct, visual sense as well as metaphorically. Similarly, the treatise notes that the placement of the crucifix set high in the church mirrors the placement of knightly shields, hung high on castle ramparts during tournaments of chivalry. Jesus's crucified body is composed of the same materials as those from which shields were made: wood, leather or skin, and color. Tactile and visual experience thus deepens the reader/viewer's experience of the crucifix; she is invited to feel the hardness of the wood and the soft vulnerability of Jesus's flesh; the shocking, violent redness of his blood accentuates the violence of his execution.

The *Tretyse*'s consistent juxtaposition of emotionalism and elegance, of grief and preciousness, is remarkably reminiscent of figure 23.3's setting of a poignant, even tragic crucifixion grouping within a ornate and elaborately gilded framework. In fact, the *Tretyse of Love* has been dubbed a Romance Gospel. Like the fourteenth-century illumination, it guides the viewer/reader toward a better appreciation of Christ's sacrifice by locating the tragedy within a pleasing and familiar frame, and depicting gospel events in accord with cultivated artistic tastes and sophisticated social mores. The goal of both literary composition and artwork is to make Christ's death and suffering meaningful to the late medieval Christian elite, who were accustomed in everyday life to equate status with grace and luxury, and to regard the failed, the naked, and the dead with revulsion. Through visual beauty and literary refinement, God has been brought to resemble an earthly beloved, that the Christian might love his or her deity all the more. And images not only helped inspire such love but also allowed for its physical, tangible expression—the now-blurred drawing of the crucifix in the bottom margin of figure 23.3, copied there so that the main image might not be similarly marred, displays the effects of generations of heartfelt kisses offered by the image's viewers.

The images and texts discussed above suggest several observations concerning how Christians interacted with Christian imagery in the Middle Ages. The most important lesson to be learned is that no single, fixed reading is provoked by an image, no matter how straightforward or explicit the image may seem. Many elements contribute to the meaning of an image: the style, material, color, size, composition, placement, and context all render apparently similar objects radically different in impact. Probably the most important factor of all is the audience—the one aspect of a medieval artwork that is most difficult for modern scholars to reconstruct. The texts discussed above make it clear that viewing the crucified body of Christ could be a very different experience for a monk in his cell, a warrior on the road, a Jew seeking sanctuary, or a noblewoman in her chamber. Depending upon who was viewing and in what situation it was viewed, a crucifix might inspire comfort about the afterlife, understanding about the Godhead, gratitude for God's sacrifice, or rage for his death; it might be meditated upon, glared at, sung to, wept over, or kissed. Rather than merely teaching a religious lesson or recalling a biblical narrative (as the usual definition of the function of medieval art has it), a devotional work of art must be seen as initiating a dialogue or conversation. The other elements contributing to this conversation include texts, both written and spoken, that were familiar to the viewer of the object, as well as the viewer's situation, society, lived experience, and store of remembered and imagined visual images. Medieval responses to and uses of the crucifix are as varied, contradictory, and creative as the images reproduced here, and as life itself.

Further Reading

David Aers, "The Humanity of Christ: Reflections on Orthodox Late Medieval Representation," in David Aers and Lynn Staley, *The Powers of the Holy Religion: Politics and Gender in Late Medieval English Culture* (Philadelphia, 1996), 15–42.

Rachel Fulton, "Praying to the Crucified Christ," in *From Judgment to Passion: Devotion to Christ and the Virgin Mary, 800–1200* (New York, 2002), 142–92.

Sara Lipton, "The Sweet Lean of His Head: Writing about Looking at the Crucifix in the High Middle Ages," *Speculum* 80 (2005): 1172–1208.

Henk van Os, *The Art of Devotion in the Late Middle Ages in Europe, 1300–1500* (Princeton, 1994).

Sixten Ringbom, "Devotional Images and Imaginative Devotions: Note on the Place of Art in Late Medieval Piety," *Gazette des beaux-arts* 73 (1969): 159–70.

Gertrude Schiller, *Iconography of Christian Art*, vol. 2, *The Passion of Christ*, trans. Janet Seligman (Greenwich, CT, 1972), 88–164.

André Vauchez, *The Laity in the Middle Ages: Religious Beliefs and Devotional Practices*, trans. Margery J. Schneider, ed. Daniel E. Bornstein (South Bend, IN, 1993).

Healing

—24—

The Old English Nine Herbs Charm

Debby Banham

Remember, mugwort, what you revealed,
what you whispered at the great denunciation.
You are called "una," oldest of plants;
you have power against three and against thirty,
you have power against poison and against what flies in,
you have power against the hateful thing that journeys through the
    land.
And you, plantain, mother of plants,
open toward the east, powerful inside,
carts have rumbled over you, queens have ridden over you,
brides have trodden over you, bulls have snorted over you;
you withstood everything then, and you crashed against everything,
so may you withstand poison and what flies in
and the hateful thing that journeys through the land.
This plant is called "crashing"; it grew on stone;
it stands against poison, it crashes at pain.
It is called "sturdy," it crashes against poison,
it drives out badness, it casts out poison.
This is the plant that fought against the worm;
this has power against poison; it has power against what flies in;
it has power against the hateful thing that journeys through the land.
Now you, betony(?), the smaller, put the greater to flight,
the greater the smaller, until there is a remedy for both of them.
Remember you, mayweed, what you revealed,
what you ended at the alder ford,
that never gave life for what flies,
since mayweed was prepared as food for it.
This is the plant that is called "curse";
the seal sent this out over the sea's ridge

as a remedy for the hatred of another poison.
These nine weapons against nine poisons.
A worm came crawling, it wounded no-one;
then Woden took nine glory-twigs,
then hit the adder, so that it flew apart into nine.
Apple and poison ended there
so that it would never turn toward a house.
Chervil [or thyme] and fennel, two great powerful ones:
the wise Lord created these plants,
holy in the heavens, when he hung,
set and sent them into seven worlds
for poor and wealthy, a remedy for all.
It stands against pain, it crashed against poison,
it has power against three and against thirty,
against the enemy's hand, against a lordly trick,
against enchantment by vile beings.
Now these nine plants have power against nine glory-flown ones,
against nine poisons, against nine things that fly in,
against the red poison, against the foul poison,
against the white poison, against the blue poison,
against the yellow poison, against the green poison,
against the pale poison, against the blue poison,
against the brown poison, against the purple poison,
against the worm blister, against the water blister,
against the thorn blister, against the [thistle] blister,
against the ice blister, against the poison blister,
if any poison come flying from the east
or any comes [flying] from the north
or any from the west over the human race.
Christ stood above the old things in a unique way.
I alone know the running streams,
and the nine adders observe;
all weeds must now grow away from plants,
the seas disperse, all salt water,
when I blow this poison away from you.

Mugwort, plantain open to the east, lamb's cress, betony(?), mayweed, nettle, wild sour apple, chervil [or thyme] and fennel, old soap; work the plants to a powder; mix with the soap and the waste from the apple. Make a lye from water and ashes; take fennel, boil it in the lye, and bathe with the mixture when he puts the salve on, both before and afterward. Sing the charm over each of the plants three times before he processes them and over the apple in the same way, and sing the same charm to the patient, into the mouth and both ears and into the wounds, before he puts the salve on.

Source: *Anglo-Saxon Remedies, Charms, and Prayers from British Library MS Harley 585: The Lacnunga*, ed. and trans. Edward Pettit (London, 2001), item no. LXXVI.

The "Nine Herbs Charm" is part of a collection called by modern scholars *Lacnunga*, meaning "treatments" in Old English. This title is somewhat misleading, since, while the collection does include plenty of straightforward medical recipes, it also contains prayers and a large proportion of the surviving Anglo-Saxon magical material, some of it medical, but some for quite different purposes, including agricultural ones. The collection is found in a single manuscript, British Library Harley 585, written around the year 1000. This manuscript also contains the Old English *Herbarium*, a widely used medical text translated from the Latin, so presumably the motivation for the compilation of the manuscript as it stands was an interest in medicine. However, it is not clear whether the *Lacnunga*, with its more miscellaneous contents, was originally assembled for this manuscript or copied from an earlier one. We know from other surviving manuscripts, such as the great Exeter Book of Old English poetry, that there was a good deal of interest in tenth-century England in preserving in writing aspects of vernacular culture that had previously been handed down orally, in much the same way as folk songs were recorded in the early twentieth century. This may also have been at least part of the reason that the *Lacnunga* was put together.

Within the *Lacnunga*, the "Nine Herbs Charm" forms a distinct, although short, section, surrounded by material that is only very loosely related. In the old edition of Grattan and Singer, the charm was divided into four sections with separate titles, but there is no authority for this in the manuscript. The only inherent distinction is between the charm itself, in verse, and the prose instructions at the end, but the instructions refer back to the plants mentioned in the body of the charm, showing that the whole belongs together.

The text is very difficult to translate in places, and the above version can only be provisional at best; where words are missing in the manuscript, those supplied are given in square brackets. Once translated, the charm is far from easy to interpret: for instance, we have no way of knowing what the "great denunciation" may have been (even if this is the correct translation), nor what mugwort may have revealed (an alternative rendition would be "betrayed") there, although we may suspect it was something to do with its medicinal or magical virtues. Similarly, we cannot tell what the name "una" might have meant; it may or may not be relevant that this is the feminine singular of "one" in Latin.

A few points are at once obvious to the reader. First, the plants mentioned in the charm do not correspond exactly with the list in the instructions. Most commentators believe quite reasonably that they are intended to correspond, in which case the following would be equivalents:

"crashing" = lamb's cress
"sturdy" = nettle
"curse" = crab apple (probably "wild sour apple")

However, if the text is taken as having any authority, *wergulu* ("curse") should be a marine plant. The charm also reads as if "crashing" and "sturdy" are meant to be alternative epithets of the same plant. One explanation for these discrepancies would be that the instructions are a later attempt to make practical a text inherited from a previous, and ill understood, age. In that case, the plant names given there may be intended to explicate unfamiliar ones in the charm itself. "Lamb's cress" remains unidentified (there is no reason to think it must be the plant known by that name in modern English), and the translation "betony" for Old English *attorlothe* ("poison hatred") is by no means certain.

Second, although the charm refers to vaguely defined poisons and flying things, the instructions make it clear that the end product is a wound salve (with accompanying lotion). This combination of a rather prosaic physical preparation with a supernatural incantation is not uncommon in medieval recipe collections, but the "Nine Herbs Charm" is probably the longest, most complex, and most obscure. There seems little possibility of refining our understanding of "what flies in" or the "hateful thing that journeys through the land"; we can only refer them and the various poisons to a generalized idea of illness coming from outside the organism. If there is a distinction between the two, "what flies in" may mean individual infection, the "hateful thing that journeys through the land" an epidemic. If these ideas are to be related to the wound salve, the charm may perhaps be intended to prevent wounds becoming infected, or lesions caused by infections, but this would be to impose modern notions of sense and order onto something that belongs to a very different habit of thought.

A third striking point is that the names of both Woden and Christ appear in the charm. We know that Woden was a pre-Christian Anglo-Saxon deity, and he appears in most of the Anglo-Saxon royal genealogies. The "Nine Herbs Charm" is not unique in containing an apparently pagan name; a charm for ensuring the fertility of land includes the phrase "*erce, erce, erce,* mother of earth," in which *erce* is assumed to be the name of a pre-Christian deity. The question is whether the compiler of the *Lacnunga*, his or her informants, or those who actually used the charm knew that Woden was a heathen god. In other words, is the charm an example of pagan practice, albeit partly christianized, surviving some four hundred years after Christianity first reached the Anglo-Saxons, or is the name Woden a mere superstition, literally left over from an extinct religion? This question is probably unanswerable, especially in view of the fact that many customs that would have horrified the modern or later medieval Church were tolerated in the early Middle Ages, so that things that might seem quite clearly pagan to us were seen as outside the realm of religion. Compare Shakespeare's references to Puck's activities in *A Midsummer Night's Dream*, which do not imply that the playwright or his audience were not Christian.

It is interesting that the passage on the Lord creating herbs, soon after the mention of Woden, although it uses the conventional vocabulary of the Christian creation myth, also refers to the Lord "hanging in the heavens." This may refer to the crucifixion, but Odin, the Norse counterpart of Woden, was believed to have received wisdom while hanging upside down. There is no other evidence that the pagan Anglo-Saxons believed the same of Woden, but this passage may suggest a folk memory of such a story.

This group of nine plants (the charm tells us there are nine, although they can be counted in various ways in the text) does not appear together anywhere else. Nor, apart from betony, which is only doubtfully part of the group, are any of them credited elsewhere with outstanding powers. It is therefore going beyond the evidence to refer to them as "sacred," as some commentators have done. It may be a mere accident of survival that we have a charm mentioning these nine; others may have been in circulation describing quite different plants in similar terms. Moreover, it is quite impossible to infer from these verses, tenth-century in their surviving form, what may have been believed in England when Woden was still worshipped as a god.

If the charm did have its roots in pre-Christian England, it must have been passed on by word of mouth for up to four hundred years before it was written down in Harley 585 or its exemplar. Texts are not fixed by oral transmission as they are when recorded in writing, but change each time they are retold. In four centuries the retellings would have been innumerable, and the changes untraceable, so that the written text might bear no resemblance to its putative pagan predecessor. Nevertheless, the charm retains its fascination as something quite different from either the orthodox Christian poetry or the conventional subclassical medicine of 1,000 years ago.

Further Reading

M. L. Cameron, *Anglo-Saxon Medicine* (Cambridge, 1993).

Alaric Hall, *Elves in Anglo-Saxon England: Matters of Belief, Health, Gender and Identity* (Woodbridge, 2007).

Karen Jolly, *Popular Religion in Late Saxon England: Elf Charms in Context* (Chapel Hill, NC, 1996).

— 25 —

Amulets and Charms

Peter Murray Jones

I. JOHN OF ARDERNE (1307–1392), AMULET FOR SPASM

This following charm against spasm has been found very useful by those who tried it in parts overseas as well as here. For it was used at Milan in Lombardy when Lord Lionel, son of the King of England, married the daughter of the Lord of Milan. The English there were troubled by spasm after drinking the powerful and hot wine of the country and too much over indulgence. At that time a certain knight and son of Lord Reginald de Gray of Shirland near Chesterfield was at Milan with Lord Lionel and had with him the following charm. A soldier was troubled by spasm so that his head was bent back almost to his arse, rather like a crossbow, who almost died for pain and anguish. The said knight saw this and took the charm written on parchment, kept in a closed purse, and put it around the neck of the patient. The bystanders said the Lord's Prayer with the Ave Maria, and as he swore to me within four or five hours he was restored to health. Afterward the knight used it to free many others from spasm, gaining great fame for the charm in that city . . .

Whoever wears this charm following on his person will be freed from spasm. Make the amulet in this way. Take a sheet of parchment and write first the sign + Thebal + Guthe + Guthanay + In the name of the father + and of the son + and of the holy spirit amen + Jesus of Nazareth + Mary + John + Michael + Gabriel + Raphael + the word was made flesh + Afterward the sheet was closed up in the form of a letter so that it could not easily be opened. Whoever wore this amulet on them honestly and acted in the name of God omnipotent and believed without any doubt, would not be harmed by spasm. This charm should be kept in reverence on account of God who gave virtue to words, to stones, and to herbs, and kept secret so that no one else can know, in order not to lose the virtue given it by God. For I have seen a woman troubled by spasm and she was given a gold ring with a beautiful peridot stone, with the three aforesaid words of the charm Thebal etc. engraved on the circle of the ring, yet it did her no good. As soon as she wore the same charm closed up on her she was delivered from spasm without delay. And so it appears that it is not right to keep the charm in the public domain, but it should be kept secret. And note that I used to write the aforesaid charm with Greek letters so that it couldn't be seen by lay folk, in the form + In the name + of the father + and of the son + and of the holy spirit amen. I am called master John Arderne the surgeon, and lived at one time in Newark near Lincoln, and later in London; I have seen and heard the aforesaid marvelous things about this charm, and a great number of other things.

Source: London, British Library, MS Sloane 56, fol. 6v.

II. THOMAS FAYREFORD (FL. 1400–1450), AMULET FOR EPILEPSY

For epileptics. Write these three names with blood taken from the little finger of the patients + Jasper + Melchior + Balthasaar + and put gold, frankincense, and myrrh into the same sheet of parchment. The patient should say three Paternosters and three Ave Marias for the souls of the fathers and mothers of the aforesaid three kings, and let the patient drink each day for a month in the morning the juice of peony with beer or wine. Write on the same parchment + ananizapta + ananizapta + ananizapta + and add to the parchment one branch of mistletoe and wear it around the neck. Three Trinity masses should be sung for the souls of the aforesaid parents of the kings. And the patient is also advised to eat each day peony root for a time, and wear the same around his neck, and he should eat also a diet suitable for that disease.

Source: London, British Library, MS Harley 2558, fol. 119r.

III. THOMAS FAYREFORD, BLESSINGS OF HERBS

When you gather herbs to help the sick, walk three times around the plant and say this: "take the herb in the name of the Father and the Son and the Holy Spirit. And I pray to my Lord God that this herb be good and effective for the medicine in which it is used." And say three Paternosters and three Ave Marias. And when you have gathered your herbs at the appropriate day or season, bless them in these words: "Almighty, you who have granted virtue to various herbs, deign to bless and sanctify all these herbs. And just as you gave to your apostles the power to trample upon serpents and scorpions, so wherever medicine from these herbs will be provided, let every infirmity and weakness be expelled and your benign grace be given to sicknesses."

Source: London, British Library, MS Harley 2558, fol. 63v.

IV. BIRTHING ROLL

Pope Innocent VIII has granted that whosoever, man or woman, wears the
length of the nails on him, and devotedly worships the three nails of our Lord
Jesus Christ with five Paternosters and five Aves and a creed, will have five
blessings. The first is that he shall not suffer sudden death or an evil death; the
second, he shall not be slain with a sword or a weapon; the third, that his enemies shall have no power to overcome him; the fourth, that no poison or false
witness shall affect him; the fifth, that he shall be sufficiently wealthy and have
a good income in this world; the sixth, that he shall not die without receiving
the Holy Sacraments of the Church; the seventh, that he shall be freed from all
wicked spirits, fevers, plagues, and other spites. And this length is the same as
that of Christ's nails, which should be regarded as relics, and worshipped devoutly by saying five Paternosters and five Aves and a Creed.

This is the measure of the blessed wounds that our Lord Jesus Christ had in
his right side, that an angel brought to Charlemagne, the noble Emperor of
Constantinople, set in a gold coffer. The inscription says that whosoever, man
or woman, shall wear this measure shall not be slain with sword or spear, nor
hurt by any shot, nor be overcome by any man in battle, nor shall fire or water
trouble him. If a woman is in childbirth, and has looked on the aforesaid measure, then she will not perish, but the child will be christened and the mother
churched. This is proven, for every man who goes to battle, who wears the
measure on him will get victory and honor over his enemies.

> Source: London, British Library, MS Rotulus Harley T 11; transcribed in
> Curt F. Bühler, "Prayers and Charms in Certain Middle English
> Scrolls," *Speculum* 39 (1964): 270–78.

V. CHARM TO FIND STOLEN OBJECTS

To find out who stole any thing from you. Write these names in virgin wax and
hold them over your head with your left hand, and in your sleep you will see
who stole the thing: + holy + cross + holy cross + Holy + cross of the Lord + In
the name of the Father and of the Son and of the Holy Spirit Amen.

> Source: London, British Library, MS Additional 34111, fol. 70v; transcribed
> and translated in part by Suzanne E. Sheldon, "Middle English and Latin
> Charms, Amulets, and Talismans from Vernacular Manuscripts,"
> PhD thesis, Tulane University, 1978.

VI. AMULET FOR FEVERS

For all fevers take three communion wafers and write on the first [Tau +] the Father is life [Tau +] and on the second the Son is virtue [Tau +] the Nazarene, and

in the third the Holy Spirit is the remedy [Tau +] King of the Jews [Tau +] and let the fasting patient eat the first wafer early in the morning once it has been dipped in holy water. Afterward take an eggshell full of wormwood and fethernoy [?] and tansey blended together with ale, and do so for three days with the wafers dipped in holy waters as before, for this is tested and found to work.

Source: Oxford, Bodleian Library, MS Ashmole 1443, p. 349.

Charms survive from the Middle Ages as incantations written down by a scribe, or as inscribed objects, which we term amulets. We must not forget too that many of the objects were supposed to work through the materials from which they were made, or through the visual signals they sent out, and very often sacralized in one way or another. It makes sense to look at the wearing of amulets as a form of ritual behavior. Charms written down in late medieval England, which make extensive use of liturgical motifs, give us a clue here. These charms require the user to say the forms of the liturgy and the names of power out loud over the patient, and sometimes also require certain specific actions from both the practitioner and the patient. They require a performance. The words and the ritual belong to a Christian cosmology in which the divine can be invoked and consequently operate in various areas of life, not confined to the context of public liturgical ritual. But many of these charms prescribe a performance of a slightly different kind—the writing down of the ritual formulas and words of power on parchment or other substances. Inscribed words and signs were thought of as enabling the transfer of power from the domain of the divine to earth, and thereby effecting changes in the physical state and status of the person who carried the inscription on them.

Amulets give objective form to the power invoked in these performative charms. Once the amulet has been sacralized, whether it be by containing a relic, touching it to a shrine, or even simply by inscribing sacred characters upon it, the wearer of the amulet has constant physical contact with an object that is a kind of material sediment left behind by that flow of divine power. The amulet also acts as a focus of devotion, for both the theological and medical perspectives on amulets stressed the importance of the state of mind of the wearer. The object is therefore also a reminder of the devotion owing to God and the saints, without whose intervention no protection will avail. As a three-dimensional object, the amulet can draw upon the meanings attached to the material from which it is made, the iconography and decoration that may be employed, as well as the inscription that invokes the power of the divine. Each of these features reinforces the others, amplifying that power the amulet has absorbed in sedimentary form.

I. Amulets were used most obviously to give protection against danger and disease, but might also be used to find lost objects, to attract a lover, or to divine some future outcome. Words of power tended to be written in Latin, even if the instructions for the amulet were in the vernacular; one writer, the English fourteenth-century surgeon John of Arderne, suggested hiding such words from the ignorant by writing them in Greek characters. He clearly believed that amulets worked better if kept secret rather than displayed on some ring or open parchment. The story Arderne tells us shows that not all charms and amulets were the province of wise women and cunning men, but that they could be owned and used by knights and lords. We know that the nobility treasured inscribed jewels and badges, and that the same words of power were featured on objects made of gold and of tin.

II. Epilepsy, like spasm, was one of the illnesses for which charms and amulets seem to have been peculiarly appropriate. Thomas Fayreford, a fifteenth-century medical practitioner in Devon and Somerset, found no difficulty in combining writing words of power in blood, the saying of masses, the use of precious substances like gold, frankincense, and myrrh, and magical herbs like mistletoe and peony, with an orthodox medical diet. There is no barrier here between academic medicine and the use of charms and amulets. The word of power he uses "ananizapta" is also found on the Middleham Jewel, a gold reliquary pendant probably worn by an English noblewoman of the mid-fifteenth century.

III. Gathering herbs to be used in medicines was a ritualized practice, and instructions for these "precationes herbarum" have survived in manuscripts. Thomas Fayreford prescribes two separate blessings, one to be performed while gathering the plant, the other to be performed once they are ready to be given to the patient. He recalls the power given to the Apostles to trample serpents, and in a structure typical of a great number of charms, invokes that same power for the purpose at hand. The distinction between Christian prayer and magic ritual is here very hard to draw, and for Fayreford and his readers would no doubt have been meaningless.

IV. A number of so-called birthing rolls have survived from the Middle Ages. The format of a scroll on which are written a series of prayers, devotions, charms, together sometimes with images, lent itself to attachment to the woman in childbirth, as well as providing convenient prompts to prayer and devotions beforehand. This roll is based upon the measure of the nails with which Christ was crucified; the nails as instruments of the passion attracted their own cult in the later Middle Ages. More than protection in childbirth is promised; the benefits include good fortune and the guarantee of the sacraments before death. Death without confession, absolution, and the sacraments was the worst evil that could befall a human being, and the roll protects both the battling soldier and the woman in the birth chamber.

V. In the margin of the charm to find stolen objects, there is a drawing of a hand with the index finger pointing to the first line of the charm. Many charms are accompanied by crudely drawn images that served not just as convenient markers for ready reference but were themselves meant to promote the function of the charm. The crosses so often found among the words of power are here drawn in Jerusalem style with dots in each of the four quadrants.

VI. Communion wafers were not just sacralized objects but surfaces available for writing and for signs like the Tau cross. The wafers could then be ingested—a symbolic as well as physical process—by the patient, who in fever cases could continue the practice as part of his medicinal diet. The formulas of "probatum est" or "experimentum est," or their vernacular equivalents, serve as warrants that a particular treatment has been tried and tested. In medical terms this established a claim to authority for a treatment with no basis in scholastic logic.

Further Reading

Peter Murray Jones and Lea T. Olsan, "Middleham Jewel: Ritual, Power, and Devotion," *Viator* 3 (2000): 249–90.

Lea T. Olsan, "Latin Charms of Medieval England: Verbal Healing in a Christian Oral Tradition," *Oral Tradition* 7 (1992): 116–42.

Suzanne Eastman Sheldon, "Middle English and Latin Charms, Amulets, and Talismans from Vernacular Manuscripts," PhD thesis, Tulane University, 1978.

Don C. Skemer, *Binding Words: Textual Amulets in the Middle Ages* (University Park, PA, 2006).

Charity

—— 26 ——

A Deaf-Mute's Story

Sharon Farmer

Louis, carriage valet of Queen Marguerite, widow of the blessed Saint Louis, was found by chance at the chateau called Orgelet when he was eight years old. That was fifteen [*sic*—the text should read twenty-five] years before the inquest concerning this miracle. And Gauchier the smith of Orgelet received him and nourished and raised him in his house for twelve years. A youth a bit older than the said Louis had led him to Orgelet and left him there; at first he was sheltered in the house of Aymon.

From the time that Louis came to the chateau of Orgelet and was found there and as long as he who is now called Louis remained there he was deaf and mute. People blasted a horn in his ear and cried to him through a horn that had been placed in his ear, but he perceived nothing and didn't hear. And at the same time they hit him and slapped him hard to prove whether or not he could speak. Nevertheless, he did not say a word, only making the signs of a mute man. And the children of Gauchier threw burning coals on Louis's bare stomach to prove whether or not he could speak and if he was truly mute, and again he did nothing except make the signs of a mute and cast the coals away. And for these reasons he was commonly held to be deaf and mute throughout the chateau. . . . After Louis had been with Gauchier, he resided, deaf and mute, with the Count and Countess of Auxerre, and, once, with John of Sorgy, the bailiff of the said count, and in the kitchen of the count. While he was living with Gauchier, when he was [still young] and not very strong, Louis blew on the fires of the smith to light the forge; and he remembers well that when he had grown stronger he assisted the smith with a hammer and helped out in other ways in the house, having been shown what to do with signs. Later, Louis went with the Countess of Auxerre to Lyon, and he was still deaf and mute. At this time, because the chamberlain of the countess did not want to give shoes to him, the said Louis followed the king, Philip of France, who was carrying the bones of his father, my lord Saint Louis, from Tunis. The deaf-mute Louis lived on the alms of the king's court and of the other nobles who were with the king. And thus he went as far as Saint Denis, where he viewed the entombing of the bones of the blessed Saint Louis, as he now recalls, now that he understands those events, because at the time he did not know what they were doing. Nor did he go there because of Saint Louis or out of devotion for him, nor out of hope to be cured and delivered there, because he did not know or understand anything about God and his saints. However, when he was with Gauchier and his wife and with the countess, he had often seen them go to church and pray there and have devotion, and kneel and raise their eyes with their hands joined together and raised to the sky. For that reason he now went to the church [of Saint Denis], but not because he knew what a church was or what devotion was. . . . And thus it happened that when the blessed king was entombed, because he saw the other men kneeling and praying at the tomb, he too knelt and joined his hands without knowing what he was doing. . . .

And when he went with the king to Saint Denis he was there for three or four days, but he did not know who the king was nor who the barons were, nor did he know one from the other, except for a valet whom he had assisted in leading a horse by the halter on the road. . . . And when he was in Saint Denis he went to the almsgiving of the abbey and thus, because of religious charity, he found enough to eat.

And the last day that he was in Saint Denis, before the hour when people are accustomed to eat, Louis was in front of the tomb in the church, and when he saw that the other men were there on their knees with their hands joined near the said tomb, [he did the same], not out of any devotion that he had nor for any intention except that he saw the others doing this. And immediately he perceived the noise of men and the footsteps of those who were going by and the sound of the bells. Nevertheless, he did not know what this was, and he was so frightened that he did not know what to do, and he very much thought that the people whom he heard speaking would run over him.

And so that same day he left Saint Denis and went toward Paris. And while he was headed in that direction he entered a field and slept there, and when he had slept he felt more secure and stronger, but he did not eat that day until the evening. And when he was in Paris he sought his means in alms, and people gave him enough. It was summertime, and he ate and slept there on the platforms that are on the public road. And ever since that hour when he says he first could hear at Saint Denis at the tomb he could hear and perceive the voices of beasts and men and the sounds of other things. . . . However, he did not understand nor did he know how to judge what it was, because he had never before heard anything. Nor could he speak, because he did not know how to speak or to form words, although from that time he had the capacity to learn to speak, if anyone would teach him.

And after that, by the same route that he had come [to Saint Denis and Paris] he returned, and he recognized the route and the places. And since he had

come from Orgelet to Lyon and from Lyon to Paris he returned from Paris to
Lyon and from Lyon to Orgelet, even though the route from Paris to Orgelet is
much shorter as he now well knows. And along the way he begged for alms as
a mute because he did not know how to talk, even though he could hear. And
at night he lay on the platforms of the towns and public roads.

And when he came to Orgelet he entered the house of Gauchier his lord and
made them understand by the best signs that he knew to show them that he could
hear. He did not know how to explain this well to the inquisitors. And those who
lived in Gauchier's hostel understood what he was indicating, and that he had returned to them as a man who could hear. And for this reason they had pity and
began to teach him in the way that very young children are taught, or even as people teach birds. And they said to Louis, "say 'bread' " and he said "bread." Or they
said to him, "say 'wine' " and he said "wine." And it was the same with the other
words that they taught him.

And afterward, when Louis had been in the house of Gauchier for several
days, the Countess of Auxerre, who was in the chateau of Saint Julien, about
three leagues from Orgelet, heard that Louis could hear. So she sent for him
and he went to her. And to facilitate his learning to speak she placed him with
her kitchen staff, so that he would be with several other people, and ordered
that he be taught to speak. And thus the kitchen staff taught him to speak by
naming certain things for him each day, and if he did not remember the names
the next day, they beat him, just as children are beaten in school when they do
not know their lessons. And from the time that he was able to hear and had
begun to speak Louis was with master John of Maynet, the former bailiff of
John, count of Auxerre, and master John taught Louis his Paternoster and his
Ave Maria, which he recited well and entirely in front of the inquisitors and
their notaries, and all the things contained in his deposition, along with those
of another layman. And while the said Louis was with the countess and master John he returned several times to the house of Gauchier, and there he
learned from Gauchier and his wife and household that they had found him at
the chateau [of Orgelet], and at what age, and he recounted those things in his
deposition.

And when the inquisitors asked Louis who had given him the name "Louis,"
he said that after he knew how to talk he told Gauchier how he had received
his hearing at the said tomb and everything that had happened to him. In response to which Gauchier said to him, "I want you to be called Louis in honor
of Louis, the king of France, who delivered you." And when they asked the
said Louis if he believed that he had received his ability to hear and to speak
through the prayers and the merits of the blessed Saint Louis, and he had responded "yes," they asked him, "why do you believe this, since at that time
you had no belief or faith or devotion for him, and you had come to the tomb
by happenstance?" And he responded that he knew of no other cause of his belief except that he was in need of this benefit. From which he believes that out

of mercy the blessed Lord Louis prayed to God for him, and thus he received his hearing, as he believes.

Source: Guillaume de St.-Pathus, confessor to Queen Marguerite,
widow of King Louis IX, reported the case in his <i>Les miracles de</i>
St. Louis, ed. Percival B. Fay (Paris, 1931), 50–55.

Sometime between May of 1282 and March of 1283 a carriage driver in his early thirties appeared before a panel of bishops in the town of St.-Denis, just north of Paris. The bishops were conducting an inquest into the miracles that were purported to have taken place at the tomb of King Louis IX of France, who had been buried there in 1271. The young man, who claimed that he had been delivered from both deafness and muteness at the tomb of Saint Louis, recounted a remarkable life story, which illustrates various forms of Christian charity in the late thirteenth century.

The story began around 1257, when, at the age of eight, the young man had been found, both deaf and mute, at a chateau called Orgelet, which was located in the Jura Mountains, not far from Lake Geneva. The chateau belonged to a prominent aristocrat, Count John of Chalon-Auxerre, for whom this was only one of several places of residence. We do not know how the deaf-mute child came to Orgelet—whether he was led there intentionally by family or friends who were eager to place him in an environment where he could find useful employment, or whether he had been abandoned, as many disabled children were in the Middle Ages. What we do know—from the account books of aristocratic households—is that noble estates and households often served as magnets for individuals with physical differences and disabilities. Dwarf retainers often held quite privileged positions in aristocratic households, and lists of stable employees and other menial workers include references to deaf or lame individuals.[1] It was probably a combination of Christian charity and a desire to collect and display exotic objects and people that motivated aristocrats to embrace these individuals as members of their household staffs.

The deaf-mute child, who later came to be known as Louis, was fortunate not only because he had come to one of the estates of a powerful aristocrat, but also because the smith there, a man named Gauchier, took him under his wing, both providing him with useful training around the forge and teaching him, through rudimentary signs, how to behave in public places, such as church. Louis's propensity to return to Gauchier's home, even after he had moved away from Orgelet as an adult, suggests that the relationship became quasi-familial, as sometimes occurred between urban masters and apprentices. Of course, there were numerous masters who treated their apprentices and employees with cruelty, but it is clear that in Gauchier's case he took his charitable obligations seriously. The narrative makes it clear that Gauchier was a pious individual. Moreover, like many medieval workers he probably recognized that he too might someday count himself among the disabled who were in need of Christian charity.

According to the surviving summary of Louis's deposition and that of his corroborating witness (whose identity we do not know), Louis remained with Gauchier for twelve years. Then, or soon thereafter, when he as about twenty, he moved to Lyon, where he apparently worked at another chateau belonging to the count and countess of Chalon-Auxerre. Not long after his arrival in Lyon, in 1271, he attached himself to the royal entourage that was transporting the body of King Louis IX, who had died in Tunis, to his burial place at the abbey of St.-Denis just north of Paris. At this point the narrative offers us an insider's view of what it was like to be deaf, living off aristocratic and ecclesiastical pious alms, and traveling across unknown terrain. It is extremely unusual in this sense—most medieval sources show us almsgiving only from the perspective of the givers. And of course, very few medieval deaf people ever arrived at the point that they could actually recount their experiences.

Once the body of King Louis had been interred at St.-Denis, devotees who were convinced of his sanctity began to visit the tomb in search of cures for their various ailments. The deaf-mute had no idea who King Louis was, or why people knelt and raised their joined hands at the tomb, but he imitated them just the same. Suddenly, we are told, the ability to hear came crashing down on him, nearly frightening him out of his wits. He fled from St.-Denis, and made his way back to Lyon, and then to Orgelet. There, he conveyed his newly found (or refound)[2] abilities both to hear and to make sounds, and Gauchier's family began the task of teaching him some words. When the Countess of Auxerre learned of his abilities, she placed him among her kitchen staff, instructing the staff to teach him to speak. By the time he appeared before the inquest in St.-Denis, around twelve years later, Louis was apparently able to recite both the Lord's Prayer and the Ave Maria. More astounding still, he evidently gave a coherent account of his sudden transformation from deaf-mute to one who was capable of both hearing and learning to speak. He had been taught to attribute that transformation to the miraculous powers of Saint Louis.

Notes

1. Countess Mahaut of Artois and Burgundy, for whom account books survive from the years 1304–1328, had at least one dwarf among her retainers, as well as a highly favored court fool who was of diminutive proportions; the list of employees in her stables included one identified as "the deaf man of the carriage" (Archives du departement de Pas-de-Calais, Series A, 316, fol. 14; Series A, 293, fol. 20). The dwarf Turold in panel five of the Bayeux tapestry provides another example of a courtly dwarf.

2. Louis must have been able to hear when he was born, since adults who have been deaf since birth cannot learn to speak, even if they do gain the ability to hear as adults.

Further Reading

John Boswell, *The Kindness of Strangers: The Abandonment of Children from Late Antiquity to the Renaissance* (New York, 1988).

Steven A. Epstein, *Wage Labor and Guilds in Medieval Europe* (Chapel Hill, NC, 1991).

Sharon Farmer, *Surviving Poverty in Medieval Paris: Gender, Ideology, and the Daily Lives of the Poor* (Ithaca, NY, 2002).

— 27 —

Bequests for the Poor

Brigitte Resl

DONATION BY GREIF, 7 JULY 1305

Since all human achievements are finite and do not last, it is fitting to protect the soul with good deeds and to confirm in writing what should be done in order to avoid any cause for conflict. Therefore I, Greif, at Our Lady's am Gestade in Vienna, announce and declare to all those who read this letter, whether seeing or hearing it, who are presently living or who will be living in future, that I have seen and considered the grief and the poverty of the sick and poor in the city's hospital outside the city of Vienna, and that I have given for God and for the benefit of my soul as well as that of my wife Percht, may God have mercy upon her, and of all our ancestors, at a time when I was able to do so, eight pounds in rents in the form of "Burgrecht," which are listed in detail in this letter, according to the old custom under which I held these rents. And I give these rents in the following manner, so that four pounds and sixty pence shall be used for two meals every year in perpetuity for the poor, one of fish and Lenten fare for two and a half pounds on the Sunday at mid-Lent, when *Letare* is sung, and one of meat for fourteen shillings on the Sunday after Easter, when *Quasi modo geniti* is sung. The remaining three pounds and 180 pence should be used to give every year in perpetuity on All Souls' Day thirteen coats of "Poltinger" cloth, four and a half yards per coat, and thirteen shirts, four yards per shirt, to those of the poor most in need and in whatever manner it can be best achieved. Along with the aforementioned meals and clothes, which should please and comfort the poor, should be commemorated my own, my wife Percht's and all our ancestors' souls. And this should take place with my, the aforementioned Greif's, knowledge, or that of my heirs, once I am no more, or that of their agent, who should watch as the aforementioned meals and clothes are bought and distributed.

In case this is not maintained every year with regard to the aforementioned meals and with the clothes, as is set out in this letter, I, or, in case I am no more, then my heirs, should take possession again of the aforementioned eight pounds in rent, and should bestow them wheresoever I, or my heirs after me, wish, to another house of God for the benefit of the Lord and of our souls.

Also, the aforementioned eight pounds in dues should remain with the aforementioned hospital for eternity and not be sold nor pawned nor exchanged for other dues nor redeemed nor alienated.

Concerning the said eight pounds in dues one pound derives from vineyards at St. Job beim Klagbaum all at the same time at Martinmas, and toward the said pound Hainrich Suechler pays forty pence from one vineyard, Heinrich Bauer eighty pence, Ulrich Potenlaimer forty pence, Dietreich Walcherinne's nephew forty, Albrecht Gravenpech forty. Of those vineyards I was in full possession and I have given them to the aforenamed hospital with all the old rights with which I and my ancestors used to possess the vineyards. And then Conrad the Heubler provides from his house at the Pleczen fountain one pound divided annually into three, at Christmas eighty pence, at St. George's Day eighty pence, at Michaelmas eighty pence; Hainrich the Smith from his house at the Peurer Castle gate one pound, also three times per year, at Christmas eighty, at St. George's Day eighty, at Michaelmas eighty. Furthermore, one provides from houses in Zieher Street and in Kärtner Street: Hainrich of Sankt Pölten from his house thirty pence at St. George's Day and twenty-four pence for chicken at three feasts per year, at Christmas, at Easter and at Whitsun, at each feast eight pence. Walter the Verber forty-five pence at St. George's Day, and thirty-six pence for chicken, at Christmas twelve, at Easter twelve, at Whitsun twelve. Hainrich the Taler thirty pence at St. George's Day, and twelve pence for chicken, at each of the aforementioned feasts four pence. Walter Fleminch thirty pence at St. George's Day and twenty-four pence for chicken, at each feast eight pence. Ötacher of Laa of the Frank's house thirty pence at St. George's Day. Seidel Sichauf twelve pence for chicken from the farm near the *Graben*, at each feast four pence. And anything lacking from the same farm shall be taken from the above-mentioned Ötacher and his house. Also, Wielant the Suester provides from his house seventy pence at St. George's Day, and six for chicken, at each feast two. Hainrich Lenk seventy pence at St. George's Day, and six for chicken, at each feast two. Herthinne Schergin one-half pound at St. George's Day, and twelve for chicken, at each feast four. Jans the Vuerer from the *Arbaiterinne's* house twelve pence and one pound three times a year, at Christmas, at St. George's Day and at Michaelmas, at each day eighty-four pence. Otto Pinter also twelve [pence] and one pound, also at the same times. *Herr* Gerlach from his house in the High Street one-half pound three times a year, at Christmas, at St. George's day, and at Michaelmas, each day forty pence. Of all the houses in Zieher Street, Kärtner Street, and of *Herr* Gerlach's house I have been in full possession. And if one of them should be leased twelve pence should be paid at the end and six pence at the beginning of the term. Furthermore, Chunigunt provides from her house in Turmfalt Street ten pence at Michaelmas; there I have also been in full possession, and six pence should be given at the end and three pence at the beginning of the term. Then Ortel of Als provides from his house at the Goldsmiths' ten pence at Michaelmas. Thus the sum is eight pounds and one penny.

And so that this donation and this my will, which I have set out honestly with the favor and good will of all my heirs, from whom I have letters and charters concerning it, may remain stable and undisturbed, for this I give my letter as a testimony to the above-mentioned city's hospital. For better security I have requested the honorable citizens from the council of the city of Vienna that they confirm my will with their letter affixed with the city's seal and with the aforementioned hospital's seal. The letter was given when there had passed since Christ's birth one thousand three hundred years and after that in the fifth year on the Wednesday after St. Ulrich's day.

Source: Bürgerspitalurkunde 26, Stadt- und Landesarchiv, Vienna.

Charity was one of the guiding Christian tenets. Giving to the poor had become a standard social obligation for the rich by the later Middle Ages. It was always considered an act of reciprocity; while the poor benefited materially in this world the rich could expect spiritual rewards for their souls in the world to come. Charitable bequests became common elements in last wills and other postmortem donations. Thus gifts to the poor formed a regular part of the "budget de l'au-delà," the budget for the otherworld, as Jacques Chiffoleau called it, together with prayers for the soul and religious services.

Charitable provisions could be linked either with funerals or with the anniversaries that were celebrated in commemoration of the donor's death. Testators had food, money, or clothes distributed to members of the poor on these occasions. Some people provided dowries for poor unmarried women or donations to poor priests. In other cases, poor people were paid to go on a pilgrimage for the sake of the testator's soul. These provisions were sometimes set out as one-off events. In other cases, donors established a fund from which money was to be given regularly to poor people on the anniversary of their death. Grants in cash or kind could be directly handed out to individuals. However, many donors preferred to make such provisions to the poor through the agency of an institution, such as parish churches, monasteries, fraternities, and hospitals. These donations have been interpreted as stages in a cycle of gift exchange. By giving something to the poor, donors hoped in return to increase the benefits to their souls. The charitable act in itself could suffice for this purpose, but the potential rewards were increased when donors asked the poor to pray for their souls. By choosing to bestow their charity upon an institution rather than individual members of the poor, testators hoped to avoid the uncertainty inherent in all testamentary dispositions and increase the chances of their wishes actually being carried out. They also sought to reach a broader audience for their good works and to increase the number of mourners or of people praying for their souls.

Charitable donations were not only made in last wills. True Christians were supposed to give during their lifetimes when they were still active and healthy.

Such acts of charity were considered more valuable. If donors set up a perpetual donation that already came into effect during their lifetime it also had the advantage that they could ensure it was carried out in the way they had intended. Apart from spiritual rewards donors may also have had other objectives in mind when they gave to the poor. Such gift-giving allowed them to display their wealth and social status, or even to enhance their positions in society, for example by choosing an institution that was supported by the local elite, thereby associating themselves with the leaders of their community.

Although the specific details of these practices varied among different regions of Europe, between town and countryside, and over time, the document translated above gives a typical and full example of one such "charitable exchange" and the various insights it can provide into contemporary attitudes and practices. The charter records a grant made by Greif ("griffin"), a member of one of the leading families in the Austrian city of Vienna around 1300. He had accumulated his wealth not only through inheritance but also as a successful businessman, and held various urban and government offices between 1268 and 1318. In July 1305 he set aside the sum of eight pounds to be given every year to the poor. Greif chose the city's hospital, the Bürgerspital, as the recipient of his charity. This hospital was the main institution in charge of social welfare in late medieval Vienna. All those inhabitants of the city who were too old or too weak to work and thus unable to support themselves were accommodated there. Greif requested that food was to be given to the poor residents of the hospital on two occasions every year. They would eat fish and other appropriate foodstuffs on a Sunday during Lent, a period when it was forbidden to eat meat, and then meat on a Sunday after Easter. Such fare would regularly feature in the diet of the rich, but it was beyond the means of the poor, and nor would it normally have been given to them in the hospital. The intention of the donation was not merely just to feed them therefore, but also to turn the two Sundays into feast days. The sums dedicated to the two meals confirm that fish was more expensive than meat. Furthermore, at the beginning of the cold season, on All Souls' Day (2 November), coats and shirts were distributed to those who needed them most.

To guarantee that these provisions would be carried out regularly in the future, the material basis of the donation had to be established in detail. Greif donated perpetual income in the form of rents from his urban properties and from vineyards outside the city. Since this material provision was essential, the charter includes very precise specifications concerning the various sources of the money and the exact sums to be paid in each case. In choosing the city's hospital as the recipient of his donation Greif was enhancing the chances that his wish would be carried out even after he could no longer oversee it himself; the institution had a long history and a proven track record both in providing for the poor and in celebrating the anniversaries of donors' deaths. Greif knew from his previous experience as administrator of the hospital that its governors exercised careful oversight over charitable income. He also took specific care to ensure the benefits of prayers for his soul be added to the rewards from charity by setting out that the poor be asked to pray for him and his family while the charitable acts were being carried out. The numbers of people resident in the hospital promised a rich harvest of prayers, something that was of particular concern to many donors.

The city's hospital was the prime beneficiary for charitable provision in late medieval Vienna, not only because of the security and volume of prayers it promised but also because of its prestige. The institution was mainly supported by the city's rich merchants, and all those who made their donations there hoped to buy into their privileged social milieu. Greif, however, had no need of such advancement by association. Not only was he the former administrator of the hospital, his family had also been closely linked to the institution from its very beginnings. Although the care with which his charitable transaction is set out gives a businesslike impression in emphasizing the donor's spiritual gains, the text reveals this personal aspect of Greif's motives. As he points out in the charter, he had himself witnessed the miserable conditions in which the poor lived in the hospital. That his actions were guided not only by professional consideration and careful planning but also by profound Christian principles and sincere concern for the welfare of the poor is further illustrated by the fact that the donation was made long before his death. Greif was still active in business as well as in public office for more than a decade after the donation was made. Similar concern for the welfare of the poor can be seen in another donation Greif made to the hospital in 1315, when he gave a wood to secure supplies of fuel to heat the hospital and a pasture for its cattle. Thus, while the introductory phrases that explain the donors' intentions in last wills and postmortem donations usually consist of highly standardized formulas preceding a core text concerned only with dry legal transactions, this document reveals something of the donor's personal compassion and feelings, rarely expressed in medieval charters.

There is a further aspect to acts of charitable exchange such as the one described in this document. While the celebrations of the anniversaries of donors' deaths by religious services, prayers, and charitable provisions benefited their souls, they also kept their memory alive. Certainly, this was another reason why prestigious institutions such as Vienna's city hospital were chosen by donors as custodians of their charitable bequests.

Further Reading

Michel Mollat, *The Poor in the Middle Ages: An Essay in Social History*, trans. Arthur Goldhammer (New Haven, CT, 1986).

Brigitte Pohl-Resl, *Rechnen mit der Ewigkeit. Das Wiener Bürgerspital im Mittelalter* (Vienna and Munich, 1996).

Miri Rubin, *Charity and Community in Medieval Cambridge* (Cambridge, 1987).

R. N. Swanson, *Religion and Devotion in Europe, c. 1215–c. 1515* (Cambridge, 1995).

The Cult of Saints and Pilgrimage

— 28 —

Translation of the Body of St. Junianus

Thomas Head

Brother Letaldus gives salutations to lord father Constantine and to the other
brothers of the monastery of Nouaillé.

The angel Gabriel was once sent by the Lord to alleviate the labors of Tobias.
The angel not only delivered him from toil but also gave him the support of the
kindness of divine piety. Then he returned to Him by whom he had been sent,
going forth from Whom does not make one absent. Gabriel first taught those
who had benefited from heavenly kindness and addressed them, saying, "It
is good to hide the secret of a king, but gloriously to reveal the works of God
[Tobias 12:7]." Therefore it is fitting that we reveal and confess the works of
Christ that are allowed to happen in our times through His most glorious confessor Junianus, both for the praise and glory of the saint's name and for the edification of those who will hear the story. All people should learn these things, for
such works as were done in the days of our fathers and are still done for us now
do not happen on account of our own merits, but through the pious kindness
and intervention of those fathers who are given to us as intercessors. They provide something for us to copy in the important correction of our own lives.

We therefore approach the task of writing this work that we have promised,
not trusting in the help of men, but supported by the aid of divine largesse,
which comes from Him who said, "Open your mouth wide and I will fill it"
[Psalms 81:10, 80:11 in the Vulgate]. Reverend fathers and brothers, you have
begged us with your prayers and you have enjoined me by your charitable command. Do not allow our rustic speech to be displeasing to you, if only so that
truth alone may bring forth the whole narrative, as it was told by you. At that
time sinners were rising up like stalks of wheat. Evil people wasted the vineyard
of the Lord just as briars and thorns choke the harvest of the land. Therefore it
pleased bishops, abbots, and other religious men that a council be held at which
the taking of booty would be prohibited and the property of the saints, which
had been unjustly stolen, would be restored. Other evils that fouled the fair
countenance of the holy church of God were also struck down by the sharp

points of anathemas. This council was held at the monastery of Charroux and a great crowd of many people [*populus*] gathered there from the areas surrounding Poitiers, Limoges, and their neighboring regions. Many bodies of saints were also brought there. The cause of religion was strengthened by their presence, and the impudence of evil people was beaten back. Convoked, as it was thought, by divine will, the council was adorned by frequent miracles through the presence of these saints. Along with these various relics of the saints honored by God, the remains of the glorious father Junianus were brought with proper honor.

Several events occurred when the relics of the holy father Junianus were being brought forth from their monastic enclosure. Not far from the monastery [of Nouaillé] those who carried the bundle containing the saint stopped and put down their holy burden. After the most holy relics departed, the faithful in their devotion erected a cross in order to memorialize and record the fact that the relics of the holy father had rested there. From that time to this, whosoever suffers from a fever and goes there is returned to their former health through the invocation of the name of Christ and the intercession of this same father Junianus. When the party came to the little village called Ruffiacus, they sought out the manor house and passed the night there in a vigil singing hymns and praise to God. The next day [the monks] resumed their journey. At the place where the relics had rested, faithful Christians erected a sort of fence from twigs, so that the place where the holy body had lain might remain safe from the approach of men and animals. Many days later a wild bull came by and wantonly struck that same fence with his horns and flanks. When suddenly he retreated from the fence, he fell down and died. In that same place, a little pool was created by placing a gutter tile to allow runoff water to be stored up. Because of the reverence for the holy relics, this pool served as an invitation for many people to wash. Among these there was a woman who suffered from elephantiasis. When she washed herself with that water, she was returned to her former health.

Source: Letaldus of Micy, "Delatio corporis s. Juniani ad synodem Karoffensem," in
Patrologia Latina, vol. CXXXVII, ed. Jean-Paul Migne (Paris, 1853), cols. 823–26.
For discussion and dating see Thomas Head, "Letaldus of Micy and
the Hagiographic Traditions of the Abbey of Nouaillé: The Context of
the *Delatio corporis s. Juniani*," *Analecta Bollandiana* 115
(1997): 253–67.

During the Middle Ages, the vast majority of the population of Europe was illiterate and rural, which is why historians are so dependent, particularly for the period before 1200, on the accounts provided by clerics for providing evidence of the religious experiences and practices of laypeople. The religious practices of rural laypeople, nobles almost as much as peasants, revolved around ever increasingly common parish churches and the monastic foundations that dotted the European countryside (and for the nobility some private family chapels or churches). Very little has survived to document parish life before 1200. The fragmentary texts that remain are largely prescriptions from the urban clergy as to how their rural colleagues ought to behave. Thus we find Bishop Theodulf of Orléans in the early years of the ninth century prohibiting the storage of "harvested crops and hay" in rural churches and warning his rural clergy against frequenting taverns, where they would not only be tempted into drunkenness but also into the unseemly networks of local gossip. More frequently we glimpse rural lay Christians when they traveled to monasteries to celebrate the feast days of local saints or to be pilgrims visiting a shrine. There they came into the vision and under the influence of monks, who sometimes recorded stories about them. Those stories, however, largely focused on their actions in or near the monastery itself. The most common stories of this type concerned miraculous cures provided for the faithful—clerical and lay, rural and urban—at the shrines of the saints. The medieval clergy were certainly not ethnographers. They did not go out from their monasteries or their urban cathedrals to learn about the religious lives of the rural faithful. They instead observed those laypeople when they came in from their villages to churches and shrines. Indeed Bishop Theodulf explicitly warned his clergy against "traveling around houses and villages out of curiosity."

The short text presented here is valuable because it provides evidence of two types of lay religious practice in the vast agricultural expanse of Europe that were not centered on a church building or a formal saint's shrine. It documents the active presence of the *rustices* (or rural agricultural workers) at ecclesiastical gatherings associated with the Peace of God and at informal shrines of their own making.

Letaldus, the author of this text, was a monk of the abbey of Micy, located near Orléans on the Loire River in central France, who lived and wrote in the years around 1000. He wrote much about Maximinus, the patron saint of Micy, and his miracles. But Letaldus is also interesting because he often journeyed outside the walls of his own monastery, not seeking to document the experiences of the laity, but in order to help other monastic communities in recording the traditions of their patron saints and shrines. One of the monasteries he so aided was Nouaillé, located in the rural hinterland to the south of the episcopal see of Poitiers. The text he composed belonged to a well-established genre of hagiographic writing, the *delatio*. This Latin word refers to taking something out on a journey, more specifically in this case the relics of a saint from his or her shrine. From time to time, monks unearthed the remains of their patron saints and took them outside the monastery walls, but the usual goal of such tours was to raise funds for the monastery. Accounts of these journeys tended to focus on events immediately around the relics themselves, where the monks established a kind of temporary shrine at each stop. And so the sort of narrative material to be found in most of these *delationes* differs relatively little from the miracle stories collected at the monasteries themselves.

In 989, however, the monks of Nouaillé removed the relics of their patron, St. Junianus, from their resting place in order to make a journey to a gathering (or synod) of bishops from all over the region of Aquitaine that was being held at the abbey of Charroux, several days distant. The synod was unusual in that it brought bishops together not in the urban setting of an episcopal see (Poitiers would have been a more traditional choice), but in the countryside, albeit in a monastery. From the statutes they promulgated in the wake of their synod, we know that the bishops sought to use religious sanctions to control violence on the part of the military aristocracy, specifically attacks against church buildings and property, the clergy themselves, and, to a lesser degree, unarmed agricultural laborers. (Indeed Letaldus's use of a metaphor at once biblical and agricultural, found in the text provided in translation here, mirrors a section of those synodal acts: "[We bishops intend] that the criminal behavior, which has time and again we know been long flowering through evil habit in our home districts—due to our long delay in calling a council—will be rooted out and more honest behavior implanted in recognition of heavenly grace." The meeting at Charroux was the first documented synod in what later became a movement widespread in the kingdom of France and known as the Peace of God. At Charroux and later synods, bishops tried to assemble large gatherings of clergy, monks, and—quite importantly—ordinary laypeople as witnesses to the injunctions they laid down against the military elite and of the oaths they demanded from such knights. The presence of laypeople at these gatherings as part of the processes of sanction employed by the bishops was an unusual move on the part of the ecclesiastical establishment to recruit the laity into the actions of the clergy. At least as important as their attendance, however, was the presence of the relics of saints. Medieval Christians believed that saints were literally present in their physical remains, and through the presence of those relics the saints themselves, as well as laypeople, bore witness to the actions of bishops and knights.

The monks of Nouaillé were not alone in bringing the relics of their patron saint to the gathering at Charroux. Many other monastic communities from the Aquitaine did likewise. But the monks of Nouaillé differed from their fellows in that several years later they asked Letaldus to write an account of the journey they had made with their relics. And so, after an introductory section (in the translation each section is parsed as a paragraph) filled with hagiographic commonplaces, Letaldus turned in the second section to the events of the synod of Charroux, providing one of the most vivid witnesses, for all its brevity, to the presence of common rural folk (the *populus*) alongside the saints and the clergy. Then in the final section he turned back to the journey itself. Stories of miracles had filtered into the monastery from the countryside. These, however, were not stories of miracles that had happened when the relics and attendant clergy were present. Rather they were stories of how the rural laity had continued to gather at and give significance to the places where the relics had rested on overnight stays during the journey. Through the erection of crosses and fences, not to mention the artful diversion of drainage ditches to create pools of water, these had become permanent shrines outside of the walls of churches, beyond the watchful oversight of the clergy. These terse accounts speak to an adaptation of the practices of the ecclesiastical establishment into the idioms, rhythms, and locales of agricultural laborers. The presence of St. Junianus in the countryside between Nouaillé and Charroux was not an evanescent one limited to the period of the passage of the monks charged with their sacred burden (despite Letaldus's slight suggestion to the contrary). Two places, at least, continued to provide the rural faithful contact with their saints and their God. The informal shrines recorded by Letaldus and physically marked by wattle fences and simple crosses hint at the existence of a large, and mostly unrecognized, geography of holy places connected to, but largely autonomous from, those overseen by clergy and monks. It is through such mirrors that we indirectly and incompletely glimpse the religious practices of laypeople during the high Middle Ages.

Further Reading

H.E.J. Cowdrey, "The Peace and the Truce of God in the Eleventh Century," *Past and Present* 46 (1970): 42–67.

For a contemporary account of a rural Christian who journeys to Letaldus's community of Micy in search of a miraculous cure, see the "Miracle of St. Maximinus" translated in Thomas Head, "The Cult of Relics in the Eleventh Century," in *Medieval Hagiography: An Anthology*, ed. Thomas Head (New York, 2000), 287–92.

Thomas Head and Richard Landes, eds., *The Peace of God: Social Violence and Religious Response in France around the Year 1000* (Ithaca, NY, 1992).

— 29 —

Pilgrimage and Spiritual Healing
in the Ninth Century

Julia M. H. Smith

We ought not to remain silent about a miracle that Jesus Christ deigned to carry out in our presence, through our lady Opportuna. The devil, the enemy of the human race, invaded a woman and alas! for grief! destroyed her limbs by flailing them around at every moment of the day and night. Loving her dearly, her husband looked after her with care and took her to many saints' shrines so that she might deserve to be freed from the oppression of the enemy by the intercession of God's elect. He heard about the reputation of the glorious virgin Opportuna, whose intercession was accustomed to driving away the power of demons, and since the enemy had been possessing his wife for an extremely long time, almost even as a hereditary vessel, and had been afflicting the most wretched woman with various tortures in a foul manner that it would be grievous to report, the pious man went with his aforementioned wife to seek out the help of the saint's power. When she had been brought into the church of St. Opportuna, the devil realized that the Lord intended to challenge his evilness through the merits of the blessed Opportuna and he began to call out through the woman's mouth thus:

"Woe, woe Opportuna, woe to me, you old crone, you who have always opposed me in Gaul and in Neustria! Your prayer often destroyed my machinations, for you were in the habit of laying traps for me during your lifetime, and now you continue to do so after your death."

I was present at this, grieving and praying among the people in the church. The devil said many insulting things to me too through the woman's mouth:

"Don't hope, you uncouth and newly appointed bishop, that by your strength I will leave the woman by my own power. I don't fear you at all, but I do dread the miracle-working power of my adversary Opportuna, and I know that you're just her useless servant!"

The priests, clergy, and common people who had gathered around this grievous spectacle did not cease praying. Without pause, they begged God with
their tears that, appeased by the request of his servant Opportuna, he would
get rid of the open hostilities of the fraudulent devil by the strength of his own
power. The woman was aspergated [i.e., sprinkled] with holy water and the
sign of the Lord's cross made over her as the prayers appropriate for the rite of
exorcism were read over her. When the place in the service was reached where
it is proclaimed: "I adjure you, serpent, in the name of the Lamb of God who
walks upon the asp and the basilisk, who crushes under foot the lion and the
dragon, . . ." the unclean spirit, who was troubling her so cruelly that she was
attacking herself with her teeth and nails, called out with a terrible voice:

"You old hag Opportuna, know that I'm going now, but I'll soon be back!"

When this was said, the woman sat down quietly, turned her eyes heavenward
and raised her hands. She then moved to the corner of the altar and vowed herself to remain in the service of the blessed Opportuna. Then she picked up the
bread of benediction and made her way to the church's guesthouse.

Several days passed and her husband, wishing to enjoy her in her health,
wanted to return home full of joy. But when they left the building and set out
on their way, behold! the devil with a multitude of his followers in the guise of
wolves and dogs attacked the woman. She cried out:

"Lady Opportuna, free your handmaid!"

Pursued by the wolves and dogs right up to the door of the church, she fled underneath the altar. She remained there praying with us for a long time and then
came out unharmed. She remained in this fortunate condition for two weeks, going wherever she wished without any disturbance and traveling without any vex-
ation. When her husband saw this, he offered money to the church's clergy to get
permission to reunite himself with his wife. When I heard this, I neither permitted nor prohibited it. If I were to have allowed it I would have seemed to go
against the will of St. Opportuna; were I to have refused it, I would have been accused by people unaware of the situation of wanting to sever the marital union of
a loving couple out of human greed.

Her husband, wishing to go to her in the usual fashion, realized that she was
vexed by the demon even more wretchedly than before, for at the instigation of
the devil, her hands and arms were shaking shamefully. Her tongue jumped
out from her throat, lolling out beyond the end of her mouth and her lips, she
appeared totally naked to men and she uttered unheard-of blasphemy against
the Almighty. Since her husband was unable to take her single-handed to the
usual church of the blessed virgin Opportuna, the townsfolk, who grieved for
the wife's afflictions and for the husband's sufferings, picked up her limp and
almost lifeless body in their own arms and put her down in front of the church
of our Lady Opportuna. She remained on the ground there for a long while, uttering no sound and seeming not to breathe, and the crowds declared that she
was dead. Prayers appropriate for a person possessed by the devil were read.

Some people began to sing the psalms that form part of the office for the dead. After many hours, the woman who was thought to be dead—because she who sins in the spirit will die—then began to shout out loud in the church:

"Lady Opportuna, help your handmaid! Because from now on I shall be your most devoted servant, resuscitate me from death of body and soul!"

When her husband talked soothingly to her and asked her what she had seen, she replied:

"My Lord Helbert," for that was his name, "Right now I was devoured by an extremely big, ferocious wolf, and when men with burning pitchforks wanted to throw me whole through the wall hangings into a sulphury pit, I was freed by the grace of God with the help of my Lady Opportuna. What sort of a wife do you want, for you cannot have me any longer."

For the rest of her life the woman, whose name was Olbiregis, served God with all the devotion of her body and with purity of mind, and for a long time many people were alive who testified that they had seen those things we want to narrate.

Source: Adalhelm of Sées, Liber miraculorum Opportunae, Acta Sanctorum
Apr. III, ed. J. Carnandet (Paris and Rome, 1866), 68–69.

This extract is an account of a moving story of affliction and love, but is also an example of the dilemmas that Christian life in the Middle Ages could pose. Although the events described here took place in the late ninth century, similar stories of release from demonic possession occur throughout the Christian Middle Ages. This one is distinctive for its moral twist, and for the lively, sympathetic narrative by an eyewitness who found himself drawn into the action in a problematic way.

Its author is Adalhelm, bishop of Sées, a small French town in what is now Normandy but was then termed Neustria. We know nothing about him, except the circumstances in which he sat down to write. He tells us that he had been appointed bishop by Emperor Charles the Fat (ruled West Francia 884–87), but that before he could be formally installed, he found himself facing opposition to his appointment. The bishop-elect made a vow to a local saint, Opportuna, that if she would help him overcome his opponents, he would write an account of her life and miracles. Adalhelm won control of his bishopric, but ignored his promise to the saint. Within a year of having been ordained bishop, he was captured by Viking pirates, whose raids on northwestern France were by this time so intense that his predecessor, Bishop Hildebrand, had had to flee for refuge to Moussy-le-Neuf (just northeast of Paris), taking the relics of Opportuna with him. After suffering hunger and cold when the Vikings shipped him abroad in chains, Adalhelm managed to return to France, only to be thrown from his horse into a stormy high tide. Unable to swim, he called again on Opportuna, who rescued him. In captivity, Adalhelm had had plenty of time to reflect upon the moral of his own travails, concluding that "it is better not to make a vow than to fail to fulfill one." Chastened by his experiences, his first act as bishop was to fulfill his vow to Opportuna, composing a lively account of her own career with an accompanying dossier of miracles, both those associated with her new place of refuge at Moussy-le-Neuf and those back in her church at Sées. This extract is Adalhelm's account of one of her posthumous miracles in Sées in which he himself became involved. Although Adelhelm recounts it with great compassion, he also keeps his eye on its moral: this too is a tale about a vow made, broken, and finally redeemed.

The story concerns Olbiregis, a married woman suffering terrible contortions and damaging her own body. A range of curative options were available in the early Middle Ages: consulting someone with professional medical training, using orally transmitted knowledge of local herbs and simples, or visiting saints' shrines in the hope of a miracle. By the late ninth century, saints' shrines were common throughout the Christian regions of Europe, many of them the shrines of locally popular figures such as Opportuna, some of them the churches of famous saints or early Christian martyrs. To travel, whether locally or long distance, in search of a cure for an ailment was extremely common, as dozens of surviving collections of ninth-century miracle stories make clear. Olbiregis's journey in search of healing was typical of such journeys, in that she was taken from church to church by her caregiver—her husband Helbert—until she was healed.

Such pilgrimages in search of a cure presumed a mentality in which physical and spiritual health were intimately linked. Demonstrations of devotion to God and his saints might restore physical well-being or, as in the case of Olbiregis, the affliction might be attributed to the work of the devil. In cases such as this, the Christian Church had had, since its earliest days, a well-developed theory of demonic possession and also efficacious strategies for dealing with it. The theological and historical foundations of both lay in Christian scripture. The psalms had predicted of Christ that "thou shalt walk upon the asp and the basilisk and thou shalt trample underfoot the lion and the dragon,"[1] and the Gospels described Jesus expelling evil spirits on many occasions,[2] linking his cures of the sick and possessed with his defeat of Satan.[3] Early Christians conducted spectacular public exorcisms, ceremonies in which Christ's name was formally invoked to command the devil to depart out of a possessed person or object, and by the end of the fourth century, entrusted them to an official ecclesiastical "exorcist" to conduct. The ability to exorcise demons was also widely attributed to saints both living and dead: Adalhelm is careful to emphasize that Olbiregis's cure was carried out by Christ through the intervention and help of St. Opportuna.

Adalhelm's account describes Olbiregis undergoing a series of cures and relapses before she was definitively released from demonic possession. There are several important aspects to note about this sequence of events. The first cure was achieved through exorcism. It involved three potent, reliable means of invoking Christ's presence and defeating the devil: holy water (water with salt added that a priest had blessed), the sign of the cross, and the traditional words of exorcism, conducted in accordance with the ceremony stipulated in surviving ninth-century liturgical books. Witnessed by a large number of bystanders, this highly charged, dramatic ceremony consisted of prayers to God and adjurations of the devil:

"God of angels, God of archangels, God of prophets, God of the apostles, God of martyrs, God of virgins, God the father of our Lord Jesus Christ, I invoke your holy name. A supplicant, I beg the clemency of your outstanding majesty, that you may deign to give me help against this most evil spirit, so that he will quickly leave and depart from wherever he is hiding when he hears your name. Let He who rules the winds and the sea and the storms rule you, O devil. Let He who ordered you to be thrown from the heights of heaven to the depths of the earth rule you. Let He who ordered you to turn backward rule you. Hear, Satan and be afraid! Go away, defeated and overthrown, in the name of our Lord Jesus Christ. . . .

"I adjure you, ancient serpent, by the judgment of the living and the dead, by the maker of the world, by Him who has the power to condemn to hell, that with fear and the army of your wrath you depart quickly out of this servant of God who flees to the fold of the Church. I adjure you not by my own weakness but in the power of the Holy Spirit that you quit this servant of God, whom almighty God made in his own image. Yield, yield, not to me but to the mysteries of Christ! . . . I adjure you, most evil snake, in the name of the immaculate Lamb, who walked upon the asp and the basilisk and who trampled underfoot the lion and the dragon, that you depart from this person, that you depart from the church of God. Tremble and flee when the name is invoked of the God whom hell fears, to whom the powers of the heavens and the powers and dominions of the world are subject, whom the cherubim and seraphim praise tirelessly. . . ."[4]

When the devil had left Olbiregis, she went to the corner of the altar in the church. This was the place where rites of passage took place that bound an individual to a different way of life. A master would free his slaves at the corner of the altar, for example, or a parent would donate a child to a monastery by wrapping the infant's hand in the altar cloth. Likewise Olbiregis: on being freed she stood at the corner of the altar and vowed herself into the service of St. Opportuna.

Helbert was a loving, caring husband who wanted to take his wife home and resume normal married life, but Opportuna had committed herself to a religious life. So when Helbert set out to take Olbiregis home, the devil returned, this time in the guise of wolves and dogs. Olbiregis's second release came after she spent time in prayer underneath the altar. Spending the night underneath a saint's tomb was a common curative strategy, but since Opportuna's relics were now in Moussy-le-Neuf for safekeeping, Olbiregis did the best she could by crawling under the altar dedicated to the saint in the Lady Opportuna's original church. Some ninth-century bishops attempted to prohibit women from going anywhere near the altar, but Adalhelm evidently accepted the suffering woman's presence here. But when, a fortnight after Olbiregis had emerged unharmed and without any renewal of her contortions, Helbert attempted to bribe the clergy to let him have his wife back, events took a new turn.

In the Carolingian period (ca. 750–ca. 900), churchmen put much effort into tightening definitions of legitimate marriage, although not until the twelfth century were their attempts to claim ecclesiastical jurisdiction over marriage fully realized. Moreover, except in a handful of high-profile cases involving Carolingian kings and counts, it is uncertain how much practical impact ecclesiastical strictures may have had; this extract suggests something of the challenges involved in the Church's efforts to win control over marriage practices.

Not only did Carolingian synods attempt to prevent marriage to close blood relatives and spiritual kinsfolk (godparents and godchildren), they also insisted upon the indissolubility of marriage. For the most part, these efforts were directed against those who sought divorce so that they might marry someone else. But what if someone wanted to end a marriage in order to enter the religious life? Early Carolingian legislation had permitted a husband or wife to separate in order for one of them to enter a monastery, although was inconsistent as to whether the remaining spouse could remarry or not. But by the end of the ninth century, a consensus was emerging that only with mutual consent could a couple separate to allow one or other to enter a religious life. Moreover, the other spouse was obligated never to remarry but to remain chaste thereafter. This view was articulated in the *Two Books concerning Synodal Cases and Ecclesiastical Discipline* (II: 109), an influential canon law collection compiled in circa 906 by Adelhelm's contemporary, Regino of Prüm (ca. 840–915). By the early eleventh century, this had become the established canon law of the Church.

Here, then, is the crux of the story. Helbert had neither been consulted about nor consented to his wife's entry into the religious life; for Olbiregis to honor her vow to Opportuna would be in violation of the canon law of marriage. This left Adalhelm in a quandary. He only tells us that he was torn between the will of the saint and upholding a loving marriage. In moral and legal terms, this means the bishop had to choose between honoring a solemn, Christian vow made at the altar or insisting upon the implementation of ecclesiastical law. No wonder he preferred not to make a decision!

The vow Olbiregis had made won out, of course, but not before Helbert had tried again to resume married life. This time, the devil seized her so badly that, after being shaken by uncontrollable convulsions, Olbiregis was carried to the church so limp she seemed dead. A full Christian funeral service was in origin reserved for monks and members of the clergy, but by the late ninth century was becoming used for members of the laity too. Some of the clergy resumed the prayers of exorcism; others began the prayers for the dead. But Olbiregis recovered—to tell of her vision of the brink of hell. Accounts of visions of the afterlife were not uncommon in the early Middle Ages, and always carried moral messages for their readers or hearers. Her vision makes the point of Adalhelm's story clear, that a vow to St. Opportuna must be fulfilled, whatever the cost.[5]

Notes

1. Ps. 90:13, Douai-Reims translation.
2. E.g., Matt. 12:22–24, 15:21–28, Mark 1:23–38, 5:1–20, Luke 7:21.
3. Mark 3:22–27.

4. *Supplementum anianense* 145: *Orationes super energumeno baptizato. Le sacramentaire grégorien: Ses principales formes d'après les plus anciens manuscrits*, ed. Jean Deshusses, 2nd ed., 3 vols. Spicilegium Friburgense 16, 24, 28 (Fribourg, 1979–82), 1: 491–94.

5. I am grateful to my St. Andrews colleague Simon MacLean for helping me understand Adalhelm.

Further Reading

James A. Brundage, *Law, Sex, and Society in Medieval Europe* (Chicago, 1987).

Eileen Gardiner, ed., *Visions of Heaven and Hell before Dante* (New York, 1989).

Patrick J. Geary, *Living with the Dead in the Middle Ages* (Ithaca, NY, 1994).

Henry Ansgar Kelly, *Towards the Death of Satan: The Growth and Decline of Christian Demonology* (London, 1968).

Julia M. H. Smith, "Religion and Lay Society," in *New Cambridge Medieval History*, vol. 2, *c. 700–c. 900*, ed. Rosamond McKitterick (Cambridge, 1995), 654–78.

Julia M. H. Smith, "Saints and Their Cults," in *Cambridge History of Christianity, III: Early Medieval Christianities, c.600–c.1100*, ed. Thomas F. T. Noble and Julia M. H. Smith (Cambridge, 2008), 581–605.

Pierre Toubert, "The Carolingian Moment (Eighth–tenth century)," in *A History of the Family*, vol. 1, *Distant Worlds, Ancient Worlds*, ed. André Burguière et al., trans. Sarah Hanbury Tenison et al. (Cambridge, 1996), 379–406.

In Pursuit of Perfection:

In the World

—— 30 ——

Interrogation of Waldensians

Peter Biller

I. EXTRACTS FROM DEPOSITIONS OF WALDENSIAN SUPPORTERS LIVING IN QUERCY, IN THE REGISTER OF THE INQUISITOR PETER CELLAN, 1241–42; WE DO NOT GET TO KNOW THE DATES OF THE ACTIONS THEY CONFESSED

[*AMONG THOSE FROM GOURDON*]

Geralda de Mailhoz had a Waldensian in her house for three days, for her husband's illness. She heard him [preaching] there and elsewhere. She gave him bread, wine, and leeks, and a Waldensian woman four loaves.

[*AMONG THOSE FROM MONTCUQ*]

Arnold Bernardi de Roset said that a Waldensian woman washed his head, and he twice sent the Waldensians some of his bread and cooked meats.

Raymonde de Bernah often received her Waldensian mother, and she gave her an ell of cloth.

[*AMONG THOSE FROM BEAUCAIRE*]

Bernarda Fabrissa rented a house to two Waldensian women, and they were there for a year. Item, these Waldensian women used to come to the house where she lived, and vice versa, and they taught her not to swear or lie. Item, she said she believed they were good men [*a male-gender inquisitor's formula is being repeated*] and she often gave the Waldensians bread and wine.

[*AMONG THOSE FROM MONTAUBAN*]

Lady Sedeira received two Waldensian women in her house, and heard their admonitions.

Lady Fauressa received the kiss of peace from Waldensian women, and often gave the Waldensians bread and wine.

Source: J. Duvernoy, *L'Inquisition en Quercy: Le registre de Pierre Cellan 1241–1231* (Castlenaud-la-Chapelle, 2001), 46–65.

II. RECORDS OF INTERROGATIONS BY THE INQUISITORS BERNARD OF CAUX AND JOHN OF ST. PIERRE IN TOULOUSE, 1245–46

[AMONG WITNESSES FROM AURIAC, JULY 1245]

Willelma Michela, sworn in as a witness, said that she lived with the Waldensians at Castelnaudary for three years. At that time the Waldensians were living openly in the region. And she dressed, ate, drank, prayed, and did other things just as they did.

[*The next day*] sworn in again, she said that she saw the Waldensians, and lived with the Waldensians Bernarda of Pomas, and Rixenda of Limoux, and Christiana for about four years. And she heard them saying that no one ought to take an oath, for the sake of the truth or a lie, nor to promise, justly or unjustly.

[AMONG WITNESSES FROM CASTELNAUDARY, MAY 1246]

Peter Simon, sworn as a witness, said that his mother Aimengarda and one of his sisters were Waldensians. However, after they became Waldensians he did not see them, except that once he saw his Waldensian sister passing through Castelnaudary. He talked to her and begged her to return to the Catholic faith; but she would not do this. This was forty years ago. . . . He saw Waldensians and heretics [=*Cathars*] living openly in Castelnaudary and in other regions . . . about forty years ago.

[AMONG WITNESSES FROM SAIX, MARCH 1245]

Raymond Boisseira . . . said that Bernarda his wife was a maid of the Waldensians. But after she became his wife she gave no charity to the Waldensians.

[AMONG WITNESSES FROM VIVIERS-LES-MAGNES, MARCH 1245]

Peter of Les Barthes . . . said that Galharda Bruna of Castres asked him to give her half a quarter of corn for the sake of charity to two poor women, which he did. And afterward he heard it said that these women were Waldensians.

Raymond Biat . . . said that on the instructions of his lord, Peter Martini, he gave a quarter of corn every year for three years to the Waldensians.

John Cochafieu . . . said that he saw the Waldensians Gaubert de les Crozes and his companion in his own house—they were there for two days, and they ate and slept there. He heard them say that no one ought to take an oath, justly or unjustly. He also said that he gave them a shirt and a basket of grapes, and this was about six years ago. Item, he saw the Waldensians in a certain hut in the wood of Saint-Affrique, and together with them Peter Martini of Viviers, who had taken him there. They then heard their preaching.

John Cochafieu's wife, Sicarda, said that she saw the Waldensians Gaubert de les Crozes and his companion in her and her husband's house. . . . Item, out of the love of God she often gave alms to the Waldensians—to the Waldensian William de Nauga and other Waldensians.

Raymond Martini said that he saw the Waldensian William Montaner in his house—the Waldensian stayed for a night, eating and sleeping there, about eight years ago. Item, he said that he and his brother, Peter Martini, went to Castres, and Galharda Martina then handed over to him and his brother the Waldensians Arnalda and Good Lady. While he and his brother were fetching these Waldensian women from Castres the women were captured. One was converted, the other burned—about six years ago.

Peter Martini . . . said that he received the Waldensians William Montaner and his companion, for two or three nights; he and his wife Raymonda and his herdsman Raymond Beadz [=*the earlier witness Raymond Biat*] were around. And he and the others ate with the Waldensians, at the same table, taking the bread they blessed. They heard them preaching—among other things, that one should not swear or kill in any circumstances, even for the sake of justice. He fed them two or three times, and this was ten years ago. Item, he saw the Waldensians Adam and Berengar in a hut in de Rassi forest, together with John Cochafieu, who came with him, about twelve years ago. Item, he and his brother Raymond fetched two Waldensian women from the house of Galharda Bruna of Castres, Arnalda and Good Lady—Galharda had asked them to do this—and when they got outside the town of Castres the Waldensian women were taken from them. One was burned; this was about eight years ago.

Arnold de Pardine said he saw two Waldensian women at Castres in Galharda Bruna's house, and he often gave them corn and bread and many other things. . . . Item, he said that he often saw the Waldensians, and in many places, and he gave them hospitality and he gave them alms so many times that he does not remember how often.

Bernard of St. Martin said that he saw the two Waldensian women Arnalda and Good Lady at Castres in the house of his sister, Galharda Bruna, and he gave them a quarter of corn and a lump of meat; this was about seven years ago.

Source: The translation is from a copy of the depositions that was made about 1260 and is now Toulouse, Bibliothèque municipale, MS 609.

III. ON THE WAY OF LIFE OF THE POOR OF LYONS; THIS DESCRIPTION OF THE WALDENSIANS PROBABLY COMES FROM SOUTHERN FRANCE IN THE LATE THIRTEENTH CENTURY.

In this sect both men and women are received, and they are called "Brothers" and "Sisters." They do not possess any immovable goods, but they renounce their own property and follow poverty. They do not work, they do not acquire or earn anything by which they could be supported, but they are supported by the goods and alms of their friends and believers. . . . They [the Brothers] live in houses and households, two or three in a hospice with two or three women [the Sisters], who pretend to be their wives or sisters.

> Source: C. Bruschi and Peter Biller, eds., *Texts and the Repression of Medieval Heresy*, York Studies in Medieval Theology 4 (Woodbridge and Rochester, NY, 2002), chap. 9.

The most perfect form of the religious life, clearly, was that followed by the Apostles; but based on which parts of the New Testament? During the twelfth century, the idea grew that the apostolic way of life was outlined by Christ's counsels in the gospels: to give everything away and wander around, begging and living off alms. Taken literally, these counsels meant improvising one's life every day, never knowing where the next bed or meal would be found. While most of the new religious movements of the period were informed to some degree by this new ideal of the apostolic life, two took it very literally. One was that of the Poor of Lyons, founded by a Lyons merchant called Valdes, and known by its opponents as the Waldensians. The other was that of the Minors, founded by an Assisi merchant's son called Francis, and known to us as the Franciscans.

Both came to form religious orders, entered by men and women, who became "Brothers" and "Sisters" after professing the three monastic vows of poverty, chastity, and obedience, and then followed a set form of religious life. But the Waldensians came to be proscribed as a "heretical sect," eventually leading a persecuted and clandestine existence, while the Franciscans became a large order within the Church. This divergence meant that their inevitable struggles and compromises with the ideal of poverty have left different traces in surviving evidence. Partisan lives of St. Francis, papal bulls, and treatises on spiritual poverty—all amply attest the conflicts that convulsed the Franciscan Order. But there is little more than a Waldensian letter about a divisive meeting held in 1218 to attest to parallel conflicts among the Waldensians. Rather, the evidence about Waldensians comes mainly from depositions given by their followers to inquisitors. These are unexpectedly informative on matters like their gifts to Waldensians, since inquisitors were interested in measuring guilt through ascertaining and weighing actions of this sort. As a result, we can learn about Franciscans *theorizing* about poverty, and Waldensians *living* poverty.

Both orders had female branches. In many cases the smaller size, poverty, and restriction of female orders in the Church in the high and later Middle Ages led to smaller documentation, and they were further "hidden from history" by male historiography until the development of women's history in the nineteenth century. The Waldensian Sisters shared this—they generated less evidence, and modern Waldensian and Protestant historians' embarrassment about a medieval past that included religious celibates helped keep them in oblivion. Trying to discern a continuity in the history of the medieval Waldensian Sisters, let alone their practice of poverty, may seem so difficult as to be almost perverse. By a minor miracle both are *just* possible through the texts presented here.

First, we glimpse something of the Sisters' overall form of life, which seems to have been in the earlier years a *bit* less restricted than that of their Franciscan equivalents, the Poor Clares. We know that Valdes's followers in the 1170s included both men and women leading a wandering life and preaching, and two inquisition witnesses with long memories who attested in 1245/46 give us a keyhole view of one little town in Languedoc about thirty years later, Castelnaudary about 1205/6. Persecution was not yet biting. Waldensians lived openly; someone could bump into a Waldsensian Sister in town. A tiny group of Waldensian Sisters lived together, Bernarda, Rixenda, and Christiana, and one person came to live with them for three years—as a young girl and novice? She did things "as they did." In other words, these things—their clothing and form of meals and prayer—were distinctive: the religious habit and regular daily life of Sisters in a Waldensian convent. We learn a bit more from the survival of inquisition evidence from 1241/42 on Waldensians who lived in the Quercy area, a little to the northwest. We again see a tiny group of Waldensian Sisters living in a house—only two this time, and only for a year. The Sisters taught doctrine and produced moral admonitions, within their own and followers' houses, but they are not seen preaching and teaching in public, as the Brothers did. Ritual included giving the kiss of peace. By the later thirteenth century, the probable date of *On the Way of Life of the Poor of Lyons*, there is no more any trace of an active life outside the house, visiting and teaching followers. 1314—when the Sister Raymonde of Castres appears in a confession to an inquisitor—is the last time we can name a Sister.

We turn to their poverty. Some of the women taking vows were making a great renunciation: of status, high marriages, and wealth. Where we get to know much about the social composition of early Waldensian followers—in Quercy—we see many from the elite, nobles, and great consular families. We should envisage at least some of our Sisters in these terms. "Lady," in the name of the Sister "Good Lady," means "noblewoman."

By the time we first glimpse them, around 1205, the Sisters, like the Franciscans, had compromised one of the elements of absolute poverty, instability of residence. Stability and another compromise can be seen in the (perhaps later) example of one year's residence in a house: its renting implies the Sisters' use of money. Then, in the late 1230s, we see two Sisters who were not living independently in a small convent, but in the house of a female follower in Castres, who eventually had to get them to go; and we see Brothers in forest huts. Provisionality had been grounded, originally, in religious poverty; now it had been brought back by persecution and fear. Later, by the time of On the Way of Life of the Poor of Lyons, some stability of residence has returned. The Sisters live in houses together with Brothers, for the sake of concealment pretending to be their wives or sisters.

The inquisition evidence documents many acts of almsgiving to the Sisters. One donor claimed not to have known they were Waldensian. This may have been a lie designed to avoid self-incrimination. But you had to be plausible when talking to an inquisitor, so this statement still shows us how the Sisters were seen by a benefactor. The gift was "for the sake of charity to two poor women." These "poor women" also carried out little jobs like washing someone. While Brothers were given clothing, a shirt, a Sister was given cloth: to be worked on. A quarter of corn: they baked. Other gifts were drink and prepared food—loaves, a lump of meat, and "many other things." "Bread and wine" may smell of the sacrament, but it was probably just a brief way of referring to the quintessential diet of Languedoc. The "Cathar" religious avoided meat for theological reasons and some Catholic religious, like the Carthusians, avoided meat for ascetic reasons. The Sisters, however, ate meat. There seems to have been nothing distinctive in their food beyond the dependence of supply on charity. The donors included nobles, one of whom we see ordering his herdsman to supply corn *regularly* to the Waldensians (gender unspecified), while one gift by a daughter to a Waldensian mother reminds us of family support. Vowed to poverty, the Sisters lived off alms, but we may be wrong to think this as normally very marginal, unstable, and irregular.

We are being given a keyhole view of the "lived religion" of these persecuted Sisters, as they tried to follow the spiritual ideal of poverty in the little communities of southern France in the thirteenth century. The keyhole is there, however, because of inquisitors' interest in individual acts of charity, and our final reflection must be this. In what ways are the texts this curiosity produced shaping this keyhole—and skewing our picture of their "lived religion"?

Further Reading

Peter Biller, The Waldenses, 1170–1530: Between a Religious Order and a Church, Variorum Collected Studies Series 676 (Aldershot, 2001), 125–58.

J. Duvernoy, L'inquisition en Quercy: Le registre des pénitences de Pierre Cellan, 1241–1242 (Castelnaud-La-Chapelle, 2001). The translation is based on their edition in Biller, The Waldenses, 155–58.

M. G. Pegg, *The Corruption of the Angels: The Great Inquisition of 1245–1246* (Princeton, 2001).

Shulamith Shahar, *Women in a Medieval Heretical Sect: Agnes and Huguette the Waldensians* (Woodbridge and Rochester, NY, 2001), 46–65.

A map and more on the inquisition are to be found in Malcolm Lambert, *Medieval Heresy* (Oxford, 3rd ed., 2002), 145, 147–48.

— 31 —

The Lives of the Beghards

Walter Simons

RULE OF THE BEGHARDS OF BRUGES,[1] 25 FEBRUARY 1292

This is the rule of the "good children" known as beghards.

1. That no one shall leave the house unless allowed to do so by the master and in the company of another beghard whom the master shall appoint.

2. No one shall engage in work other than the customary task to which he has been assigned, unless the master has ordered or allowed him to do otherwise.

3. No one shall earn a living except through work by his own hands and in a single trade; nor shall he have a journeyman work for him unless at the advice of the "children" and with permission of the mastership.

4. No one shall leave the city except with the master's permission.

5. No one shall invite a guest for a meal except with the master's permission, and that guest shall not eat in the refectory.

6. No one shall bring a woman, married or unmarried, to any house within the convent for a party, and no one shall ask [a woman] to fetch him something to drink from outside unless she were his mother, sister, or niece, and only with the master's permission.

7. No one shall divulge to outsiders what happens among the "children" inside the convent or during the chapter meeting; if someone reveals such information to an outsider, he shall once eat on the floor before all, and submit to discipline once in the chapter meeting. And if the information he divulges is weighty, he shall be expelled from the community of brethren, unless he is pardoned and allowed to stay.

8. Any beghard who does not keep the peace with his brethren in the convent, or who stays outside for three days by his own will, shall submit to one discipline in the chapter meeting, and shall eat three meals on the floor in the refectory before all; this is out of charity and will be of a certain benefit. If someone stays away

for more than three days, he shall receive discipline during the chapter meeting from every brother and eat on the floor fifteen times, and he shall not be allowed to leave the house for whatever reason until he has redeemed himself.

9. No one shall challenge the master's orders.

10. No one shall take what belongs to another unless that person gives his permission.

11. When the bell sounds, all brethren must come to the refectory and observe silence at the convent's tables, even if the master or the person who rules the convent is not present; they shall say their Blessing before the meal and Grace afterward; they cannot leave the refectory unless with permission, and shall take their meal without dawdling. Anyone who breaks the silence over meals shall receive one discipline from the master.

In the evening, when the bell is sounded for common prayer, all brethren must come to the chapel to say their prayer and remain silent afterward. Anyone who breaks the silence after evening prayer shall receive discipline for every single infraction, unless he was allowed to speak by the master.

12. No one shall go to work until after saying Matins and Prime.

13. Every brother shall attend mass three times per week.

14. Every brother shall observe the seven hours of the day.

15. Any brother who leaves the beghard convent permanently shall forfeit his contributions to the common fund and shall not make any claims to it.

16. No one shall lend [to an outsider] what is common property of the house, unless by the master's permission.

17. No one shall go and eat in the city without a companion, be it in the morning or evening.

18. Any brother who is put to work at another task must not refuse, whoever his companions are.

19. No one must go outside after sunset, whether he is in the house or elsewhere, unless by necessity.

20. No one shall sleep without underclothing.

21. Every brother shall say the daily offices in the manner of the lay brothers of the Grey Friars.

22. Chapters will be held every two weeks, and the brethren shall say confession every two weeks as well, if a confessor is available.

23. No brother shall reveal another's faults that have been redeemed nor issues that have been corrected during the chapter meeting.

24. He who strikes someone, or swears, or curses someone, or defies the master's orders and the brethren's council, shall lose his membership of the convent. A brother removed from the convent for such behavior or for other wrongdoings can make no claim to the common fund.

25. Anyone who gossips about another brother or issues threats shall eat his meal on the floor once and shall receive two disciplines during the chapter meeting.

26. You shall all be obedient to the assistant master whenever the master is not present, unless the former issues an order that runs counter to the master's; if neither the master nor his assistant is present, everyone shall obey the person appointed to rule the house in their absence in all matters that pertain to the house.

27. If a brother, because his work requires it, and with permission, is allowed to have a journeyman, anything the latter shall earn beyond the agreed work time shall benefit the common fund of the convent. If any of the brethren falls ill and, with permission, must take on a journeyman, whatever the latter shall earn beyond his usual work time shall benefit the brother who is ill, for his needs; similarly, if a brother has to leave the city for a good cause or out of necessity, he shall be entitled to the same benefits. But if he leaves the city just for relaxation, he shall not be allowed to have a journeyman to work for him.

28. If someone is ill and so weak that he cannot work or earn a living, he may, in good conscience, and with permission of the master and the community of brethren, be allowed to have a journeyman or keep the earnings of that journeyman.

29. All brethren of the convent must remain chaste. If it can be proven with utmost certainty that someone who slept outside [did not remain chaste], he shall lose his membership in the convent forever, without further contest.

30. The brethren shall lead their life so that they can receive the whole sacrament of Our Lord at the altar at least seven times per year; before each communion, they must prepare themselves by taking confession, fasting, and not eating meat for at least eight days, and by prayer and other appropriate good works, unless one is ill or too weak, and with the permission of the parish priest. The times to receive the Lord are: Easter, Pentecost, the Assumption of the Virgin Mary in mid-August, the Nativity of the Virgin in September, All Saints' Day, Christmas, and the Purification of the Virgin.

31. Every point of the rule must be observed by all brethren, unless the parish priest or the master allows for an exception by grace.

32. The master and assistant master must observe every point of the rule like the other brethren.

33. If a brother falls ill and is so weak that he has none of his own resources left, and if he has made full contributions to the house, he shall receive from the common funds forty shillings, for which he will reimburse the convent when he has recovered. A brother who has been ill in bed for eight days shall receive twenty-eight pence, and for as long as he remains ill he will be given twenty-eight pence a week by our hospital; each brother shall contribute one penny a week for as long as the masters and brethren prescribe.

34. If a brother accuses or defames someone falsely, and his accusation is not confirmed by the testimony of two other brethren, he shall be punished by the same penance or fine as the accused would be.

35. No one shall serve a meal to a woman in our convent, be she married or unmarried.

36. A new brother shall pay, upon his reception into the convent, three pounds; this is fifty shillings for lodging and ten shillings for the hospital.

37. Every brother who is present in the convent must come to say Blessing when the bell sounds; if he does not, he shall eat once on the floor and donate two shillings to the hospital, unless he was excused by the master.

38. If a brother swears an oath on Our Lord's death or on His body or on His power, or swears an indecent oath out of anger or because of some bad custom, he shall eat once on the floor in front of all.

39. The brethren of the main convent of beghards, under the jurisdiction and protection of the city, have promised to observe all of these points of the rule before their masters, and those who will be members of the convent in later years shall do the same.

40. Every brother of the convent must say one psalter of David, or seven hundred Paternosters and as many Ave Marias at the death of a fellow brother. And we wish to do the same for our masters upon their death.

41. All of the above points of the rule were declared and renewed under the mastership of Jan Hubrecht and Matheus Hoeft, aldermen of Bruges, overseers of the main beghard convent. And since we wish the above matters to be established and remain firm, we have appended the city's seal to them. This was done on the feast of St. Matthias, in the year of Our Lord MCC and XCI.

Source: From the original charter in Middle Dutch preserved at Bruges, Stadsarchief, Reeks 438, Bogardenschool, Charters, series I, no. 17, and edited by Maurits Gysseling and Willy Pijnenburg, *Corpus van Middelnederlandse teksten (tot en met het jaar 1300). I: Ambtelijke bescheiden,* vol. 3 (The Hague, 1977), 1744–47, no. 1124.

Beghards were laymen who led an informal religious life comparable to that of their female contemporaries, the beguines. They devoted themselves to the care of the poor, elderly, and sick in quasi-monastic communities without, however, being bound by perpetual, monastic vows. Sources describing the lives of the early beguines in the Low Countries in the first decades of the thirteenth century sometimes mention male followers (*beguini* in Latin) closely associated with beguines. Around 1240, these men formed their own single-sex communities, called convents or houses of "beghards" (their vernacular name). By the end of the thirteenth century beghard houses had sprung up in most major cities of the Low Countries, northern France, and Germany (including Austria).

This text is one of the oldest known rules for beghard convents. The beghard community of Bruges originated shortly before 1252, when it received official recognition from Countess Margaret of Flanders. By 1266, it consisted of two branches, independently governed: a "main" convent, whose members took confession with Franciscan friars, and a "small" convent, whose beghards confessed to Dominican friars. The division exemplifies the rivalry between Franciscans and Dominicans in their quest to extend their influence over lay devotions in urban centers. In Bruges, the beghards following Franciscan guidance thrived for several centuries, but the smaller convent under Dominican auspices disappeared from the records after 1296.

The beghards' reputation among members of the secular and ecclesiastical elites was always dubious. Virtually all beghards earned a living as weavers or worked at other tasks in the production of textiles, which, as the largest industry in late medieval cities, attracted an increasingly unruly proletariat. Moreover, from the twelfth century onward, Church leaders had expressed doubts about the orthodoxy of religious beliefs circulating in weaving workshops and often—rightly or wrongly—regarded them as breeding grounds for the Cathar and Waldensian heresies. It is not surprising then that beghards, like beguines, were subjected to measures intended to regulate their behavior and supervise their religious interests.

The present rule, written for the "main" convent at Bruges, came at a time of mounting suspicion of beguines and beghards. In 1274, the Second Council of Lyons once again focused attention on associations of religious men and women that were not officially recognized by the Church. Ecclesiastical and secular authorities in the Low Countries and the Rhineland subsequently drafted legislation to discipline them more strictly. Synodal statutes at Trier in 1277 condemned beghards who preached publicly and alleged that they spread heresies, and in the early 1280s the bishop of Liège launched local inquiries of beguines who misbehaved. In Flanders, Count Guy of Flanders ordered his bailiff in 1286 to act against "beghards of Bruges who have a bad reputation and lead a dishonest life, attempting to hide their duplicities by assuming the habit of beghards,"[2] and in 1295 he gave similar instructions to expel beguines of Ghent who disgraced the beguine way of life because of their "suspect discussions, indecent behavior, or other criminal acts."[3] These and other measures led to a rather general denunciation of an "abominable sect of malignant men known as beghards and faithless women known as beguines in the kingdom of Germany," initiated at the Council of Vienne (1311–12) and promulgated by Pope John XXII in 1317–18. Many (but certainly not all) houses of beghards and beguines were closed or converted into convents of Franciscan Tertiaries, laymen or women who observed the approved rule of the Third Order of St. Francis under close supervision by Franciscan friars.

While the rule of the Bruges beghards fits into a broader pattern of measures taken by the Church to control such laymen, it more directly reflects the growing efforts of secular authorities to regulate their professional and social life. In Bruges, the powerful weavers' guild, concerned with unfair competition from beghard weavers who pooled their resources without much regard for guild regulation, imposed restrictions on beghard weaving as early as 1269, effectively putting a cap on their production. The city government meanwhile kept a close eye on beghard finances and internal organization through two aldermen appointed as overseers (they are mentioned at the end of the document as the authorities who issued the rule and appended their seal to it). As article 41 of the rule indicates, it reiterated old regulations introduced at various times since the convent's inception, while adding new ones, which explains why the text, like many other pieces of secular and ecclesiastical legislation of the period, lacks organizational unity and sometimes shifts perspective—beghards alternatively appearing as coauthors and subjects of the regulations. Even so, the rule was certainly intended as a comprehensive standard of correct behavior that enabled the mastership to punish and expel those beghards who failed to conform.

Borrowing broadly from contemporary monastic legislation, the rule imposed a quasi-monastic regime on the beghard convent: it posited the absolute authority of the master, comparable to that of the abbot in traditional monasticism (art. 9, 26); it prescribed chapter meetings to correct irregular behavior, albeit at two-week intervals rather than daily, as was customary among monks (art. 7, 8, 22, 23); and it adopted such generic monastic practices as the *disciplina* (translated here as "discipline," a punitive flagellation, art. 7, 8, 11, 25), chastity (art. 29), silence during meals and at certain times of the day (art. 11), and the obligations to sleep wearing underclothing (art. 20) and not to inform outsiders of internal matters (art. 7). Yet beghards were not and could not be monks or clerics, and the rule acknowledged as much. Beghards were allowed to leave the convent for short periods of time (art. 1, 4, 8, 17, 19, 27) and, since they did not take solemn or perpetual vows, were free to renounce beghard life altogether whenever they wished without further sanctions, although they would lose whatever contribution they had made to the common fund (art. 15). They also earned a living individually and held private property (art. 3, 27, 28, 33). Indeed, the demands of work dominate about one-third of the rule's articles. As other sources inform us, beghards tended to gain their livelihood as weavers of wool (or linen), working at several looms arranged within a single workshop or in an adjacent room used for combing or otherwise preparing the wool. Each beghard was responsible for his own income by such manual labor, while contributing to the common fund that paid for meals taken in the refectory, the maintenance of the common buildings (including a chapel, a hospital, and, probably, a collective dormitory), and of course the support of poor, sickly, or elderly members of the community (art. 27, 28, 33).

Their religious obligations, as prescribed in the rule, placed them at midway between the secular and religious life. Observing the seven hours of the day and attending mass three times a week (art. 13, 14), beghards may have been more devoted to prayer than were most laypeople, but their engagement with liturgical services could not possibly compare to that of religious orders. Article 21 indicates that the Bruges beghards were expected to celebrate the hours in the manner of the lay brothers of the Grey Friars (the Franciscan Order), who in the thirteenth century probably recited only Paternosters and Ave Marias at these times. As was common among associations of lay religious people in the Middle Ages, the death of a fellow beghard was an occasion for intense prayer, though it was not necessarily solemnized with pomp and circumstance (art. 40).

Article 30 offers perhaps the most illuminating insight into conventional beghard spirituality. It required beghards to receive the Eucharist at least seven times a year, far more often than did ordinary laypeople, and perhaps even more frequently than did lay members of religious orders or members of the staff at thirteenth-century hospitals. Beghards shared such devotion to the Eucharist with

Low Country beguines, whose dedication to the symbol and feast of Corpus Christi is well known (in fact, a quasi-contemporary rule for the beguines at the main beguinage of Bruges allowed them to take communion up to seven times annually, although on a slightly different set of feast days[4]). It is reasonable to suppose that beguines and beghards shared this devotion to the Eucharist since the days of their common origins in the early thirteenth century. For beghards, even more than for beguines, however, the devotion to the humanity and death of Christ, coupled with the religious ideal of the *vita apostolica*, translated into a concept of a simple religious life for laymen centered upon manual labor and the duty to earn a living "working with [one's] own hands" (art. 3).

The regulations imposed on the beghards in this document were disciplinary and not doctrinal. Although one may assume that some articles of the rule (art. 3, 30) were meant to channel beghard spirituality into religious devotions that were controversial yet ultimately acceptable, and another (art. 36, imposing an entrance fee of three pounds, a substantial sum) must have limited recruitment to members of the upper and middle classes who were less likely, it was thought, to be affected by heresy, the rule did not concern itself with heretical belief per se. Naturally, such matters did not pertain to the competence of the Bruges aldermen, who issued the present rule. Their main care was to eject from the beghard community all undesirable elements, namely those who refused to submit to community discipline or work regulations imposed by the textile guilds. Sanctions against individual beghards who harbored unorthodox opinions could of course be taken up by Church authorities at any time.

Finally, the rule acknowledged without comment the peculiar name of the beghards of Bruges, who called themselves the "good children" (*goede kindre*).[5] Beghards shared this practice with local beguines of the Wijngaard beguinage, whose rule also addressed them as "children" (*kindre*). Usage of the name became highly suspicious after 1317, when ecclesiastical authorities regarded it as indicative of heretical beliefs associated with the so-called Sect of the Free Spirit. The beghards of Bruges subsequently abandoned it; they joined the Third Order of St. Francis in 1374–76.

Notes

1. The date of the document has been converted to modern style. I have silently emended (against the Gysseling and Pijnenburg edition) a few obvious scribal errors in the original, on the basis of the rule for beghards of Middelburg, the Netherlands, issued in 1331 and in large part inspired by the Bruges text; for an edition of the Middelburg Rule, see Servus Gieben, "I penitenti di San Francesco nei Paesi Bassi (secoli XIII–XIV)," in *Il movimento francescano della Penitenza nella società medioevale. Atti del 3° Convegno di Studi Francescani, Padova, 25-26-27 settembre 1979*, ed. Mariano d'Alatri (Rome, 1980), 59–80, at 73–76.

2. Edited in L. Gilliodts-van Severen, *Inventaire diplomatique des archives de l'ancienne École Bogarde à Bruges* (Bruges: Louis L. Plancke, 1899–1900), 2: 7–8, no. VIII.

3. Ghent, Beguinage of Ter Hooie, Charters, I, no. 24 (unpublished).

4. The rule for the Beguines of the Wijngaard, which is undated but must be older than 1300–1310, has been edited, with a French translation from the original Middle Dutch, by Rodolphe Hoornaert, "La plus ancienne Règle du Béguinage de Bruges," *Annales de la Société d'Émulation de Bruges* 72 (1930): 1–79.

5. Beghard convents should not be confused with the houses of so-called Bons Enfants (Good Children), institutions that in the late medieval Low Countries and France lodged young boys preparing for the priesthood.

Further Reading

Robert Lerner, *The Heresy of the Free Spirit in the Later Middle Ages* (Berkeley, 1972).

Miri Rubin, *Corpus Christi: The Eucharist in Late Medieval Culture* (Cambridge, 1991).

Walter Simons, "Beghards," in *Encyclopedia of Monasticism* I, ed. William M. Johnston (Chicago and London, 2000), 120–21.

———, *Cities of Ladies: Beguine Communities in the Medieval Low Countries, 1200–1565* (Philadelphia, 2001).

— 32 —

The Renovation of the Chapel in
the Beguinage of Lille

Penny Galloway

6 FEBRUARY 1413

Jehan de Fromont, receiver of the beguine hospital of Saint Elizabeth, outside the gate of Saint Peter in Lille, acknowledges having received from Gilles Le Maire, receiver of Lille, the sum of one hundred *sols parisis*, in Flemish money. This money the said hospital has each year collected as a duty levied at the tolls and crossing points of Lille. This year the hospital will keep the money, which will be used to convert the chapel of the said place. I received the said one hundred sols on the day of Our Lady Chandeleur, of which sum of one hundred sols aforesaid for the term of the Chandeleur just past, I am content and well paid and square with the said Gilles Lemaire, receiver of Lille and all others involved; testified by signature and my seal put here on the 6th day of February in the year 1412.

Source: Lille Archives départementales du Nord (hereafter ADN), B7759/157225.

24 JUNE 1413

Receipt given to Jacques Leclerc, mason, for work on the new chapel.

Source: ADN, B 7760.

20 AUGUST 1415

I, Jaquemont Waignon, smith, acknowledge having had and received from Jehan de Fromont, receiver of the hospital of the beguinage of Lille, the sum of seven pounds ten *sols parisis* in the old money of Flanders. This was owed to me by the hospital for seven-and-a-half-thousand lath nails used to plaster the moldings of the new chapel of the beguinage at the cost of twenty sols per thousand nails. Having received the said sum of seven pounds ten *sols parisis* I consider myself content and well paid and square with the said hospital, the said receiver and all others. Testified by my seal on the 20th day of August 1415.

Source: ADN, B7759/157228.

26 OCTOBER 1417

I, Jehan As Pois, craftsman in glass, confess to having received from sire Gille De Le Bieque, priest, for and in the name of the beguinage of Lille, the sum of seventeen pounds *parisis*, in Flemish money, for the great window sitting in the chapel of the said place containing fifty-four pieces of glass or thereabouts of my making and design. I am content with the sum of seventeen pounds in the said money, and consider myself well paid and square with the beguinage, the said sire Gille and all others. By the testimony of my seal placed on this present document on the 26th day of the month of October 1417.

Source: ADN, B 7760/157339.

12 NOVEMBER 1423

I, Mikel Cakingon, master carpenter, give notice to having received from the beguinage of Lille from the hands of sire Gilles Delebecque and Jehan de Landas, receivers of the said beguinage, the sum of thirty-nine pounds four *sols* for paneling the said chapel, and for the materials. This work was done with the consent of the ladies of that said place and at their request. As to the sum of thirty-nine pounds four *sols* I am content and well paid as testified by this present seal of my seal on this the 12th day of the month of November in the year 1423.

Source: ADN, B7759/157248.

28 NOVEMBER 1439

I, Mikiel Moreel, confess to have received from the beguinage of Lille by the hand of Jehan Ruffault, receiver there, the sum of thirteen pounds *parisis* in Flemish money, which the said beguinage gave me for four angels and two columns sold by me and designed in the livery of the beguinage to be positioned around the altar in the chapel there. I am content with the said sum of thirteen pounds as witnessed by my seal put on the 28th day of November in the year 1439.

Source: ADN, B7759/1572.

1440 ADN, B 7759/157282

Memo to the gentlemen of the Duke of Burgundy's Chamber of Accounts, who are the governors and masters of the beguinage of St. Elisabeth beside Lille, in the parish of St. Andrew, concerning the expenses incurred by the mistress and her sister beguines of that place in relation to the relics in their chapel and concerning goods and money that were inside a secret chest, the contents of which were assembled over many years previously and exist up to the present date of 1440.

First we have the aforementioned beguines paying Jehan de Navers for an iron candlestick to be placed in front of the large altar inside their chapel, fifty-five gros.

Payment to Jaquemart Waignon for extending and reshaping the metal rods on the same altar, six gros.

Paid to the said Jaquemart for reworking the iron rods that hold up the black curtain behind the said altar, four gros.

Paid for freshening two points of the priest's vestments or surplice and for reworking the memorial ornaments in the said chapel, fourteen gros.

Since the last feast day of St. John a piece of silver inside the said secret chest was used to make a cross or a crucifix, open and gilded, to contain relics of the four evangelists. The cost to make and gild this cross was nine pounds in Flemish money for which sum the aforementioned [beguines] provided and paid. The aforementioned [beguines] have for the start of payment, given to Josne, the silversmith living on the great road at Lille, seventy-two gros. Gifts made by Miss Marguerite Deleval, beguine, twenty gros, Miss de Quinchi and Miss Ysabel du Plouch, beguines, made gifts of twelve gros.

The gift made by Marguerite Waignon and Marie Lardon: payment for the two, six gros. All of these payments made by many parties were only for the manufacturing and gilding of the said cross which cost 110 Flemish gros, which left a shortfall for the said fashioning of fifty gros. This aforementioned shortfall for the manufacture and gilding of the said cross was made up by the said beguines to the said silversmith with the silver that was leftover from the original piece the said silversmith used to make the said cross. This silver he used to make a silver vessel and three silver ladles.

The aforementioned silversmith again made two other reliquaries, one of which is decorated with the image of the lamb of God that is placed on silverware from the secret chest; the other [reliquary], which used to stand inside the said chapel all broken and disordered, is remade and reguilded. For which remade reliquary and that of the Lamb of God, the silversmith has been paid, with the said silver from inside the said secret chest.

Item another payment made for the fashioning of the vermillion fabric and three white straps that serve for the aforesaid reliquaries, forty-four gros.

Payment for redye of an altar cloth five Flemish gros.

Paid to remake the wine cruet for serving at mass fifteen deniers.

All this paid by the aforementioned beguines has been paid in silver and in money that was inside the said secret chest and used for the refurbishment of the said relics of their chapel.

The rest that still remains—three gold screens and a vessel and, in money, twenty-six gros and nine deniers—is inside the secret chest.

Source: ADN, B7759/1572.

21 NOVEMBER 1443

Katheline le Blanc called Katheline de Navers, beguine and at present mistress of the hospital of Saint Elizabeth at St. Andrews near Lille and my sister beguines of the said hospital all together, are pleased that Marie de Rabecque, beguine, our companion and sister of good account, has made a contribution of relics and gifts to our chapel, particularly the image of St. Andrew in our said chapel. For all those gifts she has received, held and taken for and to the profit of our chapel since the day of the feast of Saint Peter in 1441 until the day and date of this receipt, we claim them rightfully and absolutely as we do all others received or that will be received either pertaining to the present or in the time to come and in all the time passed. Testified by the seal of the mistress of the said hospital put on this receipt on the 21st day of the month of November 1443.

Source: Lille, Archives Hospitalières de Lille (hereafter AHL), Lille Beguinage C3.

9 JUNE 1450

I, Martin Pillot, painter living at Aneulin confess to having received from the mistress and sisters of the beguinage of Lille the sum of twenty-three pounds *parisis* in Flemish, which was due for work done for the said mistress and sisters. This work consisted of having painted in gold and other fine colors five images standing in the chapel of the said beguinage: one of Our Lady and Saint John standing next to the crucifix, one of the image of St. Elizabeth with the two angels and the columns of the altar of the said St. Elizabeth, all for the said sum of twenty-three pounds of the said money, with which I am content and well paid and consider myself all square with the said mistress and sisters of the said beguinage and all others. Testified by my seal put there on the 9th day of June in the year 1450.

Source: Lille, AHL, Lille Beguinage C3.

14 APRIL 1458

Renunciation by the chapter of St. Peter at Lille of a dispute in which they had been involved with the beguines concerning the land in and around the enclosure of the beguinage on the prince's road opposite the place and manor which is called the place of le Hente . . . the only condition being that the beguines are held to always have to pay the debt they owe and are accustomed to pay annually to our said church. In return, we, the dean and chapter aforenamed, have given and give by these present letters to the said beguines rights to the aforementioned enclosure that they may hold from us and our said church in fief and homage . . . [for this privilege] the said beguines are held by us to give and to deliver for the right of relief a coronet of roses or violets or other greenery that are in accordance with the season. This coronet is to be offered and placed on the chief of the images of Our Lady in the chapel that is called le Treille in our church. 14 April 1458 after Easter.

Source: AHL, Lille Beguinage, B 29.

Beguines were laywomen who sought to lead religious lives, either alone or in groups. The first beguine communities appeared in France, Germany, and the Low Countries in the late twelfth and early thirteenth centuries and took a number of forms, ranging from small and informal groupings of women living together to virtual townships, known as beguinages, huge walled communities with their own churches, cemeteries, and streets. The beguine community at Lille was just such a beguinage. Emerging from the extraordinary spirit of religious fervor that gripped Europe in this period, beguines differed substantially from earlier groups within the western church in a number of important respects. Theirs was, above all, a women's movement, rather than the female appendage of a masculine organization. It represented a reversal of tradition. Beguines did have male counterparts, referred to variously as beguins in southern France and beghards in the Germanic lands. However, as a group they never achieved the same measure of permanence as the beguines, nor did they spread widely in northern Europe.

The most significant features of medieval monastic orders—an individual leader, a single Rule, mother house, and organizational structure—are lacking in the beguine movement. Unlike nuns, beguines took no irrevocable vows. The women could retain the use of private property and work to support themselves. They were free to leave the community at any time to marry or enter a religious order. Individual beguine houses existed independently of, and appear to have had little or no contact with, one another. In comparison with contemporary monastic orders, beguines were not part of a single movement at all, simply communities of women reacting to the conditions of urban life. This does not, however, mean that the beguines were an insubstantial force; the chronicler Matthew Paris estimated that there were "thousands of thousands" of beguines in northern Europe in the thirteenth century.[1] Clearly this figure should not be accepted as necessarily accurate, but it does demonstrate that beguines were an extremely visible and significant part of the urban population. Particularly in the thirteenth century, beguine communities gained many powerful patrons, including King Louis IX and his successors in France and the Countesses Jeanne and Marguerite of Flanders.

Not everyone, however, thought so highly of these women. The beguines were never fully accepted into the fold of the medieval Church. The established Church, as distinct from individual churchmen and women, had an ambivalent relationship with the beguines, preferring to keep them at a safe distance. This perhaps explains why so many beguine communities (including Lille's beguinage) were established on the very periphery of towns and cities, literally on the edge of the world. Many influential churchmen distrusted and criticized the beguine way of life, deriding the women as lacking in true vocation, as heretics, and even as lesbians and prostitutes. Jacques Le Goff has suggested that beguines were regarded with suspicion by many of their contemporaries because they were "somehow shady, unclassifiable, the transgressors of frontiers, neither fish nor fowl."[2] Criticisms such as those made by Bruno, bishop of Olmütz in a letter to the pope written in 1273, support this view. Bruno complains that the beguines abused their freedom to escape the yoke of obedience to priests and "the coercion of marital bonds." He even offered a potential solution: "I would have them married or thrust into an approved order."[3]

The connection between beguines and heresy, or at least allegations of heresy, has long since been noted by historians, who emphasize that, regardless of whether or not the beguines were heretics, they were perceived as heretics by many in the medieval period. The culmination of this whispering campaign was the papal decree *cum de quibusdam mulieribus*, published in the wake of the church council of Vienne in 1311. This decree stated ". . . these women promise obedience to nobody and they neither renounce their property nor profess any approved Rule . . . their way of life is to be permanently forbidden and altogether excluded from the church of God."[4] This decree led to the persecution of beguines throughout Europe, but particularly of those women who lived east of the Rhine, but it did not lead to their total elimination. The larger beguine communities in France and Flanders took advantage of the escape clause with which *cum de quibusdam mulieribus* concludes. This allows the continued existence of women who wished to live "as the Lord shall inspire them, following a life of penance and living chastely together in their hospices, even if they have taken no vow." Following this papal decree the beguinage of St. Elisabeth in Lille, like many other beguine communities, was examined by the local bishop who, in

1328, presumably satisfied as to the orthodoxy of the women, undertook to protect the community.[5]

Sources from Lille indicate that the town contained only one beguine house: the beguinage of Saint Elizabeth, situated outside the gate of St. Peter, in the parish of St. Andrew. The beguinage did not come within the city walls until the modern period. Although the exact date of its foundation is unknown, the beguinage of St. Elizabeth is mentioned in sources from 1245 on and appears to have been one of the largest beguine communities in Europe, incorporating substantial grounds including gardens and individual houses, even employing a chaplain. The beguinage survived not only the medieval period but also the French Revolution. The last four beguines resident in the beguinage at Lille did not leave until 1855.

The primary sources presented here are, in the main, no more than receipts for payments made by the beguines of Lille and their representatives (the receivers mentioned in various of the sources) for work done toward the extensive renovation of the community's chapel. Fund-raising for this project was in itself a substantial undertaking. We see from the first document that as early as 6 February 1413 Jehan de Fromont, the receiver for the beguinage of Lille, had obtained from Gilles Le Maire, receiver of Lille, a tax exemption worth the sum of one hundred *sols* in Flemish money, to help fund the renovations to the chapel. Work on the chapel appears to have started in the summer of 1413 and continued at least until June 1450. This was a lengthy and expensive process, involving a range of local craftsmen. The amount of time and care spent on the refurbishment of the chapel, over this extended period, demonstrates in itself the significance the beguines placed on their devotional atmosphere: images, relics, altars, paneling, windows, and all.

The importance that their sacred space held for the beguines may also be gauged by the amount of their own money that the women were prepared to spend on the project. Documents 6 and 7 above demonstrate that individual women in the beguine community of Lille paid for a significant part of the overhaul and decoration of their chapel, while the community as a whole used funds from its "secret chest" to meet the costs incurred. This is all the more significant because in the early years of the fifteenth century the beguinage of Lille had very little money to spare. In 1394 the mistress of the beguinage had been forced to appeal to Philip the Bold, Duke of Burgundy to save the community from financial ruin. She informed him that rents and revenues due to the beguinage had steadily diminished to such an extent that they no longer covered the community's basic needs. Nor was it possible for the beguines to pay any debts that they owed, and, in order to avoid the virtually inevitable collapse of the community, they appealed to Philip for help. As a result Philip ordered Jean de Pacy, master of his Chamber of Accounts and Jehan aux Coffres, treasurer of the church of St. Peter at Lille, to investigate the beguines' financial situation and help them to maximize their resources. The results of their inquiry, which detailed the beguines' accounts and offered potential solutions to their financial difficulties, were made public in 1400. They suggested that there should be only fourteen beguines in the community and that each beguine should pay twelve pounds *parisis* in alms to the hospital on entry to the beguinage. The fact that only ten years after this inquiry the beguines of Lille embarked on the renovation of the chapel, which was a hugely significant financial commitment, indicates how highly they prized their sacred space.

The various images and objects described in these sources are in themselves worthy of closer consideration. Martin Pillet, a painter from Annoeullin, was responsible for the chapel's most impressive features, the five gold and finely colored images found in the chapel, including an image of the Virgin and one of the beguinage's patron saint, St. Elizabeth of Hungary. Martin Pillet's image of the Virgin discussed here is symptomatic of a wider contemporary phenomenon. Representations of the Virgin in various forms appear to have been particularly popular in beguine communities from the fourteenth century onward, and the practice of veneration of the Virgin seems to have been relatively common in beguinages at the time. The women in the beguinage of St. Elizabeth in Ghent were obliged to read psalms of Our Lady every day and to fast before all of the feast days devoted to the Virgin. The form of payment that the chapter of St. Peter demanded of the beguines of Lille as a feudal duty—a crown of roses, violets, and other flowers to be placed before the image of the Virgin, which was situated in the local church of St. Peter—suggests that the wider community associated beguines particularly with the Virgin. This obligation is recorded in the final document presented above. The beguines themselves may also have felt a particular association with Mary. Like them, Mary was not a cloistered member of a religious order. She too was a laywoman, but no less holy for that. She was also a mother, as were many beguines.

It is important to note that the various representations of the Virgin were by no means the only type of sacred image owned by the beguines of Douai and Lille. Other images and devotional objects were also popular. The documents included here tell us that there were images of St. John, St. Andrew, and St. Elizabeth in the chapel of the beguinage of Lille. The chapel also seems to have contained a freestanding crucifix. Nor was this interest in devotional objects peculiar to beguines. The fourteenth century witnessed an increase in the production and use of devotional objects in general. In terms of relics, which also functioned as devotional images, the beguines of Lille were well endowed. Relics of the four evangelists, placed in two reliquaries, adorned their chapel. Accounts from beguine communities regularly contain references to the purchase of reliquaries. The communities clearly attached great significance to these items.

These images and objects served a number of purposes. They were a standard feature of devotional life in this period, used to focus the minds of the viewing public on the objects of their veneration and prayers: Christ, the Virgin, and the saints. Relics and images, such as the ones discussed here, were found in parish churches and other religious institutions across Europe, as part of typical devotional practice. However, these images could also serve another purpose for potentially vulnerable groups like the beguines. They could be used as a means of influencing how others perceived the beguines, providing their spirituality and devotional practice, which was vulnerable to accusations of heresy and lack of commitment, with the endorsement of significant religious figures, such as the Virgin, Christ, and the saints.

The documents translated here demonstrate that it was not simply the contents of the chapel that were significant for beguines. The building itself was also important to the women. We have already seen that the beguines of Lille devoted a considerable amount of money to refurbishing their chapel, taking over twenty years to complete the process. Within their chapels the beguines followed an established pattern of services and prayers. As devout members of the laity, the beguines of Lille were placed firmly under the authority of members of the local clergy and religious orders, who celebrated mass for the women, heard their confession, and offered them spiritual guidance. In all beguine houses the pattern of mass being taken in a designated church appears to have been standard and, in the larger beguine communities, which had their own chapels, the presence of the beguines seems to have been required at all the services held in their church. Communal worship may have been particularly important in beguine houses because of the absence of other communal features in the communities. Ground plans from beguinages across Flanders reveal that beguine communities often had no refectory or any communal meeting place except for the chapel. So, it served as their social and spiritual center. As our sources reveal, this was the very heart of the beguine community, with the altar the focus of their devotion and the images and relics on it the means by which these ordinary women were transported to God.

Notes

1. Matthew Paris, *Historia Anglorum* III, ed. F. Madden, Rolls Series 44 (London, 1866), 288.

2. Le Goff's statement forms part of his preface to Jean-Claude Schmitt, *Mort d'une hérésie: L'église et les clercs face aux béguines et aux béghards du Rhin supérieur du xive et xve siècle* (Paris, 1978).

3. See Herbert Grundmann, *Religious Movements in the Middle Ages: The Historical Links between Heresy, the Mendicant Orders, and the Women's Religious Movement in the Twelfth and Thirteenth Century, with the Historical Foundations of German Mysticism*, trans. S. Rowan (Notre Dame, IN, 1995), 144–45 for discussion of Bruno.

4. Norman P. Tanner, ed., *Decrees of the Ecumenical Councils* I (London, 1990), 374.

5. AHL, Lille Beguinage, C2 (letters of protection from the bishop of Tournai, 16 March 1328).

Further Reading

Penny Galloway, "Neither Miraculous nor Astonishing: The Devotional Practices of Beguine Communities in French Flanders," in *New Trends in Feminine Spirituality: The Holy*

Women of Liege and Their Impact, ed. Juliette Dor, L. Johnson, and J. Wogan-Browne (Turnhout, 1999), 107–27.

E. McDonnell, *Beguines and Beghards in Medieval Culture: With Special Emphasis on the Belgian Scene* (New Brunswick, NJ, 1954).

Walter Simons, *Cities of Ladies: Beguine Communities in the Medieval Low Countries* (Philadelphia, 2001).

J. Ziegler, *Sculpture of Compassion: The Pietà and the Beguines in the Southern Low Countries c. 1300–c. 1600* (Brussels and Rome, 1992).

—— 33 ——

The Practices of *Devotio moderna*

John Van Engen

THE LIFE OF ALBERT TER ACHTER

In the year of our Lord 1492, our beloved brother Albert ter Achter died. A native of Delden, a city in Twentia, he was our tailor and cook. After he had spent nearly his first thirty years in the world, he was brought, at the persuasion and urging of his kinsman, our brother John Delden, to hear the collations and admonitions given by our brothers on feast days. Stricken with saving remorse, he resolved to change his life and to enlist in the new soldiery of Christ, prudently judging that everything in the world is vain and transient. After he had visited our house several times, finding the ways of our brothers pleasing to his soul, he sought a place there when it happened that our house needed a tailor and a cook. Overcoming the brothers with his repeated pleas, he was received on trial. He lived among us for a while in secular garb, and then, owing to his entreaties and the uprightness of his ways, he eventually advanced, putting off his worldly tunic, being set apart for the glory of God in the vile and abject habit of the common brothers. Tested in that habit for a long time, examined by way of prolonged exercises (for greater caution is necessary in receiving laypeople), he was received into the brotherhood after about a year and one-half of probation. Eventually he accepted the full administration of the kitchen, and with funds received from the procurator he purchased meat and fish according to season and cooked what he bought. He faithfully, frugally, and carefully watched over all kitchen tasks, so the rector frequently praised him in front of guests and strangers, exulting over a layman found so faithful and provident as over some rare and precious jewel.

He was no sluggard in spiritual exercises. For immediately after his conversion he rose up against his old self, redeeming the past with work and prayer and the exercise of various virtues. On ordinary days he faithfully applied himself to kitchen tasks and the tailor's craft, but on feast days he set himself no less manfully and devoutly to the reading of the sacred writings, so that he came away not a little learned in them. To brothers coming to him for antidotes against their vices, as is the custom, he knew to provide the most fitting remedies from sayings of the saints. Moreover, at the midday collations he was asked things, and after thinking over a question answered adequately and aptly. Even to secular clerics coming to him on feast days, he delivered exhortations aloud, admonishing them very sharply, with most apt and effective materials, to flee vice, strive after virtue, and turn away from the world.

He was constant and devout in prayer at fit times, especially on feast days and during mass. Beset by an overwhelming drowsiness, he did not cease to pray. For as soon as he felt it coming on during mass or the hours, he would rise up against it: he rubbed his temples, driven by shame before the foot of the altar, or would stand up straight with the book in his hands or kneel down on his knees, so he would not be able to sleep out of sheer embarrassment. He was so burdened by this urge [to sleep], as he used to tell it, that sometimes if he tried to say his prayers after the ninth hour, when he was accustomed to go to sleep, he would sleep sitting on his knees until morning when the brothers awoke. But he did not pray only on feast days or in the chapel, but also while working. Alone or with others, he would also say the rosary or the Angelus. So that he might more fully equal in rank and merits the clerics to whom he was not equal, he was granted permission to read the hours of the Holy Cross and of the Virgin antiphonally [literally: chorally] with a lay associate, since previously laypeople were accustomed to say those hours privately among themselves. He also prefaced them with devotional materials for that day and supplemented them with petitionary prayers; for this reason rising more quickly from bed, reading them more devoutly, and gaining a more abundant reward. He also bore most patiently the exercises imposed upon him by the rector for common faults, so that if on occasion he broke a knife or had new utensils made without permission he would present himself to the brothers before table seeking forgiveness.

He was in this way fervent and zealous until about the eighth year of his conversion when dangerously high fevers raged far and wide through the land and wiped out innumerable people. As it happened, he too was infected with this plague. After he had been confined to the infirmary for a while with a raging fever, his condition worsening, he received last rites on the feast day of Saint Lawrence in the morning before high mass. Upon receiving them, he reviewed before his mental eyes things eternal and immutable, and began to ponder with inner eyes the great and awesome judgment that was soon to come. So he went back to the beginning and examined all the more diligently his inner person, and he cleansed with frequent and repeated confessions what was uncovered by that examination. But after he had surveyed repeatedly all the corners of his conscience, and—as is often the case in such a moment—thought on these things more severely than usual, he found to be true what is often said, that the conscience at such a time admits no excuse, no cloak or pretence, as often happens when his body is still healthy. He found therefore certain things by which his conscience was made not a little anxious, and especially two that—as afterward could be concluded from his words—he had not confessed properly, or as clearly as his conscience dictated, or when coming with others to the holy Eucharist. The first was that he had frequently exceeded the limits of obedience in outfitting the kitchen with benches and utensils, and especially the table on which the fish are cleaned and eviscerated. The second was that without the knowledge of his superiors he had loaned to [literally: borrowed for] his carnal father monies not altogether rightly obtained. For around the year of our Lord 149[ ], when a powerful famine attacked the people, his spirit was troubled not a little by the deep and devastating need and poverty of family and neighbors, and he fell into a panic he ought rightly to have feared. But it was a sin worthy of forgiveness, because it had not transpired by way of any evil intent (for he put down in writing what he had laid out, to be restored at an opportune time), done rather out of pity and the need of his people, especially his father with his piteous pleading.

After he had turned over in his heart the image of these and similar things, he became weak-souled and dejected, only to be lifted again by the repeated exhortations of the rector and the brothers. Now more, now less content, he made his way through the feast day of Saint Lawrence nearly to the evening. Visited by his father in the flesh and asked how he was doing, he responded, "Badly, because I will go down to hell on your account." His father answered with tears, "May it never be, dearest son. So far as possible I will restore what remains first thing. Do not ever think you will perish because of this." But he persisted in his view. Visited also by another of our brothers, the kinsman by whose admonitions he was converted, and asked about his condition, he said to him, "I wish that you were not my relative, because I—to your shame—will go down to hell."

Meanwhile, as the day drew toward evening and the time for eating had nearly come, the fever ascended to his brain, and then he seemed totally out of his senses, uttering words that were strange and full of desperation. The brothers assigned to minister to him spoke sweetly and gently to him, and they sought various remedies to relieve his pain and grief. One of them even said to him happily and kindly, "Brother Albert, be well contented. The Lord is merciful, so you should not despair. Let us rather sing some hymns in praise to God to lift your sorrow." To which he replied exasperated, "You want me to sing. So then, I will sing!" And thereupon he began to bring forth earthy and hellish words, altogether horrible and evil, more those of a devil than of a human, as it seemed, with the devil rather than he seeming to produce them from his mouth.

On hearing such words, those standing by him were shocked and dumbfounded, and they began to say to him some comforting things about the mercy and piety of God. To which he responded, "What mercy for me? I am a son of eternal death and food for an inextinguishable fire. I betrayed my neighbor, I betrayed Christ, and I ought properly to be punished with the penalties of Judas. There is no place for me with Christ. I am given over to Lucifer and his angels so that I may burn with Lucifer in hell. Woe to me that I betrayed Christ. Woe to me that I go down to destruction because of my kin." Repeating these things several times, he added, "O that bench [table], o that bench," saying it in Dutch, "I make my bed in hell with Lucifer."

. . . [The next day, after the brothers had tried to comfort him] . . . He added as well that during that night he had been led before the tribunal of the judge and accused by the demons of various sins. They went quickly through the holy scriptures to bring accusation against me, and summarized the thrust of their charges in three points, namely, true obedience, right intention, and a love not feigned, on which they brought accusation, and wherein I was found wanting. Growing silent, because I could not respond to their charges, worried and trembling, I did not know where to turn. And when I was near to despair, behold I saw a woman of indescribable beauty sitting near the judge, whom I began to suspect was the mother of the Lord. And when I remembered her great mercy and that she was greeted as the queen of mercy, my spirit was roused up to flee to her for mercy. But then as suddenly my memory turned on itself and recalled how lukewarmly, undevotedly, and negligently I used to say her hours. So fearing to hear reproach rather than mercy, I dared not draw near. She, being most kind, saw that, and turning to me her serene face, said, "About that negligence that gives you fear I have excused you, and indeed as an advocate will powerfully defend you before the judge if you persevere in trust." And this she also did. Placated therefore by the entreaties of his mother, the judge said to me, "The sins committed by you ought to be punished with a harsh severity, but because I revere my mother, never denying her anything she asks, by her grace and intercession I grant you forgiveness. Look up therefore, and take up the treasure of unfailing life, my wounds and my death, and without despairing pay the price you owe for your sins." And suddenly he who spoke came before my gaze wounded and full of blood. "Endowed with these gifts, therefore," Brother Albert said, "and refreshed and gladdened by these visions, even if there remains much that I am wary of, I do not despair."

. . . [But the fevers and the despair nonetheless return.] . . . Many hoped and thought that in his death agony he would recover his senses and voice, as had been seen in others laboring under the same condition. But this did not happen. Rather as his death agony approached in the aforesaid third hour, though he was appealed to a good deal and some sign was asked of him by way of a finger or the eyes, he was not able to give it. Thus in the presence of the brothers, now just before the hour of rising was struck, he struggled in the deepest throes of death, emitting horrible groans and his breast crying out frequently with his body in grief and pain; and thus he passed out of this world. He was buried after the matins of the Virgin in the cemetery of St. Lebwin's in the graves of our brothers. His death, most horrible in its external appearance and without any precedent we can recall, was, we sincerely hope, not injurious to him. We certainly know that it was profitable to many others, because many of our brothers, upon hearing and seeing these things, were so moved to do better that examining their consciences they made a general confession, and the least little things that before they had accounted of no weight they washed away with confession lest such a danger befall them. So they took up the resolve of living an emended life. By which may God be blessed forever. Amen.

Source: Brussels, Royal Library manuscript 8849-59, fols. 252r–254v.

This text, still unedited, appears in a manuscript copied out in the years 1497-1501, primarily in Deventer, at the founding house of the Brothers of the Common Life. Its author is not known, but the hand of the scribe is that of the house librarian, Gerrit van den Busche. He may also be its author, in part or in whole. Telling the "lives" of "sisters" or "brothers" became an accepted genre in the later Middle Ages, derived from Cistercian and Dominican models. Recently such lives have gained attention on the women's side, as fresh sources for women's history, as well because many were written by sisters as memorials to departed companions. In the movement known as the Brothers and Sisters of the Common Life, this genre flourished in the later fifteenth century, the men's lives written mostly in Latin, the women's in Middle Dutch. Brothers and Sisters lived a regular life of intense devotion in communal households, but without taking vows, under no order or rule. Their form of life rested upon "customs" and "spiritual exercises" worked out in each household, with commonalities fostered among households over time. Hence the importance of the lives. They modeled, confirmed, and encouraged a way of living that was, in the end, a personal appropriation, voluntarily adhered to by each individual. The lives have come down to us mostly in collections. But many were composed one by one, as Brothers or Sisters died, placed in individual quires or added to an ongoing codex. Their text was considered fungible, its emphasis or length subject to individual usage and taste (two copies are rarely entirely alike). Of this particular life we have only one copy surviving, made roughly five years after Albert's death and in his own house. The original was likely written immediately on his passing.

The movement originated in the 1380s in Deventer, a major market town on the Ijssel River, also important for its school (Erasmus was there as a teenager as this story unfolds). Albert entered the house about 1383/84, when the house was already a hundred years old and was taking on increasingly institutionalized features. On the social plane, men's households began as a mix of groups usually kept separate in the Middle Ages, laymen, students, clerics in minor orders, and priests. The middle group, lesser clerics, tended to predominate from the beginning, with priests kept to a small number (usually four in households of twenty to thirty members), and laymen also few, though often highly prized (Albert was not unique in getting singled out). Students and student-clerics early on got placed in households of their own. Albert entered at age thirty, a fully mature man in medieval terms. Hence the talk of "caution" in accepting him: he was not a malleable young student, not a cleric long used to disciplined forms of life. They had to make sure he could adjust to a communal household in which clerics, saying their office together, for instance, predominated. Social rank is harder to interpret, for there were people of all ranks in the movement, and some laypeople joined after gaining a certain measure of worldly success, then turning away in despair or disgust. Albert had a great heart for poor kin and neighbors (also not unusual, especially for laymen in the movement), and retained private contact with them. He came to the house by way of a relative. Frequently, where family connections can still be traced, we find patterns of families drawn to the group, even mothers and daughters.

Entering required the "resigning" of one's whole self, both person and goods, to a communal household. This was implicit, on the spiritual plane, in their heavy talk about obedience (not peculiar, in this instance, to women's households). But it was also guaranteed by documentation, a signed statement of purpose as well as a last will and testament. If a Brother or Sister left, they could, in principle, take only their daily clothes. Albert's life and goods were no longer his own but subject to decisions taken collectively or issued by elected leaders (a "rector" headed the house, a "procurator" managed its goods and affairs). Even though he was put in charge of the kitchen and the clothing, also operated in the public market and interacted regularly with townspeople, he did so as the representative of a communal household. No decision, no goods, were his. Thus, personal choices about outfitting the kitchen, a workbench or counter he regarded as "special," and most especially loans to needy relatives, came back to haunt him when he faced death and judgment. He had betrayed the commune, had not confessed, and thus betrayed companions and Christ, to whom he was lastly accountable.

This powerful communal impulse, while potentially oppressive, also worked to break down traditional medieval divisions. Clerics in the house (this not mentioned here) were also expected to engage in manual labor, especially copying manuscripts, work that produced income, and to help in the household or in gardens and fields. Albert in turn was expected to engage in spiritual exercises, indeed even, as a layman, to rise to spiritual teaching. Not only did he become wise in offering spiritual guidance, as they recall, he became a thoughtful person who drew counsel from the writings of the "fathers" in order to advise others. He was emboldened to admonish sharply local clergy who came around to the house for spiritual sustenance, this almost unthinkable in any other setting. It sprang from a carefully fostered culture of mutual guidance, but also and especially from public teaching and discussion conducted as collations in the vernacular (they could not call them "sermons" because they were mostly not given by priests). Albert was attracted to the house by way of such talks, and then engaged in them himself, even reproaching clergymen. His capacity for this depended upon, or grew out of, a strict code of personal examination, carried out daily in private prayer before God, but also in "notebooks" (*rapiaria*) of spiritual writings each Brother or Sister was encouraged to keep, a private compilation of meaningful excerpts (these, sadly, nearly all lost). Where Albert, entering at thirty, gained the literacy to join in this endeavor is not explained, but its reality is evident in others, also on the women's side.

The most compelling feature of this particular life was its horrific death scene—all the more striking in comparison with other lives where the death scene is often comforting, even beatifying. The point, after all, was to live a life of devotion in order to ease one's way into eternal well-being. The Brothers found his end striking too, and were quite unsettled by it. But the deep distress arose—apart from high fevers or a personal disorder—from tendencies implicit in the form of devotion: its unrelenting self-examination, its dependence on voluntary adherence with few framing structures to modulate individual shortcomings, its awesome sense of personal responsibility before the community and God. His father assured him the remainder of the loan would be repaid. The Brothers assured him that it was a forgivable fault, the loan in fact written down (typical for this textual community). The Virgin Mary and the bloodied Christ even assured him of forgiveness in a vision (forms of devotion common in this era). Yet it was not clear, also not to the anxious Brothers gathered around his deathbed, that he had truly heard, that he could rise above a harsh self-examination that found him guilty of betrayal, and headed for hell. That he had acted in charity held no apparent weight in his calculation—contrary to most medieval teaching. Was this the zeal of a lay convert, turned on itself? The rigors of a community relentless in its demand for "steady progress in the virtues"? Or, alternatively (since a Brother wrote this all up), a testimony to the human honesty of this community, in its piety and writing, even amid the commonplaces of intense devotional commitment?

Further Reading

My edition of this life, with many similar and related texts, is being prepared for publication in the Corpus Christianorum series.

Wybren Scheepsma, *Medieval Religious Women in the Low Countries: The "Modern Devotion," the Canonesses of Windesheim, and Their Writings*, trans. David F. Johnson (Woodbridge, 2004).

John Van Engen, *Devotio Moderna: Basic Writings* (New York, 1988).

———, "Managing the Common Life: The Brothers at Deventer and the Codex of the Household (The Hague, MS KB 70 H 75)," in *Schriftlichkeit und Lebenspraxis im Mittelalter*, ed. Hagen Keller, Christel Meier, and Thomas Scharff (Munich, 1999), 111–69.

———, "The Sayings of the Fathers: An Inside Look at the New Devout in Deventer," in *Continuity and Change: The Harvest of Late Medieval and Reformation History: Essays Presented to Heiko A. Oberman on his 70th Birthday*, ed. Robert J. Bast and Andrew Colin Gow (Leiden, 2000), 279–302, *Dicta patrum* on 303–20.

In Pursuit of Perfection: Prophecy, Revelation, and Ecstatic States

— 34 —

The Possession of Blessed Jordan of Saxony

Aviad M. Kleinberg

110. Having concluded the account of the events worthy of memory concerning the times of Master Dominic, we can now turn to certain things that happened later. When Brother Everard died near Lausanne, I continued my journey and arrived in Lombardy to take up the office assigned to me in that province. There lived there at that time a certain Bolognese brother, Bernard, who, possessed by a most savage demon, was tormented by him so horrendously day and night, and disturbed the brethren beyond measure. No doubt divine providence provided this trial to exercise the patience of His servants.

111. But let me tell you how such a scourge came to be visited upon that brother. It seems that, after he entered the Order, in great anguish because of his sins, he desired that the Lord inflict on him some kind of purifying punishment. Hence he often pondered in his heart whether he wished to be possessed by a demon, but his spirit recoiled from the thought and could not consent to it. Finally, however, after many deliberations, one day he experienced such terrible displeasure at his past offenses that he consented in spirit to have his body given over to a demon as a means of purification, as he himself has told me. Thus, with God's permission, the affair he had conceived in his heart at once came to pass.

112. The demon vomited many marvelous things from his mouth. Although the possessed was neither learned in theology nor versed in the Holy Scriptures, he sometimes produced from his mouth such profound opinions about Holy Scripture that his words would have been considered worthy of praise even had they been pronounced by St. Augustine himself. Incited by pride, however, he was exceedingly conceited, when anyone gave ear to his words.

113. I recall that, on one occasion, he proposed to me that if I were to desist from preaching he would stop altogether being a trial to the brethren. To this I replied: "God forbid that I should enter into an agreement with death, or sign a contract with hell. Your temptations will, against your will, profit the brothers, and will strengthen them in the life of grace, 'for the life of man on earth is temptation' (Job 7:1)."

114. He often tried to sow traces of his malice in our hearts, disguising his lies with deceptive words. When I noticed this, I said to him: "Why do you keep trying your deception on us?! We are not unaware of your real intentions." He answered: "I too am aware of your falsehood. At first you scorn and reject what is offered you, but after a while, you will surrender to my wickedness and gladly accept it." Let the soldiers of Christ take note of these words, whose struggle "is not against flesh and blood, but against principalities and powers, against the rulers of this darkness, against the spirits of wickedness in the high places (Eph 6:12)." May they learn from the unremitting efforts of those hosts to persevere in their fervor against them and avoid inaction and laziness in their spirit.

115. Furthermore, he would at times use such effective language by way of preaching, that his way of speaking and his piety as well as the profundity of his words drew copious tears from the hearts of his hearers. Sometimes the demon also immersed the body of the possessed brother in a wonderful way in odors sweeter than any human could concoct. He also wickedly subjected me to the same kind of temptation, pretending to be gravely tortured by those fragrances, as though they were produced by an angel from heaven, while it was entirely his own doing, striving to arouse in me in this manner a rash presumption of saintliness.

116. On another occasion he was gravely tormenting this brother in our presence, feigning being afflicted, exclaiming in an agonized voice: "here is that smell, that smell, that smell!" And shortly afterwards, as the sweetness of the odor poured upon that brother reached us, his face and voice displayed feigned horror and dismay. He said to me, "Do you know what horrified me just now? That brother's angel has come to comfort him with these odors, and the consolation he derives is a source of severe torment to me. But behold, I shall produce from my treasures the very different odors with which I am wont to accompany myself." As soon as he spoke he filled the air with a stench of sulfur, hoping to conceal the falsity of the sweetness of the first odors, by the succession of smells.

117. Hence, when he did the same to me, I was greatly perplexed with ambiguity. I doubted my own merits, and yet I hesitated, uncertain. Wherever I went I was surrounded by a wonderful fragrance. I scarcely dared take my hands out of my sleeves for fear of losing sweetness [or sanctity]of which I was not yet sure. If I held a chalice, as one does when offering the host of the Lord's body, such a wonderfully pleasant odor seemed to me to flow from it, that I could be entirely overwhelmed by the immensity of its sweetness.

118. But the Spirit of Truth did not allow the wiles of this evil spirit to continue long in its deceptions. For one day, as I was preparing for mass and was attentively reciting the psalm "O Lord! Give judgment for me against those that harm me [35:1]," a psalm very effective for warding off temptation, I reached the verse "all my bones will say, who is like unto you, O Lord," when suddenly I noticed such a strong odor of sweetness surrounding me that it suffused the

marrow of my bones. Stupified and struck by the extreme unusualness of the phenomenon, I asked the Lord, in His mercy, to reveal whether this was an artifice of the devil, and not to suffer a poor soul for whom there was no certain helper but Him to be put to shame by the powerful. Hardly had I prayed thus to the Lord—I say this in His glory—I received an inner illumination of the spirit, and such an unquestionable certainty through infused truth that no ambiguity at all was left in me, and I recognized all of this to be the enemy's deceit.

119. Now that the secret of iniquity was exposed, I notified Brother Bernard, certain as to the nature of the diabolic temptation. At once the odors ceased from both of us and from then on, he who used to utter words so full of devotion began to speak in evil and foul ways. When I asked him, "Where are your beautiful words now?" he responded: "since my evil plan has now been revealed, I wish henceforth to exercise my wickedness openly."

Source: Jordan of Saxony, "Libellus de principiis," in *Monumenta Ordinis Praedicatorum Historica* 16 (1935): 77–82.

Toward the end of his book on the beginnings of the Order of Preachers, Jordan of Saxony, St. Dominic's successor as Master General of the order, interrupts the narrative with a short personal account. In 1221 he became the Provincial of Lombardy. When he arrived at the Dominican convent in Bologna, he found there a certain simple brother, Bernard, who was possessed by a terrible demon. Day and night the demon tormented him—and the unfortunate community—"beyond measure (*supra modum*)." The screaming, howling, and rolling on the ground was certainly a nuisance, but it was what one expects of a demoniac. What was uncanny about Brother Bernard was that his possession had another side to it. Although unlettered, and untrained in either theology or Holy Scriptures he—or rather the devil—spoke most wonderfully about these matters. So much so, that "his words would have been considered worthy even had they been pronounced by Augustine himself." Worse, the demoniac preached such marvelous, such beautiful, such efficacious sermons, that, moved by the profundity and the piety of his words, his listeners shed copious and sincere tears. Finally, the possessed brother's body gave off an odor, sweet beyond the capacity of human making—the odor of sanctity?

What did this mixing of signals signify? In the neat dichotomous world of a thirteenth-century theologian, there ought to have been a clear line of demarcation between the holy and the demonic. True, the devil could imitate an angel of light, but the imitation cannot be too convincing. The devil's preaching should be empty and misleading. It should misinterpret Holy Scriptures and lead Christian souls to sin rather than sincere tears of repentance. The problem is that the road to God is straight and narrow and few are allowed to walk on it. Would a simple brother be allowed to preach in public, to offer ingenious biblical exegesis and profound theological commentary? Probably not. The devil's way, in contrast, is wide open and offers few barriers. Like all demoniacs, Bernard was a ventriloquist. He needed a belly out of which to speak. For reasons I shall discuss later he chose to speak through the devil. Bernard's demon had strong egalitarian convictions. Bernard, Jordan tells us, was very proud of his intellectual and rhetorical achievements. When ordered by Jordan to shut up, he offered his superior a deal: I will, if you will. At this point Jordan still felt confident. He responded with all the self-assured arrogance of a young theologian that he would not "enter into a pact with hell." When the demon tried to reason with him, Jordan rejected him out of hand: "We are not unaware of your intentions."

The devil did not like Jordan's attitude. He issued a threat and a challenge: "At first you scorn and reject what is offered you, but after a while, you will surrender to my wickedness and gladly accept it." At first the devil offers Jordan an explanation: good and evil can live side by side without mixing, like oil and water in a bottle. Bernard has the devil *and* an angel in him. This was supposed to have dispelled Jordan's bewilderment. The odor, the effective preaching, the theological skill were angelic, the stench, the howling, the contortions demonic. Jordan was beginning to have doubts. These doubts soon led to a deep psychological crisis. It is one thing to spurn heavenly smells in another; it is a very different thing to despise your own odor of sanctity.

It seems that for a while some sort of a truce was reached, a modus vivendi. There were now in the convent two theologians and preachers, both exuding a marvelous fragrance (though this last trait seems to have been discernible only to the two of them). One, Jordan, was the official head of the community, exercising his authority by right; the other was Bernard, exercising authority by usurpation. Who was more authoritative? Did Jordan's sermons touch people to the heart and bring them to tears? We are not told. Obviously this state of affairs could not continue much longer, especially since Jordan was undergoing a deeper and deeper moral crisis. For Jordan was internally torn, just as Bernard had been. Was he not displaying all the positive symptoms and none of the negative ones? Still very young, he became in a very short time the leader of men in an up and coming religious order; he was praised to high heaven by his peers, brought hordes of Parisian students to the order, and the odor of sanctity was about him (chapter 88). A rash presumption of his own saintliness (*temerariam sanctitatis presumptionem*) was entering his heart. Was he, then, a saint? What would that make of Bernard?

At last Jordan could not bear the ambiguity any longer. One day, when he was chanting in choir psalm 35, he was flooded by such an overwhelmingly sweet fragrance that it seemed to him that all the sinews of his bones were thoroughly steeped in it. Stunned by this extraordinary occurrence, he prayed the Lord in deep distress to take pity on him and finally regained his lost confidence (*omnino nihil ambigerem*). Good and evil were again nicely and clearly separated. Bernard's fragrance as well as all the other positive aspects of Bernard's possession stopped. The game, clearly, was over.

But was the game really over? The devil may have lost a battle, but he did not yet lose the war. The wonderful autobiographical vignette of Blessed Jordan tells us something unexpected about the role demoniacs were allowed to play in medieval society. The spiritual realm was supposedly reclassified and remapped by Christianity between the first and fourth centuries. In the old Greco-Roman religion there were no inherently evil spiritual powers. Although the gods, demigods, and *daimones* were capable of harming human beings, none of them was evil by nature. Christianity introduced a dualistic element into the Roman world. Furthermore, whereas Jews treated pagan divinities as mere fables, not worthy of consideration, Christians did not deny their reality. Instead, they have redefined them as devils *diaboloi*—malignant creatures, rebellious subjects of the one true God, bent upon harming humans. As in other aspects of its history, Christianity demonstrates here its disposition toward bricolage. Indifference would have thrown the old gods into the outer darkness, reclassification—even a hostile one—made them part of a common discourse. Thus, instead of arguing that Greco-Roman myths and stories were made up, fictive, and therefore insignificant, Christian apologists reinterpreted pagan myth as the record of a huge conspiracy. In a similar vein they have reinterpreted Jewish history as a colossal allegory.

Theoretically, then, the world was split into three realms: one wholly good, where angels and the souls of the righteous dwelled alongside God; one wholly evil, where evil souls and devils—demons—dwelled alongside Satan; and a third—the visible world—where good and evil were mixed. Whether there was in the early Church some notion of purgatory, or whether, as Jacques Le Goff has argued, a developed concept of such a place emerged only in the twelfth century, is not our concern here. What matters is that *only* human beings were capable of change, *in vita* or *postmortem*. Origen was a notable exception to this view. He claimed that demons, and even Satan himself, are capable of change and in the end will be saved. But *apokatastasis* was rejected outright by all orthodox authorities. Spiritual creatures, then, were either good *or* evil. Demons were evil.

Robert Levy, Jeannette Marie Mageo and Alan Howard have shown in a recently published book a similar reclassification-simplification of the spiritual world occurring in this century with the arrival of Christianity to Polynesia.[1] This aspect of Christianity is all the more striking, as one does not detect a similar process in the other two monotheistic religions of the Mediterranean basin, Judaism and Islam. In both, demons were far less "demonized" than in Christianity. The Jewish *maziqim* and the Muslim jinnis retained some of the moral ambiguity of the polytheistic divinities. As we shall see, the less tolerant attitude of Christianity was not without consequences.

Let us turn to spirit possession. Raymond Firth has identified three types of possession states: In the first, the alien agent is seen as a dangerous hostile invader that must be gotten rid of. In the second type, which he calls "spirit mediumship," the possessing entity enters its host in specific ritual and crisis situations, not as an evil parasite that seeks to harm him or her, but as a bearer of important messages from the invisible world. Specialists in the group are assigned to get this information and interpret it. In the third type, "shamanism," the spirits are domesticated. The

shaman can summon them more or less at will, entering trance states and participating actively in the spiritual world. As other anthropologists have noted, it is
possible to move from the first to the third type. One can begin as a victim of an in-
vading spirit and gradually learn to "domesticate" and use it, sometimes becoming
a member of a spirit cult or even a "professional" shaman.

Formal Christianity recognizes only the first type of possession. It views the
possessing entity *always* as a devil. Christian spirit possession, then, is by definition negative. This peculiar western trait has seemed so self-evident to most students of possession that they have never paused to consider its consequences. As
we know from the work of anthropologists like Lewis, Lambek, Kapferer, and
Leiris, for example, possession cults are an important means of acquiring sociocul-
tural power. While some forms of ritualized possession can be found within the
ruling elite, most mediums come from the margins. Women, serfs, the poor, ho-
mosexuals, the marginalized are much more likely to claim possession. By contact
with a powerful spiritual entity, they are invested with power beyond their "normal" powers. Even the poorest victim of demonic possession gains *parhesia*, the
right to speak in public. Sometimes, as we saw, he can gain much more. If the
possessed is allowed to keep his spirit and domesticate it, he can move with it up
the social ladder. How high? It varies. In some African societies, for example, the
progress is limited by the allocation of different possessing agents to different
groups. While charismatic-power build-up is allowed to marginals, it has its limits.
The spirits that such people are allowed to claim as "their" spirits belong to inferior
categories of spirits—sometimes the old gods of a conquered group. Thus, while
the virtuosos may acquire power, convey divine messages, and initiate ritual re-
forms, in the cultural margins they are rarely allowed to interfere with the religious
praxis of their culture's center. In other cultures, however, possession can lead all
the way to the top.

Unlike other methods of contact with the world of the spirits, possession involves very few preconditions. To speak in God's name, one must be either worthy of being God's mouthpiece—a pious, morally distinguished person—or be
willing not to claim any personal authority as a result of being God's messenger.
In a series of studies, William Christian has shown that visionaries had to be either devout Catholics with some standing in their community, preferably virgins,
or children and marginals who were seen as incapable of interpreting what they
have seen.[2] The possessed, in contrast, do not have to be pure, or particularly religious. Very often they were neither. In possession, then, the question of moral
worth does not arise at all, and the question of credibility is significantly less
acute. While what the possessed was saying might be called in doubt, the presence of a spiritual being was usually not questioned. In most cases of divine inspiration, one had to believe the visionary, that he or she was accurately reporting
what he or she has seen and heard. In possession one did not have to believe; it
was the spirit itself that was speaking, for all to hear, through the host. Moreover,
God's messages are supposed to be consistent with tradition and with theological

knowledge in the broad sense, but a spirit could be mischievous and lying. The possessed could begin his career with an unruly spirit and slowly, with the help of expert counsel and expert training domesticate the spirit—either by changing its nature or by moving to a "better" spirit. One could learn on the job.[3]

In the Middle Ages the Christian ecclesiastical hierarchy insisted that all possessing agents were evil. Attempts to claim possession by saints, the dead, and angels were dismissed as demonic fraud. In view of this strong theological position it is surprising to find an uninterrupted record of "good possession" throughout the Middle Ages. Brother Bernard was not alone, nor was he the most successful of popular ventriloquists. Writing in the same time as Jordan, Cardinal Jacques de Vitry tells, in his *Historia occidentalis*, of a demoniac from Germany who was preaching the gospel truth (*veritatem evangelii*) to good effect. Jacques does not specify who the *host of* this malignant spirit was, but clearly it was a person with no right to preach, for the question is promptly put to him, by whose authority he dares preach and teach. The demoniac's answer is significant: "My name is pen in ink (*penna in inchausto*). I am compelled by God to preach the truth in contempt *of* the dead dogs, incapable *of* barking. And, because I can speak nothing but what is and worth recording (*scribi dignum*), 'pen in ink' is my name."[4] The image is appropriate: the demonic pen dipped in human ink, held by the invisible hand of God. The emphasis on constraint serves to solve the basic problem *of* Christian "positive possession": How can an evil spirit do good? How can the Father of Lies speak the truth?

One might reasonably ask, why claim *demonic* possession at all? Why not claim to be inspired by the Holy Spirit, for example? Surely it is easier to deliver significant messages to the community, without having to explain how a benign message comes from a tainted source. The answer has to do with the nature of demonic possession as initially a psychological crisis. Whereas career as a saint or a visionary could begin in a great variety of ways, the diagnosis of a person as demoniac usually followed a more or less conventional set of symptoms. The most important *of* those were wild contortions, screams, incoherent speech, and altered states of consciousness. For us these symptoms express internal conflict. For a medieval audience they meant either madness or demonic possession. The incoherence, the violence, and the lack of control made it impossible in most cases to be identified with the forces of order. Having shown such symptoms, then, the patient, with a taste and a talent for spiritual power, had little choice but to make the best of a bad situation—accept his status as a demoniac and offer a version *of* demonic possession that could serve him best.

There were all kinds of possible variations. A common strategy was to admit the presence of an evil agent, a demon, and argue that this agent has been forced by God to become his messenger. Thus, one could eat his cake and have it too. We don't have to rely on Jacques de Vitry, whose intentions in telling about the anonymous German demoniac we may rightly suspect. Similar stories of demoniacs preaching, serving as social critics, as local healers and as clairvoyants, can be found in Walter Map, Caesarius of Heisterbach, Thomas of Cantimpre, and Etienne de Bourbon, for example. Etienne tells the story he heard from a fellow Dominican about a sermon delivered by a demoniac in Tuscany in front of a large crowd. The demoniac had promised beforehand that he will not lie, and has, then, preached a very fine sermon on the virtues of preaching. He has ended his sermon with a reiteration of his good faith and veracity: "Know," he says, "that I am a devil. I am compelled to preach the truth to you . . . on judgment day I will be able to accuse you all the more. You will be reproached for having the truth preached to you by the very devils."

Reports of "good possession," however, appear not only in such major works, laden with political undertones, but also in a host of less well-known sources: in the Lives of Aluzio, of Robert of Chaise Dieu, of Joachim of Fiore; in the miracle tales of Justus and Clemens; in the Lives of Remerius of Pisa, Osanna of Mantua, and the Dane, William of Roschild, to name but a few. The sources are reluctant to dwell on the benign activity of demoniacs. Often it is reported in negative terms: demoniacs who expose the sins of priests are seen as a nuisance, demoniacs who heal are seen as meddling in illicit magic, and demoniacs who preach are criticized as usurpers of Episcopal authority. When credence is given by the authors to the words of demons it is usually to denounce someone whom the demoniac—carefully coached—identifies as an ally of Satan—heretics and Jews, for example. This tradition goes back to the days of Jerome and Ambrose in whose presence demons were compelled to denounce Arians and Manichaeans and to identify—no questions asked—the relics of saints "discovered" by Catholics.[5] It had its heyday in the sixteenth and seventeenth centuries, when Protestant and Catholic demons joined in the general choir of mutual denunciation. But whatever their political allegiances and effects, the sources show demoniacs—that is, demons—acting successfully in ways that were theologically very dubious.

The very existence of the supposedly nonexistent class of Christian career-demoniacs alerts us to the great variety of strategies of acquiring spiritual power in medieval society, strategies that have yet to be studied systematically. For if the ecclesiastical hierarchy was able, on theological grounds, to narrow the field of action of one strategy—possession (by preventing the emergence of possession cults)—it was less successful—if that was its aim—in limiting the influence of other strategies: prophecy, visionary activity, wonder-working, and local sainthood. Not only were these not limited, they were, in the Middle Ages, the life and soul of religious life.

Notes

1. Robert I. Levy, Jeannette Marie Mageo, and Alan Howard, "Gods Spirits, and History: A Theoretical Perspective," in Spirits in Culture, History, and Mind, ed. Jeannette Marie Mageo and Alan Howard (New York, 1996), 22–23.

2. William A. Christian, Jr., Visionaries: The Spanish Republic and the Reign of Christ (Berkeley, 1996), 243–61.

3. Michel Leiris, *La possession et ses aspects theatraux chez les Ethiopiens de Gondar* (Paris, 1989), 40–42; Vincent Crapanzano, "Spirit Possession," in *Encyclopedia of Religion* XIV (New York, 1987), 12–19.

4. Jacques de Vitry, *Historia occidentalis* 2.5, edited by John F. Hinnebusch (Friburg, 1972), 86–87.

5. Paulinus, *Vita Ambrosii* 15, PL 14, col. 32.

— 35 —

On the Stigmatization
of Saint Margaret of Hungary

Gábor Klaniczay

THOMAS ANTONII DE SENIS "CAFFARINI," *LIBELLUS DE SUPPLEMENTO, LEGENDE PROLIXE VIRGINIS BEATE CATHERINE DE SENIS*

On that day the virgin Margaret was praying in tears in her oratory, which was between the sacristy and the wall of the church. And the aforementioned nun saw her rise to a height of several cubits above the ground, and she remained like this for the length of two hours. When she returned afterward to her bed, the same nun saw that she had bleeding stigmata on five places of her body, on the hands, on the feet, and on her side. The holy virgin begged that the nun would keep her secret as long as she lives. She did as she was asked, but after the death of the virgin she revealed everything.

> Source: Thomas Antonii de Senis "Caffarini," *Libellus de supplemento*, *Legende prolixe virginis beate Catherine de Senis*, ed. Iuliana Cavallini and Imelda Foralosso (Rome, 1974), 175.

SPECCHIO DELLE ANIME SEMPLICI, DALLA BEATA MARGARITA FIGLIUOLA DEL RE D'UNGARIA SCRIPTO

Once when she stood devoutly praying in her oratory, in front of the crucifix, and weeping copiously, she was ravished by the Holy Spirit, and elevated from the ground four cubits high, and she stayed this way for more than two hours. And with God's permission the rapture of this blessed virgin was seen by her companion and relative who was a member of the same order, and with whom she frequently discussed the secret revelations she had received from Christ. Turning back to her room, this companion and relative of hers saw that her cloak was soaking with blood, just like her hands and feet. She was astonished to see this, and devoutly asked the blessed Margaret what it all meant. Then the blessed Margaret asked her to give her word that she would keep it secret while she herself was alive, and she told how the crucified Christ appeared to her in the form of a seraph, and how he imprinted his holy wounds (stigmata) onto her body. And when the time came and the blessed Margaret passed away from this life, the moment came for this relative to testify to this very great miracle, which she did with the greatest reverence and devotion, telling it to everybody.

Source: Florio Banfi, *Le stimmate della B. Margherita d'Ungheria*, in *Memorie Domenicane* 50–51 (1934): 304–6; cf. also R. Guarnieri, *Il movimento del Libero Spirito: I. Dalle origini al secolo XVI; II. Il* "Miroir des simples âmes" *di Margherita Porete; and III. Appendici*, in *Archivio italiano per la storia della pietà* 4 (Rome, 1965), 351–708.

GIROLAMO ALBERTUCCI DE BORSELLI, *CRONICA MAGISTRORUM GENERALIUM ORDINIS FRATRUM PRAEDICATORUM*

How she obtained the stigmata and was marked by Jesus Christ was narrated after the death of blessed Margaret by the nun who was her confidential companion. On that day the virgin Margaret was devoutly praying, in tears, in her oratory, which she had ordered to be made for herself between the sacristy and the wall of the church. This was seen by the aforementioned nun, who was her relative and her confidante. She saw the blessed Margaret elevated from the ground to a height of four cubits or more, and she stayed this way for more than two hours. When she returned to her bed afterward, and took off her clothes, [the same nun] saw [Margaret's] hands and feet bleeding, and her cloak was also soaked with blood. She was astonished to see this. The blessed Margaret asked her to swear that she would not tell anybody what she had just seen while she herself was alive. The nun asked: "Dearest mother, tell me how this happened?" The blessed Margaret responded: "While I was meditating on his passion, the crucified Jesus Christ appeared to me and impressed his *stigmata*." After the death of blessed Margaret, the above-mentioned nun testified under oath about what she saw and had not spoken about until then.

Source: Girolamo Borselli, *Cronica magistrorum generalium Ordinis fratrum Praedicatorum* (Biblioteca universitaria di Bologna, Cod. Lat. 1999, fol. 33r). I have also edited this legend: "Borselli és Taeggio Margit-legendája Bánfi Florio apparátusával," in *Miscellanea fontium historiae Europaeae: Emlékkönyv H. Balázs Éva történészprofesszor 80. születésnapjára*, ed. Kalmár János (Budapest, 1997), 11–56.

The Hungarian princess Margaret, daughter of King Béla IV, spent her life as a Dominican nun and died in 1270, in the reputation of sanctity. The miracle of her stigmatization, described in three related versions above, however, was not mentioned in her oldest legend or in the acts of her canonization investigations carried out in 1276. This remarkable miracle was not even known in Hungary until two promoters of the canonization of Catherine of Siena (1347–1380), Raymond of Capua, and Tommaso Antonii of Siena, named Caffarini, made direct inquiries about the alleged stigmata of the Blessed Margaret. Answering a letter about this matter in 1409, Friar Gregorius, Prior Provincial of the Hungarian Dominicans, explained to Caffarini that Margaret was not stigmatized at all, but this divine favor had been bestowed upon another Hungarian Dominican nun, her *magistra*, the blessed Helen.

Even though the reputation of the stigmata of St. Margaret of Hungary was short-lived even in the Middle Ages, it provides a good illustration of the emergence of the cult of female stigmatics in the late Middle Ages. The nascent cult of this royal Dominican nun (whose canonization process was not concluded in the thirteenth century) provided the first opportunity for the Dominican Order to claim for one of its members the "unheard-of" miracle that had happened to St. Francis of Assisi: the reception of the holy wounds of Christ.

Before the date that can be assigned to these fourteenth- and fifteenth-century texts, the fame of Margaret's stigmata had spread in Italy through a number of panel paintings and frescoes. The earliest among them is the panel by the Master of the Dominican Effigies (after 1336, probably before 1350) in the sacristy of Santa Maria Novella in Florence, depicting Margaret with a crown on her head and lilies in one of her stigmatized hands, in the company of other Dominican saints or candidates for sainthood. A second image appears in a fresco in San Domenico, Perugia (ca. 1368), showing Margaret half kneeling, with her crown beside her on the ground, receiving the stigmata from Christ appearing to her in the form of a seraph.

A fresco in the church of San Nicolò, Treviso, is also from around 1370. Margaret is crowned here by two angels and adorned by two inscriptions: "*Beata Margareta regina Ungariae ordinis fratrum predicatorum*" and "*Ego enim stigmata Xti in corpore meo porto*"—a quotation adapted from St. Paul (Galatians 6:17).

In Tuscany and Lombardy, Caffarini himself saw several such pictures of Margaret with bleeding stigmata, and he quotes Antonio da Bitonto, Naples Provincial of the Dominicans, according to whom in Puglia "in every convent there was a depiction of a nun of the Dominican Order, named Margaret of Hungary, with five stigmata and a lily in each wound. . . ." This repeated image stimulated Caffarini to inquire about St. Margaret in Hungary when he was assembling data to support the campaign to canonize St. Catherine of Siena, to dissipate doubts about whether the divine favor of stigmatization could be extended to representatives of the "feeble sex." He composed the *Libellus de Supplemento* around 1417, which included not only a long *Tractatus de stigmatibus*, but also, among other hagiographic analogies to Catherine, the account of the stigmatization of Margaret of Hungary translated above, which Caffarini copied in Pisa around 1393 from a legend-version no longer extant.

Fig. 35.1. Fresco of Margaret in San Domenico, Perugia, Italy, ca. 1368.
Photo by László Bókay.

Fig. 35.2. Fresco of Margaret in the church of San Nicolò, Treviso, Italy, ca. 1370.
Photo by Péter Klaniczay.

The second, longer description is a translation from the introduction to the late fourteenth-century Italian version of the famous medieval mystic treatise *The Mirror of the Simple Souls*, written by the Vallonian Beguine Marguerite of Porète, who was executed in Paris in 1310. In its Italian translations this book—one of the masterpieces of medieval mysticism—was erroneously ascribed to the blessed Margaret, "daughter of the King of Hungary." The attribution might have been due to a wish to camouflage the original author, whose name was tainted with the accusation of heresy, but it also might have relied on the spreading fame of Margaret as a stigmatized visionary in Italy. Perhaps to reinforce the plausibility of her authorship, the introduction elaborates on this passage of her legend. In one of the preserved fifteenth-century copies of the *Specchio*, perhaps once in the possession of the Observant Franciscan inquisitor St. John Capestran, a note among the marginalia adds: "the Roman Church does not accept all this, and we know that it is reputed to be a tale and a fraud in the eyes of many."

Despite Caffarini's disappointment, and the later condemnation, and even despite the successful recognition of the sanctity of St. Catherine of Siena (1461) as the first stigmatized female (and Dominican) counterpart of St. Francis, the cult of Margaret's stigmata continued to survive. The third translated text comes from the hagiographic appendix to the *Chronicle of the Masters of the Order of the Preachers* by Girolamo Albertucci de Borselli (1432–1497), written in Bologna around 1495. It also appears in anthologies of the legends of Dominican saints by Ambrogio Taeggio (died ca. 1520) and Leandro Alberti (1479–1552). The context of these renewed and detailed accounts is the new vogue of stigmatic female visionaries—so-called living saints (*sante vive*)—associated with the Dominican Order at the end of the fifteenth century.

The fame of St. Margaret's stigmata is the first chapter in the enduring endeavor of the Dominican Order to discover within its ranks a series of female counterparts to St. Francis. The legends and images related to the stigmata of the Hungarian princess-nun have to be understood in the context of the lasting rivalry among three religious orders, the Franciscans, the Cistercians, and the Dominicans, because having stigmatized "living saints" became a special issue within these orders.

St. Francis's miraculous reception of the stigmata on Mount Alverna on September 14, in 1224, as told in his legends and represented on a mass scale by medieval painting, became a rich, enigmatic, and effective core symbol of thirteenth-century Christianity. It has received due attention in recent historiography (by André Vauchez, Chiara Frugoni, Arnold Davidson, Octavian Schmucki, and Giovanni Miccoli). Besides the abstract identification with the suffering Christ and the optimistic message that a contemporary could reach that perfection, it also carries particular hints for the Franciscan Order; it might also refer to Francis's dissatisfaction, documented in his writings and his various legends, at seeing his order growing in power and departing from its original purity and poverty. St. Francis, one might say, was crucified for his effort to build up an ascetic Christian Church in a rapidly developing urban world pregnant with the secular spirit of the Renaissance. It is thus no wonder that the stigmata of St. Francis were highly controversial even within the

Church. Among those who were most reluctant to attribute this prestigious emblem to St. Francis, thus making him an *alter Christus*, were the two rival religious orders of the age, the Cistercians and the Dominicans.

The religious message expressed by the stunning miracle of St. Francis, however, was the common treasure of a broadly diffused new spirituality of the thirteenth century. In many places of medieval Christianity from England to Italy and from the Netherlands to Central Europe, sometimes independent of one another, sometimes trying to outdo one another, many beguines, nuns, friars, and tertiaries similarly strove "to show in their outward body the composure of their inward mind," as James of Vitry said about Mary of Oignies. Stigmata could thus be interpreted as the most convincing bodily proof that the visionary was truly "beholding" Christ. From the 1260s on, after the energetic interventions of Pope Alexander IV, which seemed to settle the early controversies, we see repeated attempts to appropriate this special sign of perfection within the two religious orders that had been skeptical about the stigmatization of St. Francis. A noteworthy aspect of the new cult is that these new claims came almost exclusively from woman mystics: stigmatization became a feature of late medieval ecstatic female religiosity. A historian of stigmatization, Antoine Imbert-Gourbeyre, in 1894 counted 321 historical claims to stigmatization; approximately a third of them occurred before 1500, and the majority were by women.

The first series of noteworthy attempts came from the Cistercians. In 1267 Philip of Clairvaux, from the Cistercian abbey of Herkenrode, diocese of Liège, reported that a beguine living nearby in the village of Spalbeek, named Elizabeth, "bore most openly the *stigmata* of our Lord Jesus Christ, that is, in her hands, feet, and side, without ambiguous simulation or doubtful fraud. The visibly open, fresh wounds are bleeding frequently and especially on Fridays." Philip of Clairvaux colorfully describes how Elizabeth presented, in a series of ecstatic raptures, a meticulously precise performance of Christ's sufferings from the moment of his arrest till the deposition from the cross, and fitting this presentation, in addition, to the rhythm of the seven canonical prayers. Philip also underlines that "he himself with his companions, abbots and monks" could observe with their own eyes the blood drops or streams coming from the eyes and the wounds of the virgin. After his detailed description of the miraculous bodily signs, Abbot Philip raises the question of how the divine choice for representing "this glorious victory, this wonderful virtue" could fall upon "a representative of the feeble feminine sex," and tries to justify it with eloquent arguments.

A vehement controversy ensued, initiated by the Franciscan master of Paris, Guibert of Tournai, who raised his voice in his treatise titled *On the Scandals of the Church* (*Collectio de scandalis Ecclesiae*) against this attempt to steal the privilege of stigmatization from St. Francis. We know less detail about a similar case of stigmatization about the same time, mentioned in the life of a Cistercian nun from Val-des-Roses, Ida of Louvain (died ca. 1300).

Another very detailed description was transmitted from Germany about the stigmatization of Lukardis, a Cistercian nun of Oberweimar, Saxony. Her desire to receive Christ's wounds surfaced in 1279, during a spectacular vision of the crucifix. She was asked by Christ to alleviate his pain by "attaching her hands to His hands, her feet, to His feet and her breast to His breast." Subsequently she strove for two years to make those felt wounds real, and finally, in 1281, with further visions of Christ pressing his wounded hands to hers, the five holy wounds gradually appeared and started to bleed every Friday. The Cistercians were also involved in the strange heterodox cult around the Milanese Guglielma (died 1280/81), popularized after her death by a female "pope," Mayfreda Pirovano, and after 1300 vehemently persecuted by the Inquisition. Guglielma's principal saintly attribute was that she was held to be the incarnation of the Holy Spirit, yet, at the same time, she was also reported to have "five wounds in her body which were like the wounds of Jesus Christ."

The attempts of the Dominicans to come forward with their own stigmatized saint date from about the same time. They first proposed a male counterpart to St. Francis; one of the early histories of the order, the *Vitae Fratrum*, compiled before 1260 by Gerard Frachet (1205–1271), mentioned the case of Walter, prior of the Dominican convent in Strasbourg, who, after a deep meditation on the Passion, once felt deep pains in his body at the places of the five holy wounds of Christ (which nevertheless did not become bleeding wounds visible from the outside). It is worth noting that this episode occurred when he entered for a prayer the church of the Franciscan convent in Colmar.

In his encyclopedic survey of thirteenth-century religiosity, in *De bonum universale de apibus*, written between 1256 and 1263, the Dominican Thomas of Cantimpré made several allusions to stigmatized nuns and beguines, expressing the order's attention to lay female religiosity—a stigmatized woman he referred to in his work may have been Lutgarde of Ayvières (died 1246), whose biography he himself recorded, or perhaps Ida of Louvain. In the same decades, near Cologne, another stigmatized beguine, Christina of Stommeln (1242–ca. 1290) was discovered and enthusiastically accompanied by a Swedish student of theology in the Dominican *studium generale* in Cologne, later prior of the convent in Visby, Petrus of Dacia (1230/40–1289). He recorded that Christine had "the *stigmata* at the hands, the feet, the front, and the side at the age of fifteen, for she desired to have something that reminded her of the passion of Christ." These bleeding wounds, and some blood-dripping nails extracted from them (perhaps causing them?), had been personally observed and described by Petrus of Dacia in 1267.

One does not really know when the fame of Helen of Hungary, a stigmatized Dominican nun from Veszprém, started to spread. Her legend, preserved in the manuscript sent to Venice from Hungary in connection with Caffarini's inquiry originally related to St. Margaret, tells that "she bore *stigmata* on both of her hands and feet, and on her chest." Her first wound appeared on the right hand "during the night of the feast of St. Francis" (she protested out of modesty: "My Lord, this should not happen . . ."), and the second one "on the day of the Apostles Peter and Paul, at noon." From the wound in her right hand "there grew a golden hair, . . . then a resplendent golden lily." The leaves of these lilies were torn out and subsequently collected by the nuns of her convent. As the death of Helen is described by her legend as preceding the Tartar invasion (+1241), she certainly could not have been the *magistra* of Margaret of Hungary, as Prior George asserted in his letter to Caffarini in 1409. Suspicions about her real historical existence could arise from the fact that her name is mentioned nowhere in the legends and canonization investigations of St. Margaret of Hungary, which gave an extensive presentation of the Dominican milieu of that period in Hungary. The reputation of Helen as a stigmatized saint is first attested around 1330, when her name and stigmata were mentioned in a Paris controversy on St. Francis in the *quodlibet* of the Franciscan master Pierre Thomas.

St. Margaret's stigmatized representations and the related passages inserted into her legends also date from the fourteenth century and testify to an intensification of the claim of the Dominicans to have their own stigmatized saint. Her fame was subsequently superseded by the attribution of this sign of divine election to the most popular contemporary living saint of the Dominican Order, St. Catherine of Siena. According to her confessor, Raymond of Capua, Catherine revealed to him that in 1375, in Pisa, she received the wounds of Christ during an ecstatic prayer, but they remained invisible upon her request and were kept secret until her death in 1380.

As already mentioned, even for St. Catherine of Siena, the widely acclaimed visionary, it took several decades to have her sanctity and, related to it, her stigmata recognized. After her canonization by Pope Pius II in 1461, a new debate erupted concerning the representation of her stigmata during the pontificate of Sixtus IV, who came from the Franciscan Order. In the bull *Spectat ad Romani* of September 1472 he prohibited the representations of St. Catherine "with Christ's *stigmata* . . . and like St. Francis," and even the mention of them in sermons. This prohibition was repeated in 1475 and 1478. A counterreaction of the Dominican Order to the prohibition soon reappeared after the death of this pope (1484). During the papacy of Alexander VI (1492–1502), several paintings depicting the stigmatization of St. Catherine bear witness to this trend, and we know that the pope, although he did not rescind the ruling of his predecessor, allowed the depiction of the stigmata. The reappearance of a new series of stigmatized representations of St. Margaret of Hungary in various Italian churches and convents around 1494–95 is to be regarded as a part of the conscious effort of the Dominicans to strengthen the claim of St. Catherine's stigmata, as Caffarini did a century earlier, by making explicit reference to other stigmatized Dominican nuns.

Precisely in the middle of the 1490s, Girolamo Borselli and Ambrogio Taeggio took a special interest in a renewed formulation of the legend of St. Margaret of Hungary's stigmata, and the same interest must have contributed to the appearance of a series of "new St. Catherines." Shortly after having finished his Chronicle, and presumably his legend with the stigmatization account of St. Margaret, in 1495, Borselli gave a series of Lenten sermons in Rome in Santa Maria Novella. He met on this occasion the new celebrity of the Dominican Order, Lucia Broccadelli, more commonly known as Lucia da Narni (1476–1544), who in 1494 entered the Dominican third order, who in that period lived in Rome, in the same house where her ideal, Catherine of Siena, had lived more than a century earlier. Borselli described their meeting and mentioned the account Lucia da Narni gave of a vision in which she saw "the bleeding Christ on the cross." A year later, in 1496, during a meditation on the Passion on Good Friday, the bleeding stigmata appeared on the hands, feet, and side of Lucia da Narni in the city of Viterbo. This spectacular repetition of the miracle of St. Catherine, as she took care to point out, was meant to provide tangible proof of the veracity of the stigmata of her ideal.

How much the phenomena of stigmatized saints preoccupied and seduced the religious imagination of the last decade of the fifteenth century is demonstrated by the fact that other stigmatized *sante vive* appeared simultaneously. The prominent reincarnation of St. Catherine, Columba da Rieti (+1501), was asked to give an opinion on the stigmata of Lucia da Narni. Caterina da Racconigi (1486–1547), the protégée of Gianfrancesco Pico della Mirandola, equally claimed to have the stigmata of Christ, like St. Catherine, professing out of modesty, to have invisible aching wounds. Other stigmatized female Dominican saints included Stefana Quinzani di Soncino (1457–1530), who periodically experienced in her body the pains of the crucifixion, shown also by bleeding wounds. Osanna Andreasi (+1505), another Dominican nun, regularly felt the pain of the crown of thorns, the side wound, and the wounds in the legs.

This new series of late medieval female stigmatizations, together with the persistence of the fame of St. Margaret of Hungary's stigmata, illustrate that the appearance of such miraculous religious phenomena is closely bound to the intensive efforts to represent and narrate the archetype and the possible repetitions of this event. It is precisely this context of representations, images, and texts that created the heated visionary atmosphere where such phenomena could become, or at least could be accepted to have become, bloody bodily realities.

Further Reading

Caroline Walker Bynum, "The Female Body and Religious Practice in the Later Middle Ages," in *Zone 3: Fragments for a History of the Human Body* I, ed. Michel Feher et al. (New York, 1989), 160–219.

Nancy Caciola, *Discerning Spirits: Divine and Demonic Possession in the Middle Ages* (Ithaca, NY and London, 2003).

Arnold Davidson, "Miracles of Bodily Transformation; or, How St. Francis Received the Stigmata," in *Picturing Science, Producing Art*, ed. Caroline A. Jones, Peter Galison, and Amy Slaton (London, 1998), 101–24.

Gábor Klaniczay, *Holy Rulers and Blessed Princesses: Dynastic Cults in Medieval Central Europe*, trans. Éva Pálmai (Cambridge, 2002).

———, "Le stigmate di santa Margherita d'Ungheria: Immagini e testi," *Iconographica: Rivista di iconografia medievale e moderna* 1 (2002): 16–31.

Tibor Klaniczay, "La fortuna di Santa Margherita d'Ungheria in Italia," in *Spiritualità e lettere nella cultura italiana e ungherese del basso medioevo*, ed. Sante Graciotti and Cesare Vasoli (Florence, 1995), 3–27.

Aviad M. Kleinberg, *Prophets in Their Own Country: Living Saints and the Making of Sainthood in the Later Middle Ages* (Chicago, 1992).

O. Schmucki, *The Stigmata of St. Francis of Assisi: A Critical Investigation in the Light of Thirteenth-Century Sources* (New York, 1991).

Walter Simons and J. E. Ziegler, "Phenomenal Religion in the Thirteenth Century and Its Image: Elisabeth Spalbeek and the Passion Cult," *Studies in Church History* (1990): 117–26.

André Vauchez, "Les stigmates de Saint François et leurs détracteurs dans les derniers siècles du moyen âge," *Mélanges de l'École Française de Rome* 80 (1968): 595–625.

Gabriella Zarri, *Le sante vive: Cultura e religiosità femminile nella prima età moderna*? (Turin, 1990).

— 36 —

Eschatological Prophecy:
"Woe to the World in
One Hundred Years"

Robert E. Lerner

Woe to the world in one hundred years, for it has departed from my virtue.

The inhabitants of Syria, who were torn apart by profane studies, will be driven in my indignation by a profane people from the shore of the sea. And there will be desolation in the land until a new David comes to restore the ark of Zion.

The concubine Greece will be exposed to plunder again and through the art of the western bat be led back to the house of the bride.

Headstrong Sicily will be threshed until after the consumption of the bees the bride will have been reformed. And when the talons of the eagle cease to crush Sicily, it will not escape, for the double storm will swallow it.

The heights of the Romans will demolish the unicorn, and they will be consumed by their own fits of rage. To them the collapse of the bridge and the sinking of the ass offer signs of approaching destruction. When, at the suggestion of the two-tongued one, the idol is forged, the truth, while the people will be shameless and the elders will keep silent, will be refused a hearing, and it will be concealed in the command of the statues of Egypt.

Italy, the nest of wild asses, will be bitten by the lions and the wolves, who were born in its own forest. And after the groin is lacerated, the blood will flow to the big toes. But after it has become acquainted with the earthly abyss, it will learn to recognize a remedy for its thirst.

Germany will be tortured by pain in its entrails, and after the necks are broken, it will drink with gigantic priests from the cup of wrath. In its sea a multitude of beasts will rouse a storm, plunging the seafarers into peril. For through the disorder of the princes the peace of the populace is endangered.

The mangy-red arms of the king of the bees, squeezing together the sides of the subjects and stretching from sea to sea, are folded together through immoderate repulsion. For the king, who gulped down the menstruations of the bride, stran-gled by his own cord, falls from the throne by the hate of his neighbors, while the vines of the feigned alliance dry up. Nor will the ambitious pollution of the blood go unpunished because the children mourn.

The nest, moreover, of Aristotle, wasting away gradually, will be emptied, because the detestable gabbling of the chicks will cover up the truth, deriding its ministers.

The caves of the Irish will be confused by the tumult of their own sea, and, so that the untamed ferocity will become milder, one king will bring an undis-tinguished people under pressure.

Spain, the nurse of Mohammedan depravity, will be lacerated by reciprocal raging. For its kingdoms will fight against one another. And when the young ox has lived three times seven years, the consuming fire will be multiplied un-til the bat devours the mosquitoes of Spain, and, by subjugating Africa and trampling on the head of the beast, will receive the monarchy and subse-quently humiliate the dwellers of the Nile.

And afterward the son of perdition will rise up in a sudden attack in order to sieve humanity, so that with a very sharp sword he will separate the sons of Jerusalem from the sons of Babylon. And the dragon will come to the end of his last angry rage and remain mocked and defeated for eternity. Amen.

Source: In Josep Perarnau i Espelt, "El text primitiu del *De mysterio cymbalorum ecclesiae* d'Arnau de Vilanova," *Arxiu de textos catalans antics* 7–8 (1988–89): 7–169, at 102–3.

The anonymous prophecy "Woe to the World in One Hundred Years" belongs to a class of texts that cropped up in western Europe between roughly 1150 and the end of the Middle Ages. All of these prophetic texts foretold events that were supposed to occur very soon, just before the imminent coming of Antichrist. All were said to have descended from on high, either by means of writing by angels or by visionary knowledge vouchsafed to God's chosen by supernatural revelation. All were written in opaque language full of strange imagery. Because their language was so opaque the question arises as to whether many phrases in these prophecies were meant in-tentionally to be incomprehensible. Whatever the case, these prophecies invariably began with sufficiently clear "predictions" of events that had already happened in order to enhance their impression of infallibility. And all presented a succession of events that were sufficiently clear in their general outlines.

"Woe to the World" differs in one noteworthy respect from most of the other prophecies in the genre: although we do not know who wrote it, we do know the circumstances by which it was introduced to the reading public. The prophecy was first disseminated in 1301 in a treatise concerning the coming of Antichrist by the Catalan physician Arnald of Villanova. Arnald's purpose was to prove that Antichrist would arrive later in the fourteenth century. After he "demonstrated" this by inter-preting scripture, he clinched his proof by citing verbatim the "Woe to the World" prophecy, which he said had recently been communicated to him. The supposed author was an unnamed person who was nearly illiterate, and who as a result of a transport wrote down a supernatural message in "eloquent Latin." Scholars are unsure as to whether Arnald was dissembling and actually wrote the prophecy himself or else accepted it from the actual author who knew he would be a dupe. Either way the main lines of the text certainly suited Arnald's purposes and also reflected a Catalan political stance that would have been congenial to him.

The prophecy ostensibly reports the Lord speaking and promising imminent chastisements. Punishments will be visited on "the world" in a counterclockwise circuit around the Mediterranean, starting with Syria and Greece, moving thence through Europe to the Straits of Gibraltar, and returning through North Africa to the Holy Land. Where the Lord goes He will punish: the inhabitants of Syria will be driven from the shore by a profane people; Greece will be plundered; obstinate Sicily will be threshed; Italy will be bitten by lions and wolves. And so forth. Only when the circuit reaches Spain is there a respite from successive afflictions with the appearance of a messianic hero—a "bat" who will devour the Spanish "mosquitoes," and then "subjugate Africa, trample the head of the beast, receive the monarchy, and humiliate the dwellers of the Nile." But the hero's work will merely be a clearing of the slate for the "son of perdition," who will "separate the sons of Jerusalem from the sons of Babylon." Finally, however, "the dragon" will be defeated for eternity.

It hardly needs saying that the syntax of the prophecy is tortuous and the language quintessentially obscure. Because the voice uttering the prophecy is supposed to be that of the Lord, the words are meant to sound oracular and portentous, as if heard from a whirlwind. Because some passages may never have been meant to be comprehensible to anyone, the modern scholar who attempts to explicate every phrase runs the risk of seeming ridiculous, like Humpty Dumpty explicating the nonsensical Jabberwocky poem. ("Well, *toves* are something like badgers—they're something like lizards—and they're something like corkscrews.") Nevertheless, a few explanations can still be offered with reasonable confidence. For one, the opening passage about the expulsion of the "dwellers in Syria" is classic "prediction after the fact," for it clearly alludes to the fall of the last Christian outposts in the Holy Land, Tripoli and Acre, in 1289 and 1291. (This securely dates the text to sometime between 1291 and 1301.) The Spanish "bat" that appears toward the end must be a coming king of Aragon because a bat was an Aragonese heraldic emblem. And the "mosquitoes" the bat devours are Muslims, as can be told not only from the context but also because Catalan for mosquitoes puns on Catalan for mosques (mosquits/mesquitas). Since Arnald of Villanova was a Catalan who worked in the service of the king of Aragon it makes sense that he would disseminate a prophecy that propagandized in favor of the Aragonese crown.

Evidently the Spanish bat that marches victoriously to Egypt is also the "new David" and "western bat," who appear at the beginning of the prophecy. Since the "new David" will restore the Holy Land and the "western bat" will bring back "Greece" (meaning the schismatic Greek Church) to "the house of the bride" (the Roman Church), the messianic Spanish bat accomplishes everything that devout western Christians had hoped for in the later Middle Ages: the destruction of Islam, the winning back of the Holy Land, and the reuniting of the Greek and Latin Churches. But the bat's victory according to the prophecy will occur at the edge of time, for it will be followed by the appearance of the "son of perdition," a sobriquet taken from the New Testament (II Thessalonians 2:3) as a designation for Antichrist. And directly after the appearance of Antichrist comes the Last Judgment, described here as the "separation of the sons of Jerusalem from the sons of Babylon" and the defeat of the "dragon" (Satan) for eternity.

Like most other puzzling and portentous short eschatological prophecies, "Woe to the World" had a long life. It was copied and recopied frequently from the time of its first appearance in 1301 until the end of the Middle Ages. Many readers copied it without explaining why they did so or indicating how they interpreted it. It can only be inferred that they believed it might offer them useful insights into the future, just as many people today ponder the obscure prophecies of Nostradamus. Moreover, since so much of the language was ambiguous, readers could easily interpret particular passages as foretelling events that they feared or desired. Since the prediction of the coming of Antichrist was sufficiently clear, that must have reinforced the beliefs of many who took a dim view of the present world condition.

Fortunately it is possible to go further in estimating the reception of the "Woe to the World" prophecy in specific cases. Two extensive fourteenth-century commentaries on the prophecy survive. One was written around 1332 by an Italian Augustinian Friar, Gentile of Foligno, who, oddly enough, read it optimistically. Gentile recognized that most of the text promised imminent punishments, but he also knew of other prophecies that promised the imminent arrival of holy popes who would reform Christianity before Antichrist's coming. Thus in his commentary on "Woe to the World" he merged the two themes: on the one hand, there would be punishments, especially for the Church, along the lines foretold by the prophecy, but these would end with the appearance of three successive holy popes who would reign between 1345 and the coming of Antichrist around 1389. On the other hand, Gentile made no mention of the messianic Spanish bat of "Woe to the World," probably because he had no independent belief in the coming of a messianic secular ruler.

A second extensive commentary on "Woe to the World" was written in 1354 by the southern French Franciscan, John of Rupescissa, who was then imprisoned in a papal dungeon. Rupescissa had been held captive for many years because he sympathized with the ideals of heretical Spiritual Franciscans. Given his situation, he was bitterly pessimistic about the future of the Church and hence was entirely receptive to the truculent prophecies of doom in "Woe to the World." Indeed he had such a bleak view of the current state of the world that he was unable to imagine the possibility of an imminent messianic hero. Unlike Gentile, he grappled with the culminating lines about the advent of a messianic Spanish bat and properly recognized that the bat had to be a king of Aragon. But in his view the bat would be a chastising scourge and its triumph over Islam a case of evil triumphing over evil. Thus the victory of the bat coincided very naturally with the ultimate advent of Antichrist.

Rupescissa's identification of the bat of "Woe to the World" as a king of Aragon itself had some remarkable fortunes. Because Rupescissa adhered to a French and papal tradition that was hostile to the house of Aragon, he viewed the bat negatively. But readers of the prophecy in the lands of the crown of Aragon were apt to see the bat as a longed-for hero. Already in the later fourteenth century two different Catalan readers who had learned of Rupescissa's identification of the bat as a king of Aragon wrote jubilantly of how the bat would "obtain the monarchy and destroy the African beast, namely the Mohammedan."

When no heroic bat appeared in Aragon by the end of the prophecy's allotted "hundred years"—roughly the end of the fourteenth century—the prophecy lost its interest. But it was resuscitated toward the end of the fifteenth century as a result of its applicability to a real ruler with heroic potential, Ferdinand of Aragon. Given a desire to apply the prophecy to him, the supposed date of the prophecy's issuance was set back one hundred years so that it could apply to the present. Accordingly, during the time when Ferdinand was preparing his military campaign against the Kingdom of Granada (the last Muslim territory in Spain), royal propagandists brought forth the "Woe to the World" prophecy to prove that he would succeed. Then, when Ferdinand actually conquered Granada in 1492, propagandists wrote on the basis of the same prophecy that he would soon cross the Straits of Gibraltar to conquer the Muslims of North Africa. Because the prophetic text had foretold certain events that had actually transpired, the entire narrative seemed all the more ineluctable.

Of greatest interest is the fact that none other than Christopher Columbus drew on the prophecy's prediction of a "New David" who would "come to restore the ark of Zion" in order to ingratiate himself with Ferdinand of Aragon in letters of 1501 and 1503. Implicitly Columbus was assuming that Ferdinand was the hero who would conquer North Africa all the way to Egypt and then restore the Holy Land. Accordingly it may be that his voyages, which he claimed were meant to aid in the conversion of the world, were connected with the Aragonese messianic aspect of the "Woe to the World" prophecy.

Further Reading

Mathias Kaup and Robert E. Lerner, "Gentile of Foligno Interprets the Prophecy 'Woe to the World,'" *Traditio* 56 (2001): 149–211.

Robert E. Lerner, "Medieval Prophecy and Politics," *Annali dell'Istituto storico italo-germanico in Trento* 25 (1999): 417–32.

———, "Medieval Prophecy and Religious Dissent," *Past and Present* 72 (1976): 3–24.

———, *The Powers of Prophecy* (Berkeley, 1983).

37

Raymond de Sabanac, Preface to
Constance de Rabastens,
The Revelations

Renate Blumenfeld-Kosinski

The Holy Fathers and the Doctors of the Church say that a person who has visions has to be examined in such a manner that one can know whether she is a spiritual person or whether she is worldly or secular; whether she lives under discipline or a special obedience in a continuous spiritual [tradition] of some saint or ancient father, spiritual, discreet, mature, virtuous, Catholic, and approved, or whether she lives according to her own will. And further, whether she has submitted the temptations and the visions that contain them to the examination and judgment of her spiritual father or other aged spiritual fathers, with humility, all the while being afraid to be led astray or deceived; or whether she has shown and submitted these visions to some examination and judgment; or whether, based on [these visions] she has perhaps arrogated to herself some vanity or vainglory, or whether she shows disdain toward others. And it must also be examined if this person who has these visions follows them by true acts of obedience, humility, charity, and steadfastness, or rather by acts of concern for her reputation, of boasting, and of arrogance; or whether she shows a demonstrable and growing appetite for human praise, a neglect of prayers, or a desire for honors and dignities; and further one should examine whether this person has a reputation among people of being a true Catholic, faithful and obedient. And whether she has persevered with humility in having visions for a long time or whether she is a novice and persevering. And whether the person having these visions has a good and true natural and spiritual understanding and whether she has discreet judgment in reason and spirit or whether she has flighty judgment, with too much imagination or fantasy. For Saint Gregory says in his book the *Dialogues* that saintly men can distinguish by intuition and by the voice of their hearts whether visions showing the same images are illusions or revelations, and for this reason they know what is sent by a good spirit and what by a bad spirit.

And one has to know whether this person will be examined at other times concerning the merit and circumstances of her visions by knowledgeable, literate, spiritual, approved men or not. All these things must be considered during the examination of this person.

As far as the manner of seeing or hearing spiritually, as well as of receiving revelations and visions, is concerned, the Holy Fathers and Doctors of the Holy Church say that one must examine with great subtlety whether this person who has seen visions and has heard words was awake or asleep or dreaming, and whether she had a corporeal, imaginative, or spiritual vision, or whether this was by chance an intellectual supernatural vision. Or whether in a new mental ravishment called ecstasy, that is, an elevation of her thoughts, she has seen or felt things that come from divine love or not. . . . [words missing in manuscript]. Or whether she has seen any mysteries speaking of spiritual . . . [words missing], or in which species or semblance these persons see things, and whether they feel an illumination or an elucidation of a supernatural intelligence and a manifestation of divine truth.

As far as the things seen or not seen, the quality of the person and the subject matter of these visions are concerned, it has to be examined whether these visions accord with the Holy Scriptures or whether they disagree with or contradict them; and whether these visions are a delight for humans leading to virtuous actions and the salvation of souls or whether they lead to error in the Catholic faith; and also whether they demonstrate anything monstrous or superfluous or new in nature, and whether they lead to anything that does not agree with or is far from reason; or whether they keep us away from good, virtuous, and humble deeds; and whether these visions are always true or sometimes false and lying, that is, one has to know whether these things sometimes reveal themselves to be true and sometimes false and whether they indicate to us future honors, riches, and human praise or humility in all things, or whether they rather lead us to an exaltation of arrogance . . . [words missing]. . . . and we are admonished to obey the pure, spiritual persons . . . and our prelates.

For reasons of brevity I finally say that perfection in this matter comes from the quality of the visionary and from the quality and manner of seeing, and also from the quality and manner of the visions and from the manner of knowing the spirits shown, inspired, and administered by these visions; and whether these are good or bad spirits will be demonstrated in the *Book of Revelations*.

For without such a subtle examination made beforehand, a dangerous error could ensue if we were to approve or disapprove without reflection or with abrupt haste the person who sees visions of these things or revelations. Suppose that by chance a hasty, indiscreet, and thoughtless person, given to fantasies, would approve of such a visionary and her visions, she would receive false things as true and would perilously reject the true as false, and thus the good and true visions and divine pronouncements would be disdained, they would not be believed and one would not obey them . . . [passages missing in manuscript]. And even today such error occurs because of a lack of a discreet and careful examination. So, all these above-mentioned matters that we have considered in theory, are visible in the things she has seen and in the quality of this person, that is, in this spouse of Christ, the blessed Constance from the place called Rabastens in the county of Toulouse. It would be suitable to recount her life in order to demonstrate the truth of all the above-mentioned points, for her reputation would thus be manifest in all the world. But for this very reason I will not write her life. Rather, the marvelous visions contained in this book which our lord God has revealed to her will clearly tell of and demonstrate her virtues, for her life and her noble deeds and the explanation of all this would take too long to recount. If, however, it becomes necessary that such a particular book should circulate, it will appear at that time.

Source: Raymond de Sabanac, *Les Révélations de Constance de Rabastens,*
ed. A. Pagès and Noël Valois, in *Annales du Midi* 8 (1896): 241–78.

Constance de Rabastens, a simple woman from the region between Toulouse and Albi, began to have visions in 1384, around the time of her husband's death. She was probably in her early forties then. Little is known of her life, for, as the end of our text indicates, out of concern for Constance's humility her spiritual director, Raymond de Sabanac (a law professor at the University of Toulouse), decided not to write her life but only to record her visions. This text, including the preface translated here, today exists in only one manuscript preserved at the Bibliothèque nationale de France in Paris (BnF, latin 5055). The text is written in a fourteenth-century hand in medieval Catalan (though the original was most likely in a Languedoc dialect), while all the other texts in the manuscript are in Latin. Constance's visions were at first quite personal, focusing on penitence and her relation with Christ, but then became more and more political, centering on one of the most pressing problems of her times: the Great Schism of the Western Church that had begun by a double papal election in 1378 and was to last until the Council of Constance in 1417. Because Constance's revelations strongly condemned Clement VII, the Avignon pope accepted in France (including the region of Languedoc) and instead favored the Roman pope Urban VI, she was finally ordered by the archbishop of Toulouse to stop having visions and when she did not, she was imprisoned. Thus her spiritual director had taken on a rather dangerous mission: to disseminate an account of visions by an unknown laywoman who challenged the position of both the secular and ecclesiastical authorities of her region.

In order to establish the authenticity and orthodoxy of Constance's revelations—and undoubtedly also to protect himself—Raymond composed a preface that uses the method of discernment of spirits (*discretio spirituum*). Distinguishing good from bad and true from false spirits was a concern evident already in the New Testament. In John's first letter we find this exhortation: "Beloved, believe not every spirit, but test the spirits to see whether they are of God: because many false prophets have gone out in the world" (1 John 4:1). But how can one test these spirits? Over the centuries the Church Fathers and later theologians developed a practice of discerning spirits, basically a series of questions that had to be answered satisfactorily for the visionary to receive approval for his or her visions. Saint Augustine (354–430) reflected at length on how to ascertain that a vision is of divine origin, as did Gregory the Great (540–604), whose *Dialogues* Raymond de Sabanac mentions as an authority in his preface. One of the great systematizers of the method was the famous theologian and chancellor of the University of Paris Jean Gerson (1363–1429), who between 1402 and 1423 wrote a series of texts on testing the spirits. In a kind of ditty from his 1415 *De probatione spirituum* (on the testing of spirits) he summarized the questions one needs to ask of visionaries and their visions: "Tu, quis, quid, quare, cui, qualiter, unde, require," that is, ask who is the person who receives the revelation, what does it mean and refer to, what is its reason for being, to whom was the vision shown, what kind of life does the visionary lead, and where does the vision come from?

These concerns echo those laid out by Alfonso of Jaén (also known as Alfonso of Pecha), born around 1330, who had been the confidant of Saint Birgitta of Sweden (1303–1373) and after her death became the organizer of the campaign for her canonization and the editor of her *Revelations*. He composed a defense of the orthodoxy of her visions, the *Epistola solitarii ad reges* (the solitary's letter to kings) in which he drew on a large number of authorities on the practice of discerning spirits, from Cassian (360–435), Augustine, Jerome (342–420), and Gregory the Great to Thomas Aquinas (1225–1274) and Nicholas of Lyra (1270–1340). The *Epistola* was the principal text, then, that began to circulate just when Raymond de Sabanac needed to buttress Constance's orthodoxy.

Like most confessors and spiritual directors who wrote the lives or visions of holy women Raymond emphasizes his subject's humility and obedience. Women like Constance are not allowed to live according to their own will but must submit to their spiritual fathers. Raymond uses various expressions for pride, vainglory, or arrogance—vices that must be eschewed by visionary women but to which they were said to be particularly prone since they were singled out by God for special revelations. The visionary must be a true Catholic, not neglect her prayers, and not give herself over to fantastic imaginings. This cautionary note introduced by Raymond reminds us of the thin borderline between claims to sanctity and witchcraft, a line that became more and more blurred as we approach the early modern era. In order to ensure the visionary's orthodoxy and bar the threat of demonic influences one thus has to examine whether the revelations are in accordance with the teachings of the Scriptures. Several of Constance's visions reflect imagery from the *Book of Revelation* or the *Apocalypse*, and in one of auditory revelations—when Christ speaks directly to her—he charges her with being the exegete of the Scriptures he entrusts her with. In fact, Christ chooses her to explain the Holy Scriptures to the inquisitor of Toulouse. While this may sound thoroughly orthodox and confirm the divine origin of her visions, this command obviously challenged the authority of the Church leaders. No wonder, then, that

Raymond was at times afraid to transcribe the visions of this forceful woman. Could he protect her and himself by the claims made in the preface that he subjected her to a thorough examination before accepting to be her scribe? In the middle of the text Raymond clearly felt a further need for justification; he describes how he asked the Lord for a sign whether he should continue writing and in response was struck by an illness that forced him to wear glasses henceforth. Although not part of the official discourse on the discernment of spirits, this claim to a divine order to write has its own force.

Another important criterion for determining the orthodoxy of visions was the "fruits" they would bear in the visionary's life. Raymond asks whether acts of obedience, humility, charity, and steadfastness follow Constance's visions. Looking at her *Revelations* we certainly see examples of the latter virtue, but as far as the first three are concerned, there are no specific examples of her charity, and no one from the official Church hierarchy in the Toulousain could consider Constance humble or obedient. On the contrary, she continued to proclaim her revelations against the archbishop's explicit order. She saw, for instance, the Avignon pope Clement VII in a temple filled with smoke, being menaced by a sword-wielding angel; or as a limping man bringing down the ship of the Church. Her humility was compromised by the forceful accusations she flung in the direction of Toulouse and Avignon and by the fact that she let herself be consulted on political questions by noblemen of the region.

Raymond's preface, despite its strict adherence to the practice of discerning spirits, could not be accepted as valid by the adherents of the Avignon papacy since Constance's revelations ran counter to everything they supported. Unlike Alfonso of Jaén and Raymond of Capua, whose careful managing of the careers and writings about their powerful saintly women, saints Birgitta of Sweden and Catherine of Siena (1347–1380) respectively, resulted in their eventual canonization, Raymond was most likely forced to abandon his charge to an inquisitorial prison. Although he mastered the discernment method, Constance refused to fit into the mold of the perfect visionary, and her headstrong support of the "wrong" pope, Urban VI, could not be considered orthodox in spite of Raymond's best efforts.

Further Reading

R. Blumenfeld-Kosinski, "Constance de Rabastens: Politics and Visionary Experience in the Time of the Great Schism," *Mystics Quarterly* 25 (1999): 147–68.

P. Boland, *The Concept of* discretio spirituum *in Jean Gerson's* De probatione spirituum *and* De distinctione verarum visionum a falsis (Washington, DC, 1959).

N. Caciola, *Discerning Spirits: Divine and Demonic Possession in the Middle Ages* (Ithaca, NY and London, 2003).

W. A. Christian, *Apparitions in Late Medieval and Renaissance Spain* (Princeton, 1981), 188–203.

Dyan Elliott, *Proving Woman: Female Spirituality and Inquisitorial Culture in the Late Middle Ages* (Princeton, 2004).

D. Elliott, "Seeing Double: John Gerson, the Discernment of Spirits, and Joan of Arc," *American Historical Review* 107 (2002): 26–54.

R. Voaden, *God's Words, Women's Voices: The Discernment of Spirits in the Writing of Late-Medieval Women Visionaries* (Woodbridge, 1999).

In Pursuit of Perfection:

At the Edge of the World

— 38 —

The Life of the Hermit Stephen of Obazine

György Geréby and Piroska Nagy

After, however he had been elevated to the grace of priesthood by God's dispensation, he totally abandoned the life of the world and what he previously despised in spirit, he now renounced in deeds and habit. For now laughter and those jests of old turned into mourning, and happiness into grief, the chasing of wild animals into the capturing of souls. For now the cultivation of precious garments was abandoned, and the acquisition of sweet dishes was looked down upon. Instead of a soft shirt he wore a rough hair shirt on his flesh, and instead of pleasant food he *took his bread with tears and his drink with crying* (Ps. 79:6). What is more, he treated his body with such severity that he nearly killed it with cold and fasting (2 Cor. 11:27). Since indeed in the middle of the winter, when icy cold and frost put fetters on everything, he broke the ice open with a hatchet, and submerged himself in the water up to his head, and stayed there so long that the power of the cold penetrated all his body, saying with the psalmist: *For I have become like a wineskin in the frost, yet I have not forgotten thy statutes* (Ps. 118(119):83 sec. Vulg.). He excelled in fasting, was steadfast in the vigils, and was always ready for prayers, which he did not offer for the divine ears by the arrangement of words but by the devotion of tears. His talk, seasoned with deep good sense and burning with charity inflamed his listeners with divine love and provided them with the seasoning of wisdom.

He was assiduous in reading the divine Scriptures, especially the commentators of the gospels, and by this both for himself in reading and for his audience in listening he provided eternal bliss. And while he found there much on the contempt of the world and on the glory of the coming world, his mind was mightily inflamed with the disdain of things present and with the desire of the future things, saying with the prophet: *When shall I go and arrive before the face of God* (Ps. 41:3 sec. Vulg.), or again, *Mine eyes fail for thy speech, saying, When wilt thou comfort me?* (Ps. 118(119):82).

Inflamed by such desires, he disposed himself daily for the rejection of the world, casting off all earthly cares (Luke 21:34), in order that he might continuously follow the poor Christ in poverty and nakedness, and destitute, with free and easy strides. He also wanted to avoid, however, being regarded as doing it rashly and without counsel, so he went to see a certain devout and honorable man, Stephen Mercoeur, who was a disciple of Saint Robert, as he was reared by him, and whose saintly fame was celebrated throughout the region. And when he came to him, and explained to him the desire of his soul, not being uncertain about resolution but asking for counsel, the venerable old man answered him thus: "Beloved, it is necessary that you not defer your divinely inspired desire, nor postpone it from day to day, for you should know that postponement is always harmful for the ready. On the contrary, as soon as you conceived in your mind, cast off all worldly cares (Luke 21:34) and follow joyfully in the footsteps of Christ, so that many other may be converted to God by your example." By this answer he became reinforced in his design as if by a divinely inspired oracle, and returned happily to his place.

[After this, Stephen says goodbye to the world and leaves for eremitic life together with his friend, another priest, Peter.]

(6) He constructed a little wooden hut next to a convenient tree, with a shabby roof, in which he pursued continual prayer and the singing of psalms heedless of day or night, together with his venerable friend. It was in this place that after modest rest, by which they refreshed their weary limbs somewhat, they got up to sing the divine lauds. As soon as they felt themselves being weighed down by sleep, they grabbed a handful of sticks and exposing their sides smote and flogged each other. In this way the flesh, exhausted by fasting and wearied under the weight of vigils and works, was also gashed by the protracted flagellations, and by this coerced into servitude brought forth not carnal, but spiritual fruits. It was indeed, what the Apostle said: *but I pummel my body and subdue it, lest after preaching to others I myself should be disqualified* (I Cor. 9:27).

[Somewhat later they decided to ask for the authorization of the bishop of Limoges to celebrate the mass and they constructed a monastery.]

(7) Already some disciples, converts to God, came to his school and submitted themselves to the yoke of his discipline. They led a very hard and austere life in his company. That's why there were so few who imitated them; only those followed them who, enemies of their own flesh, thought no longer of this present life.

They received wisely what came from God, and refused what came from the world, and as for the offices, they conformed themselves to the canonical rules, and they led an eremitic style of life.

After prime, prostrate on the ground and having sung the seven psalms with the litany, they immediately followed with the holy mass, unless, as often happened, [the footwear was missing]. Having finished the singing of the mass, the monks went out for manual work, while the man of God often stayed behind in the monastery.

In such a place who would dare to imagine by what kind of tears, by what kind of sighs or afflictions he satisfied his fervent desire, when being alone he had to fear neither a witness nor a judge? Who could possibly think that at this moment he would sleep or he would take a rest, he who growing impatient even of a short rest, spent the night praying and recited the psalms in advance of the nocturna? But we will refrain from describing these things that happened in concealment in more detail, lest we would be found writing about opinions rather than facts.

(18) After compline, the friars having returned for sleep as usual, the man of God frequently stayed behind in the oratory and spent the whole night in sedulous vigils, prayer, and tears.

The grace of tears was conceded to him from heaven in such a way that whenever he prostrated himself for prayer, torrents of tears streamed out of his eyes. I remember that one day, I entered one of his churches without him noticing it, not for prayer, but to do some private business. He was then speaking about some affair with secular people outside, in the porch of the church. Then, suddenly he left them and entered the church; knowing that there was no one, he prosternated for prayer in front of a lectern with such simplicity and modesty that no one present could have heard him properly. Then, as he lowered his head toward the earth, tears started to pour out of his eyes, not drop by drop, but so to say, like a stream. I was witness to this scene and I saw the tears pour from his nose to the earth, without any sound of a cough, spitting, or sighing.

I stayed there, like a thief, as if bound to the place, full of fear and trembling, him not noticing me, as long as after having wept enough, he left with the same simplicity as was his worship, of which no one took notice, as if he would have had to go out because of natural necessity. We relate this in order to make the ardor he put in prayer publicly acknowledged, and so that one could judge the intensity of the grace he had when he was free of all occupation, such that he was not kept away even by so many burdens from the spiritual exercises.

Source: *Vita sancti Stephani Obazinensis* (BHL 7916): *Vie de saint Etienne d'Obazine*, Texte établi et traduit par M. Aubrun (Publications de l'Institut d'Etudes du Massif Central, fasc. VI) Institut d'Etudes du Massif Central, Clermont-Ferrand, 1970, 46–48, 52–54, 56, 72.

The life of St. Stephen of Obazine (d. 1159) leads us to the world of the twelfth-century eremitic movement in the central region of France. Coming from a well-to-do family, Stephen was ordained a priest after his studies. Following the call to a more perfect life, he became a hermit. Finally, as adherents gathered around him, he became a monastic founder: the solitary was transformed into an abbot, the leader of his own community. This story contradicts the genuine image of a medieval hermit whom one would imagine as a solitary man living a remote, lonely, and saintly life, near only to God, who disappears almost without any trace. Some medieval hermits were well known, even celebrated; that is why their *fama sanctitatis* has survived. Recent historiography shows well that a hermit whose *fama* has come down to us was certainly not an isolated, unknown man.[1] As it is the case of Stephen, his links with society were sufficiently strong to attract disciples, lay worshippers, and even a biographer to record and commemorate his life.

Although Stephen of Obazine was far less famous than some other reputed hermits of his time, such as Robert d'Arbrissel or Saint Romuald, he did not remain in the shadows. His biography, written by an unknown contemporary monk who knew the saint but wrote the vita at the command of his superiors, was a typical product of the medieval hagiographic tradition. People of the twelfth century were not fond of originality; instead, they wanted saints' lives to conform to the rules tested and witnessed by the "authorities," that is to say, the Bible, the writings of the Church Fathers, and some later Christian authors. In his biography Stephen is described according to the conventional pattern of the hagiographic genre. The style of the rather sober *Vita*, considered as a literary text, is far less sensational concerning the deeds of Stephen than many other vitae of the period. However, it provides not only a report on the life and deeds of a saintly hermit and a monastic founder but also a model of religious life. It emphasizes the leadership qualities of its hero, according to the prescriptions for an abbot in the *Rule* of St. Benedict; and it spells out the life of Stephen as a true follower of Christ.[2] Thus, this vita, the only source we have for him, does certainly not depict a nuanced image of the *real* Stephen, the one we will never know, but at least it gives us a picture of how his contemporaries expected an ascetic religious leader to live.

Stephen's vocation can best be understood in the framework of a large movement that we can describe as the "desert call," which touched many men (and women)[3] in the eleventh and twelfth centuries. This movement referred back to the tradition of the Desert Fathers, to the legendary heroes of Eastern Christian asceticism between the third and the sixth centuries. Their model emerged in this period as one that offered an alternative to older religious patterns. They had never been forgotten; but their memory survived rather as a number of texts[4] than as a living paradigm, while the West produced its native Fathers, saints, and religious heroes. From the fifth century on, western monasticism gradually took an autonomous form, and received a long-lasting framework in the *Rule* of Benedict of Nursia (d. 547/560), which had a delayed but decisive impact. Benedictine monasticism became dominant only in the ninth century, with the Carolingian reform of the Church, which imposed its *Rule* on all the monasteries of the Empire. This model, oriented toward a strictly coenobitic form of monasticism, fitted for very different conditions, but left little place for personal religious experience. While it alluded to the eremitic form as a more perfect way and texts on the Desert Fathers, like those of John Cassian (d. 430/5), were prescribed readings for the monks, this alternative remained marginal and unregulated. It is mostly known as an eremitism *prope monasterium*, near to the monastery, a place for short-period retreats of monks.

The revival of the desert model, providing an individualist religious lifestyle and focusing on personal spiritual and ascetic achievement, was a response to the general economic development of the West, and not only to what Carolingian-style monasticism had become in the tenth century, as contemporary propaganda would have wanted. It was certainly true that far from the monastic ideal of separation from the world in order to worship God, lay donations and the imbrication in power networks made monasteries strong and wealthy institutions that Carolingian power used as its intermediaries. The end of the Carolingian order brought a mixture of institutional heaviness, local lay influence, and decline of discipline in the monastic world that was pointed out by contemporaries, for whom mainstream Benedictine monasticism no longer represented the choice of real spiritual retreat. But at the same time, society started to change; the revival of towns and new economic dynamism starting around the millennium rendered the contrast between the apostolic ideal of poverty and everyday reality even sharper. Monasteries and especially new orders like that of Cluny, which the landowning aristocracy both protected and benefited, were also the first to get involved in money economy and to get extremely rich. This context helps explain the widespread and striking return to the sources of monasticism and of Christianity. The ideal of evangelic poverty that had characterized the life of the apostles and of the first ascetic monks seemed to belong to a faraway past; it was known only from the books of the Fathers. In this period of dissatisfaction with existing institutional forms and of spiritual quest, the model of the Desert Fathers inspired many individual vocations, independent or not of existing institutions, and explains the remarkably great number of new monastic foundations and the birth of new orders in the eleventh and twelfth centuries.

While in the earlier Middle Ages hermits were rather rare in the West (with the notable exception of Ireland, southern Italy, and the Byzantine frontier), we observe a movement of eremitic revival from the tenth century on, spreading from Italy and the Empire to Central Europe, France, and the British Isles. In Central Europe, the first wave of eremitic movement was frequently linked to the conversion of this area of Europe to Christianity; its main figures were missionary bishops who finished in martyrdom, like Bruno of Querfurt (d. 1009), Adalbert of Prague (d. 997), and Gerhard, bishop of Csanád (d. 1048). In older areas of Christendom most of the hermits we are fortunate to know frequently became monastic founders: Saint Romuald (d. 1027); Robert of Tourlande, founder of La-Chaise-Dieu (1043); Johannes Gualberto, founder of a community in Vallombrosa (1037); Bruno of Cologne, founder of the Chartreuse in 1084, or the well-known founder of Cîteaux, Robert of Molesme (d. 1111). These new communities and orders were all characterized by an austere life compared to traditional Benedictine monasticism. The communities of Camaldoli and Vallombrosa, as well as Carthusians and Grandmontains, can be described as communities of quasi-hermits, while the Cistercians wanting to import austerity into the very life of the community also projected themselves as *fratres eremitae*.

Thus, even if eremitic retreat and lifestyle had some precedents in western and central France, the location of Stephen and his area of Obazine, the inspiration and life of Stephen could have occurred elsewhere in western Christendom during the eleventh and twelfth centuries. But this new eremitic movement no longer resembled its ancestor and model of the desert hermits, known or imagined by medieval monks from their early literary texts. The eremitic ideal proved to be a powerful moving force for shaping social and institutional alternatives to older religious patterns.

An important motive in Stephen's time was the *fuga mundi* (escape from the distractions of the world), to allow subjection to the sole reign of God. As we learn from his vita, Stephen converted his life to God gradually, in two steps. At the beginning of his path, when consecrated as a priest, Stephen had already renounced the lay way of life. His conversion from profane life to that of a secular cleric is in fact described according to the rules of a monastic conversion: a way to abandon "totally" the "life of the world." Such a conversion is twofold, and contains an exterior and an interior side: it is a substantial transformation of habits and feelings. From luxury, hunting, and garments, Stephen turned to an austere, penitential life, and abandoned pleasures and happiness for grief. Fasting, vigils, tears, and icy baths instead of the pleasures of earthly life express the choice of the *contemptus mundi* (detesting the world), the choice of a life focused on the world to come instead of this one, despised; these penitential practices helped him merit salvation. This lifestyle is notable in the case of a priest; according to his biographer, Stephen already lived a quasi-eremitic life. Thus, we are not so surprised to read in the vita that dissatisfied with the life of a secular cleric, Stephen underwent a second conversion: leaving all the burdens of the world and becoming a real hermit.

Why this second conversion? The text tells us clearly: although the "capturing of souls" replaced the aristocratic habit of hunting in Stephen's life, and although his reading of the Scriptures and its commentaries made him a highly esteemed preacher, he remained dissatisfied. His longing for the world to come and the "disdain of things present" became too pressing for him to stay between two worlds. His retreat had to become perfect. We find a clue to this shift in the writings of Peter Damian (d. 1072), the great reformer of the eleventh century, himself first a hermit whose model was Romuald, then a cardinal-bishop who lived and worked near the pope. As he explains, even the cure of souls is still a way to remain in the world, not to consecrate oneself entirely to otherworldly thoughts. Stephen wanted to "follow poor and naked the steps of the poor and naked Christ," to be entirely destitute of any cares that linked him to this world. This choice implies at the same time a continuity and a clear distinction between two ecclesiastical *status*, that of the secular cleric and that of those who renounced the secular world, becoming monks or hermits.

A Hermit in Society: Seeking Authority

Leaving his secular clerical ties, Stephen's move was not toward total isolation. On the one hand, he tried to loosen the links that belong to this world; on the other, as a hermit, he sought an institutional framework. First, he needed a guide, the support of a spiritual authority. That was why Stephen went to meet Stephen of Mercoeur, an abbot of great authority of the monastery of La-Chaise-Dieu, a recent monastic foundation of eremitic style. To make the next, decisive step of his life, he needed the urging of the abbot to finish his conversion and not to waste any time. With his support, Stephen left the priesthood and, together with his friend Peter, another priest, he said farewell to the secular clerical life. After a preparatory stay of ten months with Bertrand, an established (but unknown) hermit, they went their own way, and wandered in search of a place to settle. They were looking for a religious order that would be "perfect enough" as the vita says, to serve God; and this kind of order did not exist in their region. That is how they arrived in the dark forests of the region of Obazine, where they decided to stay.

Now, one could ask, what kind of retreat, eremitism, and *fuga mundi* are we speaking about if Stephen never stayed alone? Leaving the priesthood he went wandering with a friend. Together they associated with the hermit Bertrand. In the woods of Obazine, people living nearby helped them; their strict eremitic life seems rather populous. It was their asceticism and devotion, however, that attracted so many disciples so quickly, who settled with them. This spontaneous group became the foundation of a new, eremitic community, shaped and led by Stephen. The converts came from every social background, rich and poor, male and female, as the vita says. Based on his experience, the hermit became a leader and an authority, a kind of public institution in the region, close to Peter Brown's late antique "holy men." Later, his monastery became a shelter for the population in times of famine or war; the vita records that the intercession of Stephen saved the countryside from a military conflict.

The foundation of a new community, for a man like Stephen, appears then as a logical consequence of his successful conversion. The life of the new community had to be given rhythm and regulated. Stephen obtained the authorization of Eustorgius, bishop of Limoges, to construct a monastery-like establishment so they could live their eremitic, disciplined life according to the daily rhythm of canonical liturgy. However, the choice of the rule and lifestyle involved a theological and political decision. Seeking the answer, Stephen wandered again and thought to join the Carthusians. The Carthusian prior, Guigues, advised Stephen on the most appropriate way for him to follow: the "royal way" introduced by the recently founded Cistercians.[5] Rather than a refusal, this advice can be explained by the fact the Carthusians had strict rules limiting the number of members allowed in their eremitical communities; the community of Stephen was probably already too large. The choice had to be monastic, not eremitic any more. Carthusians were also rather opposed to the penitential self-castigation that characterized Stephen's life. Thus the original spirit of the eremitic movement continued to shape the lifestyle of the new community that remained near to eremitism, centering on work, loneliness, and on a personal and affective relationship to God. This evolution of religious forms informs monasticism and more widely, the renewal of western religious experience to which the eremitic communities contributed in the period.

As an end of the process of institutionalization, in 1142 Stephen again asked and received the authorization and help of Bishop Gerald of Limoges for the consecration of a double monastery, male and female, of Obazine and Coyroux. We see here Stephen's constant care for the orthodoxy of his community—a care that can be well understood in a period when wandering hermits and preachers, easily suspected of heretical opinions or beliefs, mushroomed all over Europe. Stephen lived in a region very close to several centers of heresy in southwest France. In any case, the Church was also constantly concerned with keeping its new communities under control. The integration of the two monasteries into the Cistercian Order was a final important gesture chosen by Stephen to embrace the institution of the Church.

The vita informs us not only about the life of Stephen and his community but also about his ascetic achievements and spirituality. The rhythm of the life of the community was maintained by the singing of psalms and prayers. At an early date the bishop of Limoges authorized them to live according to the hours of the offices and to celebrate a daily mass. This rhythm was reinforced by fasting and vigils. Purification of the soul was also served by self-castigating practices like flagellation. It was the responsibility of the hermit who cared about his soul to concentrate his efforts on the constant war against laxity: "the spirit is willing, but the flesh is weak" (Matt. 26:41, Mark 14:38). Ceaseless prayer (1 Thess. 5:7, cf. Luke 22:46) had to lead to focus on the work of self-salvation, expressed in the denial of earthly cares, including the most fundamental needs like food, shelter, and sleep.

The text is clear on the aims: it was to "coerce" their flesh "into servitude," in order to bring "not carnal but spiritual fruits." The ascetic regime of Stephen required following the strict advice of the Scriptures, like plunging into icy water according to 2 Cor. 11:27: "in weariness and painfulness, in vigils, in self-afflicted hunger and thirst, in fastings, in cold and nakedness." Hard asceticism[6] characterized his spiritual choice, which was not a universal feature of eremitic life. Some could argue, like Carthusians, against physical hardships as an obstacle to the worship of God. But Stephen concentrated in this way on what we can consider as the first goal of all eremitic austerity: to subjugate the flesh, that is, first, bodily needs. This fight against oneself, one's bodily nature, is linked to a conception of the flesh (caro) that is not exactly the same as the body as such, since according to Saint Paul, and stressed by Augustine, flesh is the name given to the desiring part of the self.[7] Conversion to God implies uprooting earthly desires for the sake of desiring God. This interior, psychological process is the inner side of the conversion of life, present in monastic conversion, that can be found at the very center of the individual spiritual choice of an eremitic lifestyle. Just like his ascetic practices aimed at "killing the flesh," Stephen's frequent weeping aimed to achieve his conversion by the spiritualization of the self.

Stephen's spirituality was characterized by his "gift of tears."[8] This motif is intended to describe the success of his inner conversion and the purity of his desire, entirely turned to God. We see Stephen weep four times in our text: at the very beginning of the vita the quotation of Psalm 79:6 shows his daily practice of weeping, accompanying his meals, as a part of his conversion. We have already seen what it meant that Stephen, according to a topos, left joy and laughter for eschatological longing and tears. These tears show that his thoughts went beyond this world, to the yet unattainable happiness of heaven. The next tears in the text accompany his prayers, addressed to God not "by the arrangement of words, but by the devotion of tears." Praying in tears was considered to be more sincere than praying with words; while words, rationally formed, can imitate a feeling and be fake, tears in medieval Christian anthropology come directly from the heart. The basis of this belief is the Bible itself, which recommends praying to God not openly so everyone can see, but in the secret of the heart (in abscondito—Matt. 6:6; occulta cordis—1 Cor. 14:15). Therefore, the weeping of Stephen in prayer illustrates the purest sincerity and depth of his devotion to God.

Two other passages show him crying outside the daily order of the new community's life. He wept instead of working, and instead of sleeping as well. Before it became a hagiographic topos, it was an old religious ideal and even a habit to pray instead of sleeping—one of the most difficult hardships a man could endure. Stephen's tears are described in the vita as lonely tears wept in the church out of the needs of his heart, as a way of "satisfying his fervent desire," a kind of "natural necessity," that replaced other bodily needs. Religious tears were reputed to have an important spiritual effect: the pure water of tears was reputed to wash one's heart of sins, and thus to contribute to preparation for salvation. Although this idea had been well known in the monastic world since the patristic times, frequent weeping came back in fashion in the eleventh century as a special feature of eremitic spirituality. Around the millennium, Saint Romuald and Nilus of Rossano wept a lot, and from their model, religious weeping and the highly praised gift of tears was diffused in the West by the eremitic movement and monastic reform. According to their models, as for the earlier Desert Fathers, weeping was a way to spiritualize their flesh. Beyond the frequent metaphor of washing away sins by tears that we find in any medieval religious text, there was an anthropological conviction coming from antiquity concerning the composition of the man, body, and soul. In ancient and early Christian times, a clear certitude established the circulation of the liquids of the body, which could leave the body in different sinful or sinless ways.[9] The natural product of earthly desire was sperm; but with renouncing sexuality when pronouncing the monastic vows—a vow that was so much in evidence in the twelfth century that our text does not even mention it—the production of sperm and involuntary erection became a difficult moral problem for monks, who had to fight the "will" of their sinful flesh.[10] In the anthropological logic that sustains our text, the formation of salvific tears could also prevent the formation of sperm, as spiritual desire replaced sexual desire.

But to weep or not was not only a question of devotion or will. While one has to ask in prayer and with tears the grace of God as Stephen did, God's grace is manifest when those holy tears that relieve the suffering of spiritual desire appear. Stephen's tears are described as a divine gift: the "grace of tears," given by God to help his transformation. Thereby God personally assisted Stephen on his spiritual path. His grace of tears, a personal gift given to some of the saints, was a special sign of God's intervention in one's heart, a sign of His personal love. This feature of Stephen's spirituality was especially venerated by his contemporaries, who saw there a sign of his divine election, a real charisma. The gift of tears, sought by many at least from the eleventh century onward in the West, was only found by the most perfect of them. This gift indicated the onset of a new kind of spirituality and relationship to God. In the same way as Stephen, not content with coenobitic, collective, and ritual devotional forms, they wanted to follow "naked the path of the naked Christ" in a very personal way. Stephen's tears showed the personal guidance of God in his life, his special relationship to God. The beautiful weeping scene of the vita witnessed by the biographer by chance, shows that the tears of Stephen did not belong to his public devotion, but to a private space between him and God, in the loneliness of prayer in an abandoned church—in a way conforming to the injunction of the Gospel already referred to (Matt. 6:6). It was his love for God, and God's love for him, that is expressed in those tears—a love that had its equivalent in the strong affective relationship that linked him to his friend Peter, who accompanied him everywhere. So, if the companionship of Peter seems at first a limitation on Stephen's eremitism, it can also be understood as an earthly manifestation, a double, of his love relationship with God and of his spiritual affectivity.

The eremitic movement must have been so widespread and attractive that fraudulent hermits were also frequent; their bad reputation explains the local suspicion of Stephen's community.[11] Such success as contemporary evidence frequently records can only be explained by the wide range of social functions offered by eremitism. The lives of the above-mentioned hermits invariably point to the fact that the eremitic choice was not a life-long decision, but appears rather as an intermediary element in an ecclesiastical career. In this rarely appreciated sense, eremitism was not only an alternative for radicals to reform the establishment of the Church but also a training ground for missionaries and organizers. Rather than a short solitary retreat to repose oneself in the middle of long monastic years, it functioned as a template for experiments, institutional alternatives, or protest according to the personality and invention of the hermits themselves, who ventured on an individualist religious lifestyle and focused on personal spiritual and ascetic achievement. The choice of eremitism resembles other social schemes of the period. It appears as a time of abandoning the social order and its boundaries, a probationary and preparatory period, aiming at the acquisition of the spiritual authority necessary to found a community of one's own. A sentence from the

Vita of Stephen suggests such an aim: "God omnipotent did not want them [Stephen and his follower] to be subjected to anyone's authority, in order to fulfil what he had predestined for the holy man."[12] With such a reference to personal divine election, an eremitic search could serve as an elitist answer to the Church establishment, as a school for would-be monastic leaders, and a place for exercising unlimited innovative leadership. Interpreting the developments in this way there seems to remain no room to consider the transformation in coenobitic communities of the eremitic foundations of the period (like Arbrissel, Tiron, or for that matter, Obazine) as falling away from the "original purity" of eremitism.

Looking at this typical pattern of his career, the wandering of Stephen as a hermit, we can discover that it corresponds in an interesting way to the long period of the life of a knight described by Georges Duby as a period of "wandering."[13] His analysis concerns the same period, the twelfth century, and roughly the same region of Christendom. The young knight-errant already dubbed had to prove his chivalrous qualities before settling as a lord on his own property. For some ten to fifteen years, the young knight-errant wandered, seeking experience, adventure, fortune, and a wife, together with his companions, the members of his *meisnie*, young knights of the same condition and age. He arrived home when his reputation as a warrior was established, and he was considered as a man accepted in his social category, the aristocracy. In the same way, when Stephen established his own monastery of Obazine, his reputation, his spiritual authority, was recognized. Thus, in both cases, wandering with a *socius*, a friend of the same condition, has the same goal and social meaning: a kind of test or proof in order to attain social recognition upon which to establish one's own power—as a landlord in one case, as a monastic leader in the other.

Notes

1. Cf. for the Late Antiquity, the famous article of Peter Brown, "The Rise and Function of the Holy Man in Late Antiquity," in *Society and the Holy in Late Antiquity* (Princeton, 1982), 103–52.

2. This program is reflected in events imitating the events of the life of Christ, in direct reference to the Gospels. Stephen is a model follower of Christ and produces miracles according to the topoi of the miracles of Christ. He tells the heavy stone to move, and it moves, in accordance with the saying of Christ: "If ye shall say unto this mountain, be thou removed, and be thou cast into the sea; it shall be done" (Matt. 17:19, 21:21; Mark 11:23). Stephen meets Martha and Mary (Luke 10:38–39 (*Vita* 2, chap. 42, 164); finds water, in allusion to the water of life (Rev 22:1; 17); makes peace (Matt. 5:9; *Vita* 2, chap. 39, 158); gives food to the poor (Matt. 25:35; *Vita* 2, chap. 26, 142–43): and cures the sick (e.g., Acts 28:9; *Vita* 2, chap. 43, 166).

3. See notes 11 and 13 below. One of the interesting features of the foundation of Stephen was that it was a double monastery, one for monks and one for nuns in Obazine and Coyroux, like Fontevrault (1100), which was originally also a "double monastery." There were similar developments among the Premonstratensians, or in the movement of Gilbert of Sempringham (around 1150).

4. The memory of the Desert Fathers was known in the West mainly through the compilations of various collections of the *Apophtegmata patrum*, which circulated under the title *Vitas patrum*.

5. *Vita* 1, 26, 80–82.

6. E.g., *Vita* 1, 16, 68.

7. Peter Brown, *The Body and Society: Men, Women, and Sexual Renunciation in Early* Christianity (New York, 1988).

8. Piroska Nagy, *Le don des larmes au Moyen Age: Un instrument spirituel en quête d'institution (Ve–XI-IIe s.)* (Paris, 2000).

9. Cf. Richard Broxton Onians, *The Origins of European Thought about the Body, the Mind, the Soul, the World, Time, and Fate* (Cambridge, 1951), part II, chap. 6; Nagy, *Le don des larmes*, 62–74.

10. See Brown, *The Body and Society*.

11. Cf. *Vita* 1, 5, 52.

12. "Sed Deus omnipotens noluit eos tunc alicujus magisterio subdi, ut quod de beato viro predestinaverat adimpleret," *Vita* 1, 3, 48.

13. Georges Duby, "Les 'jeunes' dans la société aristocratique dans la France du Nord-Ouest au XIIe siècle," in *Hommes et structures du Moyen Age* (Paris, 1973), 213–26.

The authors thank Judith Rasson and Matthew Suff for their help in correcting our English; Damien Boquet and Barbara Rosenwein for their valuable comments.

Further Reading

Introduction to *Vie de saint Etienne d'Obazin*, ed. and trans. M. Aubrun, Publications de l'Institut d'Etudes du Massif Central, fasc. VI (Clermont-Ferrand, 1970), 7–31; a bibliography on Stephen and Obazine can be found on 32–35.

Ermites de France et d'Italie (xie–xve siècles), ed. André Vauchez (Rome: École Française de Rome, 2003).

Henrietta Leyser, *Hermits and New Monasticism: A Study of Religious Communities in Western Europe, 1000–1150* (London, 1984).

Giovanni Tabacco, "Eremo e cenobio, in *Spiritualità e cultura nel Medioevo: Dodici percorsi nei territori del potere e della fede* (Napoli, 1993), 159–66.

— 39 —

Creating an Anchorhold

Alexandra Barratt

This indented writing, made at Whalley, the sixteenth day of December, in the thirty-fourth year of the reign of King Edward, the Third since the Conquest, between Henry, Duke of Lancaster, Earl of Derby, of Lincoln and of Leicester, and steward of England, on the one part, and the Abbot and Convent of Whalley, on the other part, WITNESSES that the said Duke, by special license obtained from our Lord the King concerning this, has given and granted, and by this present document confirms, to the said Abbot and Convent and to their successors for ever, two cottages, seven acres of land, one hundred and ninety-three acres of pasture land, and two hundred acres of woodland, with appurtenances, called Ramsgrove in his chase by Blackburn. And also that the said Duke by the same license has GRANTED that two holdings, one hundred and twenty-six acres of land, twenty-six acres of meadow, and one hundred and thirty acres of pasture, with the appurtenances called Standen, Hulcroft, and Greenlech, in the towns of Pendleton and Clitheroe together with the sheepfold of Standen, and the foldage of the same sheepfold, with all the profits coming from them and the attachments of the said foldage as was the custom formerly. Which holdings, land, meadow, pasture, sheepfold, and foldage with their appurtenances and the profits of the above mentioned, William of Yves holds for his whole life, by the lease and grant of the said Duke, and which, after the death of that same William, were to revert to the said Duke and to his heirs, SHOULD ENTIRELY REMAIN, after the death of the said William, to the said Abbot and Convent and to their successors for ever, to have and to hold the aforesaid holdings, cottages, lands, meadows, pastures, woodlands, sheepfold, and foldage, with all their easements, franchises, severalties, commons and with all other profits and appurtenances, to the said Abbot and Convent and their successors, from the said Duke and from his heirs, for ever, also should be as entirely, freely, fully, quit of rent, and peaceably in all points and in all profits as the said Duke and his ancestors at any time formerly held or had them: for the services that follow:

TO WIT, to provide adequate and appropriate sustenance for a female recluse dwelling in a place within the cemetery of the parish church of Whalley, and to her successor recluses there dwelling for ever, and for two women, chosen as their servants by the said recluses, and for each one of them who for the time being shall be there, praying in perpetuity for the said Duke, his ancestors and his heirs. TO WIT, to pay to the said recluses and to their successor recluses there for ever in the said abbey, each week of the year, from one year to the next, seventeen loaves of convent standard, each loaf weighing fifty shillings sterling, seven loaves of the second sort of the same weight, eight gallons of the superior convent ale, and three pennies for relish to be eaten with bread; and to provide and pay, from year to year in perpetuity, at the Feast of All Saints, in the said Abbey, for the same recluses and their successors there, ten hard fish called stockfish, ten fatty fish, ten ling, one bushel of oat flour for their soup and one bushel of rye, two gallons of oil to light their lamps, one weight of tallow for candles and also six cartloads of peat and one cartload of wood for fuel, transported by the said Abbot and his successors to the place of the said recluses; and thus to recover, repair, and maintain all the houses and the enclosures which ever are there erected for the dwelling of the said recluse and her successors, and to repair them whenever it is necessary in a way appropriate to the standing of the said recluses, forever in the manner in which the said houses are at present constructed, at the expense of the said Abbot and Convent, and their successors.

AND to provide there a monk chaplain from the same Abbey, of a chaste way of life, and a clerk to serve him at mass, to sing masses every day in perpetuity, in the chapel of the said recluse and her successor recluses there, for the said Duke, his ancestors and his heirs for ever. And let the said monk, as to the offices of the mass and the hour of singing, be governed according to the disposition of the said recluse, and of each of her successor recluses, who shall be there for the time being. And at a time of the absence of each recluse, whether by death or by whatever other reasons so that there is no recluse there, the said Abbot and Convent and their successors shall receive and welcome another on the nomination, command or order of the said Duke or of his heirs, without gainsaying or counterclaim; so that she shall dwell there as a recluse and shall take, receive, and have as to all things as is said above. And the said monk, every day during the absence of a recluse, shall sing mass in the said chapel, as was said above. And the same Abbot and Convent and their successors shall provide vestments, chalice, bread, wine, lighting, and other altar furniture necessary for the said mass, in perpetuity.

AND IN ADDITION, they shall pay the said Duke and his heirs an annual rent from the lands and tenements aforesaid, to wit, one rose a year, at the Feast of the Nativity of Saint John the Baptist, for the entire life of the said William of Yves; and after the death of the said William, sixty-six shillings and eight pence in sterling annually for ever, twice yearly: that is to say on the said Nativity and on the Feast of St. Martin in winter, in equal portions. And to carry out for ever well and loyally in every detail all these things aforenamed, with respect to the said recluse and her successor recluses there and with respect to the said Duke and his heirs, THE AFORESAID ABBOT AND CONVENT bind themselves and their successors and their abbey for ever.

AND FURTHERMORE they agree on behalf of themselves and their successors at whatever hour that fault or failure in any way should be detected after this time in them or in any of their successors, in any detail or item above mentioned, voluntarily to yield to the said Duke, and to his heirs, and to their bailiffs in all the aforesaid lands and tenements and in each part of them, in the hands of whomsoever they may come, that they may distrain and pursue, carry off, and retain what has been distrained, so that satisfaction may be made to the said recluse, and to her successor recluses there, of all things aforenamed that appertain to them; and also to the said Duke and to his heirs concerning what appertains to him and to his heirs, and so that all the said things aforenamed, by the said Abbot and Convent and by their successors, may be fully performed and carried out in all details as was said earlier. And on this the aforesaid Duke and his heirs WARRANT, acquit, and defend against all people for ever all the holdings, cottages, lands, meadows, pastures, woodlands, sheepfolds, and foldages itemized above, with all their liberties, profits, and appurtenances, in the manner above mentioned, to the said Abbot and Convent and to their successors.

IN WITNESS of this, on the portion of this indented writing remaining with the aforesaid Abbot and Convent, the said Duke has set his seal; and on the other portion of the same indented writing remaining with the aforesaid Duke the said Abbot and Convent have set their common seal, with these witnesses: William of Dacre, Adam of Hoghton, Roger of Pilkington, and Nicholas le Botiller, knights; Richard of Radecliffe, steward and master forester of Black-burnshire and of Bowland; William of Radecliff, sheriff of Lancaster; Robert of Clitherhoe; Gilbert of the Leigh; John of Bailey; and many others.

GIVEN at Whalley, the second day of January, the tenth year of the said Duke's dukedom.

Source: William Dugdale, *Monasticon Anglicanum* V (London, 1825), 645–46.

Retreat into the desert to pursue a life of prayer and asceticism was the earliest form of Christian monasticism. The eleventh and twelfth centuries in western Europe saw a revival of this eremitic ideal. In England, however, there was no desert, and remote places were unsafe, particularly for women who were more often attracted to this way of life than men. Instead, there developed forms of the so-called anchoritic life.

An anchorite is, literally, someone who "goes away" or "withdraws." Medieval anchorites, sometimes also called "recluses," did this by secluding ("enclosing") themselves in a cell or anchorhold. This often consisted of several rooms and might be attached physically to, or in the grounds of, a monastic or parish church; the remains of some still survive in England. Although often called "solitaries," anchorites did not live entirely alone. They would need at least one servant who constituted a link with the outside world and performed domestic duties such as fetching water and fuel, obtaining food, and cooking. Some anchorites lived in groups of two or three, informal groupings that might eventually develop into full-fledged religious communities.

Some anchorites were, technically, "religious," that is, they had already taken vows as monks, friars, or nuns. On discovering a vocation to a more rigorous form of life, with the permission of their religious superiors they might leave their original community and enter an anchorhold. More often, an anchoress or female anchorite was a laywoman. Her desire to adopt this way of life would be conveyed to the local bishop who would examine her suitability. If approved, she would be enclosed in a solemn ceremony conducted by the bishop or his representative, versions of which are found in the Sarum Manual and the late medieval York and Exeter Pontificals. The candidate for enclosure was clothed in special garments, vowed chastity, obedience, and stability (that is, she promised to remain in the anchorhold until death), and was then conducted to her cell, which would be blessed. In the Exeter ceremony, after the candidate had entered she received the sacrament of extreme unction (the anointing of the sick), otherwise given only to those on the point of death. She was sprinkled with dust, and the door of the anchorhold was blocked up. All this dramatically emphasized her "death to the world" and her "burial" in the cell.

The anchoritic life was one of prayer and contemplation, combined with asceticism or the practice of various kinds of physical deprivation designed to subdue the sinful flesh. If the anchoress was literate, that is, able to read (if not understand) Latin, she was expected to recite the Divine Office, the round of seven services during the day and one at night that structured religious community life. She might also engage in spiritual reading (a number of treatises were written in or translated into English specifically for anchoresses), in meditation and, eventually perhaps, achieve contemplative prayer. If she could not read, she would simply repeat the Lord's Prayer, the Hail Mary, and other simple devotions from memory. In the spirit of Christian monasticism, all were expected to do some manual labor, whether spinning, weaving, or sewing. Some might exercise a wider ministry: Julian of Norwich (fl. 1373), the best known of all medieval English anchoresses, wrote two accounts of her spiritual experiences and was known as a spiritual director. Margery Kempe, wife, mother, and visionary, a peripatetic holy woman, went to visit Julian to consult her about her own revelations, "to know if there was any deceit in them. For the anchoress was expert in such things and knew how to give good advice."

But the contemplative ideal meant that, theoretically, the anchorite should have no involvement whatsoever in the secular world. She must not, for instance, keep a school, make money from needlework or embroidery, provide safe-deposit facilities for her neighbors, let alone sit at her window gossiping with other women or, God forbid, priests! Even owning a cow was impermissible as the animal might stray, be impounded, and thus embroil its unfortunate owner in awkward dealings with secular society. The anchoress could not be active in good works and was not even supposed to give alms to the poor in person. As she often lived in the heart of the village or town and might well have relatives nearby, this was a hard ideal to achieve.

In spite of the anchoress's personal asceticism, then, her way of life was costly in financial terms. The indenture (a duplicate copy of a contract between two parties) translated here highlights this most practical aspect of the anchoritic life. No more than Anna Karenina could the anchoress "live on air": she needed an endowment (or source of capital), the income from which would support her financially so that she would not need to resort to attempts to generate external income that might distract her from her life of prayer or pose moral dangers.

Not everyone, however, agreed that an anchorite should be fully funded in advance of her enclosure. Aelred, the abbot of the Cistercian monastery of Rievaulx, writing to his sister, a recluse, in the mid-twelfth century, considered that "if possible, she should live by the labor of her own hands, for this is more perfect." But even he recognized that this was often impractical and conceded as second best that "before she is enclosed she should seek out certain people from whom she should humbly receive every day what is sufficient for one day." It is unlikely that he had in mind the kind of permanent provision (however modest) made by Duke Henry. The anonymous author of *The Cloud of Unknowing*, a fourteenth-century English treatise on mystical prayer, remarked that if his correspondent is faithful to his life of prayer, God

> will stir other men in spirit to give us our necessities that belong to this life, such as food and clothing with all these other things, if he sees that we will not leave this work of love [i.e., prayer] for preoccupation concerning them. And this I say in confutation of the error of those who say that it is not lawful for men to determine to serve God in the contemplative life unless they are sure in advance of their bodily necessities. For they say, God sends the cow, but not by the horn!

But this writer was probably a Carthusian monk, not an astute magnate to whom such an approach of simple trust in God would seem merely an excuse for hand-to-mouth financial arrangements.

Clearly Henry of Grosmont, Duke of Lancaster (1310–1361), father of Blanche, whose death Geoffrey Chaucer commemorated in his early dream poem *The Book of the Duchess*, and father-in-law of John of Gaunt, believed that anchorholds should be financially secure. In other words, he was determined to tie the cow God sent securely by her horn. He therefore makes over 670 acres of land of various kinds to the Cistercian abbot and convent (community) at Whalley. To do so, as a tenant-in-chief that held his own lands direct from the Crown, he had to obtain permission from the king. The monastery was to farm or rent the land and, out of the income derived from it, was to maintain an anchorhold in the grounds of the local parish church of St. Wilfred, and provide for the needs of the anchoress, her servants, and their successors in perpetuity.

These needs are partly physical. The monks must keep the building in which she lives, apparently already in existence, in good repair. They must provide her with food and drink, mainly in kind, from the monastery's own kitchen, but there is also some provision for cash payments, presumably so that the anchoress's servant could buy some local fresh produce. They must also provide means of adequate heating and lighting, an important consideration in Lancashire in the north of England. The food and drink is to be fetched from the abbey, presumably by one of her servants, while the fuel is to be home delivered. But the abbot and convent must also provide a chaplain from the monastery (by now, most choir monks were ordained priests) to sing Mass daily, a clerk to assist him, and everything necessary for the proper celebration of the liturgy in the anchoress's own chapel. Any financial surplus after the income has been used for these purposes remains with the monastery, of course. One can only hope that the monks of Whalley did not stint their resident anchoress in order to maximize their profit.

We may also note what the endowment does not provide. There is no reference to furnishings for the anchorhold or kitchen equipment, to books, whether service books or books of devotion, or to the anchoress's clothing or bedding. It is quite likely that the anchoress and her family contributed these items themselves; young women who entered a convent were expected to bring such items with them.

Although the document records a contract between two parties, there are actually three involved in this transaction. What was in it for each of them? Firstly, we can only speculate about the motives of the anchoress (or those of her servants, who were probably unpaid, receiving only food and lodging, but who might later succeed to the anchorhold). In general, life as an anchoress, and the anchorites themselves, were surprisingly popular. Judging by the number of bequests made to them in medieval wills, we may conclude that there was much social support for such a choice of vocation and that most anchorites were perceived as faithful to their vocation and spiritual assets to the community.

Secondly, the abbot and convent of Whalley gained the use of additional lands and income (although they did have to pay the Duke some rent). No doubt Duke Henry's financial advisers had carefully calibrated the scale of the endowment, but it must have been adequate, even more than adequate, to fund the anchorhold and its occupant, or the monastery would never have agreed to the arrangement. Six hundred and seventy acres seems generous, even if some of the land was uncultivated or marginal, especially in medieval England where one or two acres could provide subsistence living for a peasant family. The monks therefore stood to make at least a modest profit from the transaction, and could also improve the lands made over to them by the Duke to make them more productive.

Finally, what of the Duke himself? This indenture was written only a few months before his death, but Duke Henry had been a very great man indeed, a grandson of King Henry III, a successful soldier, and a politically powerful figure in fourteenth-century England. He was also a devout Christian of genuine personal piety. He was the first layperson, as far as we know, to write a devotional treatise in Anglo-Norman: his *Livre des Seyntz Medicines*, written in 1354, survives in two manuscript copies. He had made many other, far grander, religious benefactions (he was cofounder of Corpus Christi College, Cambridge) and in particular on his death provided for the foundation of a college of secular canons in Leicester. He no doubt genuinely valued the prayers of the Whalley anchoress, whom incidentally he had the right to nominate, thus exercising benign patronage over a deserving holy woman. But probably even more he treasured the daily Masses that would be offered in her chapel for himself and his family. To use one's financial resources in order to ensure prayers in perpetuity for oneself, one's forebears, and one's children may seem a strange, even immoral, investment to the modern mind. To the medievals it was simply the equivalent of private health insurance. Yes, salvation (or medical care) is or should be freely available to all—but why take the risk of having to rely on the public system if you can afford to go private? Most people in the fourteenth century would have applauded Duke Henry's actions as prudent, pious, and farsighted.

But the best laid plans of mice, men, and even dukes. . . . In 1443, less than a century after its establishment in 1361, a complaint was made to King Henry VI that the current anchoress, Isole de Heton, who had been enclosed in 1436, had broken her enclosure, apparently to attend to the financial affairs of her young son (she was a widow). Clearly the selection procedures on this occasion had been remiss, as anchoritism is one career that does not combine well with motherhood. There were also complaints about the immoral behavior of the anchoress's servants, who were said to have become pregnant. As a result the anchorhold was dissolved, and no trace of it now remains.

Further Reading

Aelred of Rievaulx's letter of advice to his sister, *De Institutione Inclusarum*, is translated as "A Rule of Life for a Recluse," by Mary Paul MacPherson, *Aelred of Rievaulx: Treatises and the Pastoral Prayer*, Cistercian Fathers 2 (Kalamazoo, MI, 1971).

Rotha M. Clay, *The Hermits and Anchorites of England* (London, 1914).

The Cloud of Unknowing, advice on contemplation for a young male recluse, is available in a modern translation by James Walsh, Classics of Western Spirituality (New York, 1985).

Kenneth Fowler, *The King's Lieutenant: Henry of Grosmont, First Duke of Lancaster, 1310–1361* (London, 1969).

Bella Millett, *Ancrene Wisse, The Katherine Group, and the Wooing Group*, Annotated Bibliographies of Old and Middle English II (Cambridge, 1996).

Anne Savage and Nicholas Watson, eds. and trans., *Anchoritic Spirituality: "Ancrene Wisse" and Associated Works*, Classics of Western Spirituality (New York and Mahwah, NJ, 1991).

Of the numerous modern translations of Julian of Norwich (many of them of little scholarly value) the most recent is by Elizabeth Spearing, *Julian of Norwich: Revelations of Divine Love (Short Text and Long Text)* (London, 1998).

Ann K. Warren, *Anchorites and Their Patrons in Medieval England* (Berkeley, 1985).

— 40 —

The Ritual for the Ordination of Nuns

Nancy Bradley Warren

**THIS IS THE FORM FOR HOW A NOVICE SHALL BE
MADE AND RECEIVED INTO RELIGION**

In the beginning, when she has made her petition and asked to enter the
house, and the prioress and convent have granted her petition, then she shall
come to the prioress and kneel down before her. And the prioress shall take her
[the postulant's] hand in hers and kiss her, and she shall be received. Afterward, when it pleases the convent to bring her to the chapter to be examined,
they shall dress her in the clothing that she shall wear for the first year until
she is professed.

When the chapter is finished, the one who will be her mistress shall say to the
prioress, "There is a novice to be examined." Then the prioress shall give the
command to bring her in, and her mistress shall bring her in. When she comes
to the place where the nuns prostrate themselves in their acts of supplication,
she shall prostrate herself, and the prioress shall ask her, "Dear daughter, what
do you request?" Lying still, she shall say, "The mercy of God and your mercy."
The prioress shall tell her to rise, and say to her, "What is that mercy for which
you ask?" And she shall say, "To dwell in this place in the habit of religion to
serve God, and to punish my sins, and to amend my life, and finally to save my
soul."

Then the prioress shall say again, "Dear daughter, this thing that you ask is a
hard and strait thing. Nevertheless, to those whom God inspires and gives the
grace, will, and power to fulfill it, and who stand stable in the purpose in which
they began, it is easy, meritorious, beneficial, and leads to everlasting life. But at
the beginning of this spiritual life, there are three things that behoove you. The
first is to forsake your own will and live under obedience, being obedient princi-
pally to the prioress and to the elder sisters in the order in all lawful and honest
things. The second is to live in willful poverty, owning nothing without the
knowledge of the prioress and taking nothing from any of your friends (neither

gold, nor silver, nor any other gift) unless the prioress sees it, and it is fully disposed to her will; thus you avoid being proprietary and falling into extremely serious danger concerning your religion. For whosoever has any thing, gold, silver, jewels, or any property, without the knowledge and permission of her prioress, she stands cursed. The third is to live chaste and take God as your spouse and forsake all your desires and the pleasure of the flesh. It also behooves you to be abstinent and fast when other people eat, to rise for divine service when other people sleep, to give yourself to prayer and devotion to purchase grace, and to stand firmly in the purpose that you undertake. And, daughter, if you say that you may fulfill with God's grace all these points that I have rehearsed, speak your will now here before the convent."

And then the novice, if she will abide in her purpose, shall say, "I shall fulfill the good purpose I have undertaken to the end of my life through the grace of God and through your good instruction." Then the prioress shall say to her, "Dear daughter, may God of His great grace give you good perseverance. Go with the mistress in the name of Christ."

Her mistress shall take her to the area reserved for the novices and teach her according to the religion. When two months have passed, her mistress shall expound to her the rule and all the points and the strictness of religion, concealing nothing from her. When her mistress has expounded the rule to her, she shall be examined again in the chapter in the same fashion described previously. And if she stands in her purpose after six months, the rule shall be expounded to her again by her mistress. Then she shall be examined in the chapter in the aforesaid manner. And if she still abides fully in her purpose until the end of the year of proof, then on the day that she shall be professed, she shall be brought into the chapter, and be examined for a fourth time, in the manner rehearsed before.

Then, if she fully agrees to the behest that she made previously, after the Gospel on the day that she shall be professed, her mistress shall come to her and lead her to the steps before the altar. There she shall read her profession, with the priest who sings the mass standing at the right corner of the altar. When she has read her profession, her mistress shall give her a pen with ink, and the novice shall make a cross on the book of her profession, and so go up to the high altar, her mistress with her, and lay it on the right end of the altar. She shall kiss the altar, and bow devoutly, and going again to the steps sing there three times: "Suscipe me, domine, etc." The convent, always standing in their stalls, shall rehearse the same again three times, and "Gloria patri."

Then the novice shall prostrate herself before the steps when "Kyrie Eleison, Christe Eleison, Kyrie Eleison" is sung. And then the priest shall say or sing, "Et ne nos." The choir shall answer, "Sed libera nos." Then the cantor shall begin this psalm, "Misere mei, deus," each side of the choir saying the verses in turn, and also "Gloria Patri" and "Sicut." When the psalm has ended, the priest shall come from the high altar to the steps and say, "Salvam fac ancillam tuam." The choir shall answer, "Deus meus, sperantem in te." And afterward the priest shall sing,

"Mitte ei, domine, auxilium de sancto." The choir shall answer, "Et de sion tuere eam." The priest: "Esto ei, domine, turris fortitudinis." The choir: "A facie inimic." And then the priest: "Nichil proficiet inimicus in ea." And then the choir: "Et filius iniquitatis non apponet nocere ei." The priest: "Domine, deus virtutum, converte nos." The choir: "Et ostende faciem." The priest: "Dominus vobiscum." The choir: "Et cum." The priest: "Oremus."

And then the priest shall say four collects over the novice, who is lying prostrate by the steps, while the prioress and the convent are standing in their stalls, their faces turned toward the altar. When the collects are all ended, the novice shall stand while the veil is consecrated. When the priest has consecrated it, three or four of her sisters with her mistress shall come out of the choir and stand around her as her veil is being put on, while all the rest of the convent stand in their stalls. When she is arrayed in her veil, the cantor shall begin this hymn solemnly, "Veni, creator," with each side of the choir singing verses in turn.

And in the meantime, the novice shall make her profession to the prioress, kneeling and saying in this manner, "Promitto tibi obedienciam secundum regulam sancti benedicti." [I promise you obedience according to the rule of St. Benedict.] And the prioress shall say then, "Det tibi Deus vitam eternam." [God grant you eternal life.] She shall do in the same fashion to all of the nuns on the prioress's side of the choir, and then likewise to those on the other side. When she shall come again before the altar, she shall prostrate herself there, and the priest that made her a nun shall say or sing, "Salvam fac," "Nichil proficient" (as is said before), and three collects: "Deus, qui caritatis," "Acciones nostras," "Fidelium dues." When the three collects are said, the prioress shall come and lead her to the stall where she shall stand, and the priest shall go to the altar and begin the creed. And at this point she ought to receive the Eucharist.

Source: Ernst A. Kock, ed., *Three Middle English Versions of the Rule of St. Benedict and Two Contemporary Rituals for the Ordination of Nuns*, Early English Text Society o.s. 120 (London, 1902).

The historian Roger Chartier has said that practices "are the visible indices of demonstrated or desired identities."[1] Indeed, the practices found in the *Northern Lansdowne Ritual for the Ordination of Nuns* indicate that identities of late medieval Benedictine nuns were complex mixtures of elements. The practices evident in the *Ritual* reflect that which was desired and demonstrated *for* the nuns by the ecclesiastical authorities who designed the ritual and regulated their lives; they also reveal that which was desired and demonstrated *by* the nuns themselves. Competing loci of authority as well as tensions between empowerment and constraint shaped what it meant to become, and live as, a nun in the Benedictine Order during the fifteenth century.

The practice of examination dominated the process of entry into religious life. The repeated examinations of the candidate, both at her admittance and throughout her probationary year, have a double valence that resonates with the larger tensions between empowerment and constraint. Examination is, on the one hand, a process designed to ensure conformity; monastic life is, after all, by definition a life governed by rule, and it was essential to ensure that a prospective member of the community understood, and was committed to follow, that rule. On the other hand, the examinations open up opportunities for the exercise of autonomy, both communal and individual. The prioress and the convent had to agree to the postulant's petition, and the convent decided corporately when to bring the candidate to the chapter for her first examination. So, the *Ritual* allows the community a significant degree of self-determination in shaping its composition, although the community's right to decide who might enter the novitiate could be overridden by a founder's or monarch's right to nominate a member. The *Ritual* also gives the postulant several opportunities to exercise her own autonomy during the probationary period; at each of her examinations, she was faced with the choice whether to continue in the process of adopting the identity of Benedictine nun.

Alongside the opportunities to demonstrate communal and personal autonomy evident in the opening section of the *Ritual*, female authority, especially female authority to teach and instruct, emerges quite strikingly. In the formalized dialogue that takes place at the postulant's petition for entry, the prioress is directed to instruct the postulant in the three substantial vows of poverty, chastity, and obedience, the fundamental vows that define monastic life for both genders across all religious orders. The prioress addresses the postulant as "daughter," evoking her maternal authority within the community. The prioress's authority is dramatically emphasized at the end of the ceremony of profession, when the newly professed nun must kneel and promise obedience to her. The postulant, once received as a novice, also underwent extended instruction by the mistress of the novices, another key female authority figure within the community.

The novice's instruction, as well as the service of profession, suggests the possibility of another kind of autonomy—textual autonomy—among later medieval Benedictine nuns. The repeated process of the novice mistress's "expounding" the rule implies a situation in which nuns would have engaged in textual interpretation, whether or not all of them were actually literate and able to read the text for themselves. Similarly, the retention of Latin phrases in the text of the profession service suggest at least the possibility that some nuns could have known some Latin, a possibility that many scholars long discounted. Evidence from manuscripts and records of nunnery libraries demonstrates that some Benedictine communities (for instance, Barking) owned large numbers of volumes and had rich textual cultures.[2] In the fifteenth century, the era to which the *Ritual* belongs, women's access to religious texts and any interpretations of such texts women might offer were often regarded with suspicion or outright hostility. For instance, the poet Thomas Hoccleve, in his "Remonstrance against Oldcastle," admonished women not to make arguments concerning Holy Scripture but rather to sit down, spin, and "cackle" about other things. So the opportunities for textual autonomy possible for women in nunneries were potentially quite liberating.

Aspects of the *Ritual*, then, suggest that the processes of transformation from secular to religious life, and, subsequently, the process of actually living the religious life, are ongoing processes of "becoming," in which education, development, and self-determination are central. In the description of the profession service itself, though, the emphasis changes; the profession service stresses the process of "making." This idea shows up in the heading of the *Ritual* ("the form for how a novice shall be *made* and received into religion") and in the service, where there is a striking reference to "the priest that made her a nun."

The distinction between participating in active "becoming" and being a passive subject that is "made" informs one of the key differences in religious life for monks and nuns, a difference set up at the entry into religion—that is, the meaning of the vow of chastity. Monks and nuns both took vows of chastity, but, as the *Ritual* illustrates, that vow had a gendered significance for nuns, who became brides of Christ upon their profession. In instructing the novice, for instance, the abbess tells her "to live chaste and take God as your spouse." A newly professed bride of Christ had, unlike the monk who undertook a quest for spiritual perfection, "the dubious advantage of being in the same state where she would ideally end."[3]

The nun's chief aim was to preserve her soul by preserving her chastity, the virtue into which all others more or less collapsed. The absolute value of chastity for nuns is emphasized in the profession service during the formal ceremony in which the veil is consecrated and given to the new nun. The veil hides the nun's hair, long a symbol of female sexuality, and it has ancient associations with both virginity and marriage. It is, significantly, the only piece of the nun's habit that is consecrated during the ceremony. Indeed, it is the only piece of her habit given particular mention, appropriately since it is, as Penelope Johnson has observed, "the outward sign of inward chastity" and the one "distinctively female part of a nun's habit."[4]

The extraordinarily high value placed on nuns' chastity, evident in the *Ritual*, had practical as well as spiritual implications. Although all monastics, male and female alike, were subject to supervision by clerics outside of their immediate community, nuns were typically under much closer ecclesiastical control than monks, largely because their chastity was seen always to be at risk, either from attack from outside or from internal female fallibility. The desire to control nuns closely and protect their chastity had practical as well as symbolic or spiritual implications, since it resulted in much more stringent attempts to enforce claustration for nuns than monks.

The supervisory paradigm is grounded in the ceremony marking the entry into religious life. In the profession service, the priest who "makes" the nun acts "symbolically as parent and spouse" in addition to officiating as a cleric.[5] His multivalent role foregrounds the extent to which nuns were subject to clerical authority. Priests performed necessary sacramental work from which nuns—like all women—were excluded, as we see at the end of the *Ritual* when the new nun is directed to receive the Eucharist. They also, though, served as stand-ins for the nuns' divine spouse, and, as such, as guardians of their valuable chastity with rights like those of human husbands to regulate wives' conduct.

Notes

1. Roger Chartier, *Cultural History between Practices and Representations*, trans. Lydia G. Cochrane (Ithaca, NY, 1988), 10.

2. On the libraries of nunneries see David N. Bell, *What Nuns Read: Books and Libraries in Medieval English Nunneries*, Cistercian Studies Series 158 (Kalamazoo, MI, 1995); for information on Barking's books, see 107–20.

3. Barbara Newman, *From Virile Woman to WomanChrist: Studies in Medieval Religion and Literature* (Philadelphia, 1995), 44.

4. Penelope D. Johnson, *Equal in Monastic Profession: Religious Women in Medieval France* (Chicago, 1991), 236.

5. Ibid., 64.

Further Reading

Elizabeth Makowski, *Canon Law and Cloistered Women: Periculoso and Its Commentators, 1298–1545* (Washington, DC, 1997).

Jo Ann Kay McNamara, *Sisters in Arms: Catholic Nuns through Two Millennia* (Cambridge, MA, 1996).

Marilyn Oliva, *The Convent and the Community in Late Medieval England: Female Monasteries in the Diocese of Norwich, 1350–1540* (Woodbridge, 1998).

Eileen Power, *Medieval English Nunneries c. 1275–1535* (Cambridge, 1922).

Nancy Bradley Warren, *Spiritual Economies: Female Monasticism in Later Medieval England* (Philadelphia, 2001).

Rituals of Power

— 41 —

An Anglo-Saxon Queen's Consecration

Janet L. Nelson

THE KING'S CONSECRATION ENDS

The queen's consecration follows. For the purpose of honoring her, the top of her head is to be anointed by a bishop with the oil of unction, and she is to be blessed and consecrated with due honor in royal highness in [the] church in the presence of the great men, as set forth on the following page, into partnership of the royal bedchamber. We also decree that she should be adorned with a ring to signify the integrity of faith and with a crown to signify the glory of eternity.

THE QUEEN'S CONSECRATION BEGINS

This must be said by the bishop or by [the] priest:

In the name of the Father, the Son and the Holy Spirit. May this anointing be of benefit to you for honor and eternal strengthening.

Let this prayer follow:

Almighty eternal God, looking kindly on our prayer pour the flowing spirit of your blessing on this thy woman-servant, so that she who by the laying-on of our hand is instituted queen this day may remain worthy and chosen through thy sanctifying, and may she never more be separated from thy grace as unworthy. Through [our Lord . . .]

Then the highest of the bishops taking the ring and placing it on her finger must say:

Receive this ring of faith, the sign of the Holy Trinity, by which you may be able to shun all heretical depravities, and to also by the strength supplied to you to summon barbarian peoples to acknowledgment of the truth.

Let this prayer follow:

God to whom belongs all power and all office, give to this thy woman-servant by the sign of faith a prosperous carrying-out of her office, and may she always remain firm in it and strive continually to please thee. Through . . .

Then let her be crowned:

Receive the crown of glory, the honor of joy, in which you may shine forth in splendor and be crowned in eternal exultation. Through . . .

Let this prayer follow:

Lord, the fount and origin of all good things, and giver of all benefits, grant, we beg, to this thy woman-servant that she may perform well this office she has received, and strengthen in her with good works the glory conferred on her by thee and for thee. Through . . .

Source: Leopold G. Wickham Legg, ed., *English Coronation Records*
(Westminster, 1901), 21–23.

The earliest Anglo-Saxon kings to be consecrated lived in the eighth and ninth centuries. The beginnings are obscure, varying between the different Anglo-Saxon kingdoms; but once the innovation had been used in one kingdom, others followed, and once begun, so far as we can tell, usage continued. Anglo-Saxon queenly consecration is both later, and discontinuous. It has its own history, and its own refracted light to throw on Anglo-Saxon kingship.

Kingly consecrations in western Europe are first documented in seventh-century Spain, before the Muslim conquest, then in eighth-century Francia, and in England.[1] They signify change and growth in three dimensions: kingship as an institution, the aristocracy's interest in kingship, and the entente between kings and churchmen. While kings often succeeded their fathers, the notion that they had to be elected, or chosen, by the leading men in the kingdom was widely and strongly held. Both elective and hereditary elements were always present in medieval kingship. Election emphasized the merits of the individual king, inheritance the importance of descent. On a sliding scale, the Spanish kingdom was most elective, the Anglo-Saxon kingdoms increasingly dynastic, and the Frankish kingdom the most strongly dynastic. At a change of dynasty in 751 the Franks adopted an ecclesiastical ritual of royal consecration. By 800, two Anglo-Saxon kingdoms had followed suit, and a third did so in the ninth century.

The Christian Church had used anointing rituals long before this to identify objects and people in special categories of holiness or danger or spiritual transformation: altars were anointed, for instance, following biblical precedent; the sick and the dying were anointed; and so were children in baptism and confirmation. The idea of anointing kings and priests with holy oil came from the Bible. The relevant passages were read figuratively until the seventh/eighth century, but from then on, very sporadically at first, literal interpretations began to be applied in ritual practice. The *Anglo-Saxon Chronicle* for 787 says that a king of Mercia was *gehalgod*, "hallowed." It can be inferred that a literal anointing was performed by clergy on that occasion, because there is much more explicit evidence for this from Francia in the decades just preceding, and because the Franks and Mercians were in frequent contact with each other. Franks and Mercians were also in touch with a series of popes who actively promoted royal anointings as a means to fostering aristocratic and ecclesiastical cooperation with kingship. In 796 a Northumbrian king was "consecrated."[2]

A century later, both the Mercian and the Northumbrian kingdoms had been dismembered and were under new management. Kingly anointing had already taken root in Wessex, and one West Saxon queen had been consecrated, in 856. For that queen, a Frankish princess, a Frankish bishop compiled the earliest extant and clearly dated royal consecration service, drawing on an earlier Anglo-Saxon king-making rite.[3] The 856 service is highly unusual in surviving as an ad hoc set of instructions, explicitly associated with one well-documented event. The vast majority of comparable data consist of model services that were generally normative rather than bound to any particular occasion. They are known in the plural as *ordines* (singular *ordo*, in this context a liturgical service). They survive, along with many other rites, in pontificals, liturgical books for bishops' use, and they consist largely of prayer texts with only very brief indications of the gestures intended to accompany them. Directions for the ritual's choreography would be written in a separate parchment-roll. Bishops were professionally concerned with action *inside* the church. Rituals preceding or following the ecclesiastical part, performed in secular space and probably managed by lay officials, are occasionally recorded in more or less contemporary narrative sources: such accounts, though problematic in other ways, reveal how much the royal *ordines* could leave out.[4] Nevertheless, despite these limitations, the new genre brings new evidence for the rituals whereby rulers were made and, hence, for the symbolic representation of rulership. The ninth and tenth centuries were the crucial formative period for these *ordines*, as for other major rites such as episcopal ordination and church consecration.

The first pair of royal *ordines* for king and queen survives in a Frankish pontifical written perhaps circa 875. It can be deduced that copies of these *ordines* reached England either in the late 890s or early in the tenth century. The king's *ordo*, together with small borrowings from two other West Frankish king's *ordines*, was spliced with an earlier Anglo-Saxon one to form what historians have labeled the Second English *Ordo*. It survives in three slightly different versions. In all three, the accompanying queen's *ordo* is straightforwardly that of the Frankish model, implying that there was no indigenous Anglo-Saxon queen's *ordo* (hence, that the rite composed for 856 had not been used subsequently). The manuscript evidence is Continental and later, but the following proposed reconstruction is convincing.[5] The Second *Ordo* (in its first or second versions) was used for the consecrations of tenth-century English kings, perhaps from Edward the Elder (900) onward, or perhaps Athelstan (925) onward, down to Edgar in ?960. The (Frankish) queen's *ordo* was presumably alongside it in some English pontificals during those decades (no manuscript with that arrangement survives from so early); if so, it was copied out routinely, as frequently happened in liturgical books, without actually being used. There is no clear evidence for the consecration of any English queen between 900 and the mid-960s, inclusive. Why this silence? Two of the six kings in question (Athelstan and Eadred) were unmarried, arguably as part of family agreements on brothers' or nephews' succession. Two others of the six (Edward and Edgar) were serial monogamists, that is, took a new wife when one consort died or was repudiated. Three of these kings (Edward, Eadwig, and Edgar) consorted with women not recognized as wives and not endowed with marital property, presumably because the unions were seen as provisional. The king's wife was not marked out as the mother of a future king. What all this amounted to was the keeping of options open where the royal succession was concerned, and a series of unstable alliances between the kings and the aristocratic families from which all tenth-century English royal consorts were drawn. This fluid situation was compatible with powerful monarchy, yet in contemporary western European context, it marks out Anglo-Saxon kingship as distinctly unusual.

When the silence is broken, there is a strong note of something new needing the most public kind of justification and the loud support of the leading men of the kingdom, laymen as well as ecclesiastics. The same queen's *ordo*, associated with the third and latest form of the Second English [king's] *Ordo*, survives in no fewer than five Anglo-Saxon pontificals of the later tenth and early eleventh centuries, and in all of these, a remarkable passage acts both as a bridge between the paired *ordines* and as a preface to the queen's *ordo*; this is the first section of the text given above. The purpose of this preface is evidently to justify the queen's consecration by reference to a "decree," which sounds like the decision of an assembly.[6] The reference to a "following page" seems to refer not to a book but to a document (a parchment-roll?) prepared for a particular liturgical occasion in which "the great men" of the kingdom are to share. While "partnership of the royal bedchamber" could in theory refer to a royal marriage, no combined marriage-with-consecrations is known for the period; and in any case I think here it is an ornate allusion to queenship. For queenship is the theme of this *ordo*.

"Almighty eternal God," the prayer immediately following the queen's anointing, is taken almost word for word from the prayer over a newly ordained abbot *or abbess* (the pronouns would be changed in performance as necessary).[7] The Frankish author of this prayer had found an appropriate model in the highest ecclesiastical office to which a woman could aspire. Just as abbesses played important political roles in the late Carolingian world, so they also did at some periods in Anglo-Saxon history.[8] The prayer for the conferring of the ring evoked a leadership role for the queen in correct religious observance and specifically in the context of acculturating barbarians; again this had a wider context in queenly attributes from late antiquity through to the main tenth-century kingdoms of western Christendom. While the coronation prayer itself is a little exercise in biblically inspired rhetoric, it is framed by two *orationes pro adepta dignitate*, "prayers for an office taken on," again to be found in the abbess's ordination rite. In both these, *dignitas* means office, something with an *effectus*, that is, something to carry out and perform well (*bene gerere*). The parallel king's *ordo* borrows these prayers from the abbot's ordination, but no incongruity was seen in borrowing from the abbess's ordination for the queen. *Consortium* can have a political as well as a sexual sense, just as the royal bedchamber was the setting for acts of power (as when Charlemagne heard cases while getting dressed in the morning)[9] as well as of love. We have a window here on patrimonial government, where the political is also personal and familial.

In this case the liturgical books for once retain an echo of a particular political event: the decision to consecrate a queen. The queen, I think, was Ælfthryth, wife of Edgar, and the occasion their double consecration at Bath on Whitsunday (11 May) 973.[10] The queen's participation here was far from ancillary. If Edgar was displaying his own "imperial" power, Ælfthryth presented other aspects of this new Ottonian-style monarchy: its reform-centeredness, and its dynasticism. Reform involved women as well as men. Abbesses' heightened authority as heads of reformed houses coincided with Ælfthryth's (self-)representation as legal supervisor and protector of convents.[11] If abbesses failed to maintain the high profile reflected in their attestation of royal charters in the mid-tenth century, Ælfthryth, as queen, did sometimes attest her husband's charters and was literally at home in the court. She, unlike Edgar's earlier partners, was the king's lawful wedded wife and had been properly dowered.[12] Her offspring, unlike their elder half-siblings, were *legitimi*. Ælfthryth's subsequent consecration further enhanced not just her own status but also that of her surviving son, making him a credible counterclaimant to his older half-brother Edward. Ælfthryth pointed toward a new way of narrowing the royal descent line, thus identifying a sole lineal successor and inhibiting fraternal contest. Edgar's death, only two years after his consecration, temporarily foreclosed that outcome. Ælfthryth's own consecration, nevertheless, was not just a portent of her own future influence as queen mother; it was a sign of the formation of English queenship in theory and practice with all that entailed, in tandem, for the future of English kingship.

Notes

1. J. L. Nelson, *Politics and Ritual in Early Medieval Europe* (London, 1986).

2. Dorothy Whitelock, trans., *The Anglo-Saxon Chronicle* (London, 1979), 182.

3. Nelson, *Politics and Ritual in Early Medieval Europe*, 341–60.

4. Ibid., 283–308, 329–40.

5. Nelson, *Politics and Ritual in Early Medieval Europe*.

6. For the three manuscripts containing this word see ibid., 372.

7. J. L. Nelson, "Early Medieval Rites of Queen-Making and the Shaping of Medieval Queenship," in *Queens and Queenship in Medieval Europe*, ed. A. Duggan (Woodbridge, 1997), 301–15, at 309–10, 314.

8. Pauline Stafford, *Queen Emma and Queen Edith: Queenship and Women's Power in Eleventh-Century England* (Oxford, 1997), 197–98.

9. Einhard, *The Life of Charlemagne*, trans. L. Thorpe (Harmondsworth, 1969), chap. 24, 78.

10. Nelson, *Politics and Ritual*, 296–304.

11. Pauline Stafford, 'Queens, Nunneries, and Reforming Churchmen: Gender, Religious Status, and Reform in Tenth- and Eleventh-Century England," *Past and Present* 163 (1999): 3–35, 140; Barbara Yorke, *Nunneries and the Anglo-Saxon Royal Houses* (London, 2003), 169.

12. Stafford, "Queens, Nunneries," 70–71.

Further Reading

The Second English Queen's *Ordo*, from the Sampson Pontifical, Cambridge, Corpus Christi College 146, ed. Wickham Legg (London, 1901), 21–22.

English Historical Documents, 500-1042, ed. Dorothy Whitelock (London, 2nd ed., 1979).

Leopold G. Wickham Legg, *English Coronation Records* (Westminster, 1901).

Secondary Work

S. MacLean, "Queenship, Nunneries and Royal Widowhood in Carolingian Europe," *Past and Present* 178 (2003), 3–38.

J. L. Nelson, "The Lord's Anointed and the People's Choice: Carolingian Royal Ritual," in *Rituals of Royalty: Power and Ceremonial in Traditional Societies*, ed. D. Cannadine and S. Price (Cambridge, 1987), 137–80.

— 42 —

Mass at the Election of the Mayor
of London, 1406

Caroline Barron

On Wednesday, the Feast of the Translation of St. Edward the King and Confessor (13 October), in the eighth year, etc., John Wodecok, Mayor of the City of London, considering that upon the same day he and all the Aldermen of the said city, and as many as possible of the wealthier and more substantial Commoners of the same city, ought to meet at the Guildhall, as the usage is, to elect a new Mayor for the ensuing year, ordered that a Mass of the Holy Spirit should be celebrated, with solemn music, in the Chapel annexed to the said Guildhall; to the end that the same Commonalty, by the grace of the Holy Spirit, might be able peacefully and amicably to nominate two able and proper persons to be Mayor of the said city for the ensuing year, by favor of the clemency of Our Savior, according to the customs of the said city.

Which Mass having in the said Chapel been solemnly celebrated, there being present the said John Wodecok, the Mayor; John Prestone, Recorder; Nicholas Wottone and Geoffrey Broke, Sheriffs; the Prior of the Holy Trinity, John Hadlee, William Staundone, Richard Whytyngtone, Drew Barentyn, Thomas Knolles, John Shadworth, William Askham, William Bramptone, John Warner, William Walderne, William Venour, Robert Chychely, Thomas Fauconer, Thomas Polle, William Louthe, William Crowmere, Henry Bartone, and Henry Pountfreyt, Aldermen; and many reputable Commoners of the City aforesaid; the same Mayor, Recorder, Sheriffs, Aldermen, and Commoners entered the Guildhall, where the precept of the said Mayor and Aldermen, as to the cause of the said congregation, was becomingly set forth and declared by the said Recorder to the Commoners aforesaid; to the end that such Commoners should nominate unto the said Mayor and Aldermen such able and proper persons as had before filled the office of Sheriff in the City aforesaid; it being for the said Commoners to take no care which one of the persons so to be nominated should be chosen by the Mayor and Aldermen to be Mayor for the ensuing year. Which being done, the said Mayor,

Recorder, Sheriffs, and Aldermen, went up into the Chamber of the Mayor's Court, within the Guildhall aforesaid, there to await the nomination of such two persons. Whereupon, the Commoners peacefully and amicably, without any clamor or discussion, did becomingly nominate Richard Whytyngtone, mercer, and Drew Barentyn, goldsmith, through John Westone, Common Counter of the said city, and presented the same.

And hereupon, the Mayor and Aldermen, with closed doors, in the said Chamber chose Richard Whytngtone aforesaid, by guidance of the Holy Spirit, to be Mayor of the City for the ensuing year; after which, the Mayor and Aldermen, coming down from the Chamber into the Hall, to the Commoners there assembled, as the custom is, notified by the Recorder unto the same Commoners, how that, by Divine inspiration, the lot had fallen upon the said Richard Whytngtone, as above stated.

And further, the said Commoners unanimously entreated the Mayor and Aldermen, that they would ordain that in every future year, on the Day of the Translation of St. Edward, a Mass of the Holy Spirit, for the reasons before stated, should be celebrated, before the election of the Mayor, in the Chapel aforesaid. And hereupon, the Mayor and Aldermen, considering the entreaty of the said Commoners to be fair, reasonable, and consonant with right, and especially to the glory and laud of God, and to the honor of the said city, by assent and consent of the said Commoners, did ordain and decree that every year in the future a solemn Mass with music shall be celebrated in the presence of the Mayor and aldermen; the same Mass, by ordinance of the Chamberlain for the time being, to be solemnly chaunted by the finest singers in the Chapel aforesaid, and upon that Feast.

Source: Corporation of London Record Office, Letter Book I, fol. 54; this translation is from H. T. Riley, ed., *Memorials of London and London Life . . . 1276–1419* (London, 1868), 565–66.

This decision to commence the proceedings for the annual election of a new mayor with a mass of the Holy Spirit, which was apparently taken on the sole authority of the outgoing mayor, marks a significant stage in the sanctification of civic office-holding in London. In 1406 the mayor was clearly attempting to harness religious practice to the cause of law and order. The mercer John Woodcock seems to have been not only an exceptionally successful and wealthy merchant but also a particularly pious man.[1] When he died in 1409, his will was laden with pious bequests. By the time of his mayoralty he was probably one of the wealthiest men in London who, like his successor Richard Whittington, had been able to move effortlessly from supplying the household of Richard II to becoming a leading financier and purveyor of luxury goods for the wardrobe of Henry IV. Thus Woodcock had the financial and moral authority to introduce changes into the mayoral election procedures.

Unlike most other English towns where the governing group often took the form of a religious fraternity that had taken upon itself the secular government of the civic community (for example at Westminster), in London self-government had grown from completely secular roots. The sheriffs and then the mayor, since their first appearance in the twelfth century, had been seen simply as royal or civic officers. The mayor was chosen every year on the feast day of St. Edward the Confessor (13 October) by the freemen (or citizens—probably about a third of the adult males in a total population of some 40,000) gathered at Guildhall. Originally all freemen could attend but, in the course of the fourteenth century, the election came to be limited to those who were especially summoned, that is, the *probi homines*, the respectable or most worthy citizens. Doubtless the elections in the thirteenth century had been rowdy affairs and the attempt to restrict attendance at the Guildhall to the respectable (i.e., law-abiding) citizens was part of the continuing attempt by the city's rulers to govern London in such a way that the king would not feel it necessary to intervene.

It is clear from this account that it was the city's common countor or, as he was more usually known, the city's common sergeant, John Weston, who organized the meeting in the Guildhall and presented the citizens' two candidates to the mayor and aldermen, and it was the city's recorder, John Prestone, who announced the aldermen's choice to the assembled citizens. John Carpenter, the city's common clerk (in office 1417–38), when writing his account of how the mayor should be elected in his compendium of civic custom known as the *Liber Albus*, noted that the mayor and aldermen voted for the mayor and that the votes were counted by the common clerk under the supervision of the recorder.[2]

The government of London had centered on the Guildhall since the early thirteenth century. In 1406 the mayor, aldermen, and citizens were still using the thirteenth-century hall and chapel although these were both soon to be rebuilt (1411–55). The fact that there had been a chapel adjacent to the London Guildhall since the end of the thirteenth century may suggest that the government of London already had a religious dimension. The chapel, dedicated to the Virgin Mary, St. Mary Magdalene, and All Saints, at the beginning of the fourteenth century seems to have been the meeting place of the Society of the Pui, a kind of medieval glee club that encouraged good fellowship and harmony among its members by composing and performing songs. But the emphasis here was secular rather than religious, although the intention of the founders was, like that of Woodcock, to foster "mirthfulness, peace, honesty, joyousness, gaiety and good love" among men.[3] In the middle of the fourteenth century, when Guildhall College was founded next door, the five chaplains were expected to maintain the services in the chapel.[4] Woodcock's focus on the need for good harmony and order in the election of the mayor, and the great civic rebuilding schemes of the middle years of the fifteenth century indicate, perhaps, a growing awareness of the importance of the ceremonial (which in this period must almost always have been religious) in fostering acceptance of, indeed enthusiasm for, civic government.[5]

Woodcock's choice of a mass of the Holy Spirit to precede the mayor's election was an obvious one for, as the Holy Spirit inspired the Apostles at Pentecost, so the same inspiration might illuminate the London citizens as they set about their task. The mass of the Holy Spirit was often celebrated before important clerical elections, and so Woodcock is here borrowing a clerical practice for use in a secular election. He may have been influenced by a recent innovation in Bruges where John de Waghenare as recently as 1395 had endowed a mass of the Holy Spirit in the church of St. Donatian to be celebrated on the day of the renewal of the magistracy in the city. It is clear from this account in the city's Letter Book that Woodcock's innovation "caught on." Not only did twenty-three of the twenty-four aldermen attend the mass, together with "many reputable Commoners of the City," but it was also decided that a solemn mass of the Holy Spirit should always in future be celebrated in the Guildhall chapel on the feast of St. Edward, that is, on the day of the annual election of the mayor. So Woodcock's innovation continued until the Reformation when the mass was replaced by a solemn sermon.

It is significant that Woodcock specified that the mass was to be celebrated "with solemn music." At this time it was becoming fashionable for services in collegiate and parish churches to be accompanied by polyphonic music. This was expensive because it required the presence of a number of singers and, in particular, boys to provide the treble parts. By his will drawn up in 1442 the common clerk John Carpenter provided an endowment for the mayor and commonalty that was to fund the feeding, clothing, and education of four London boys to be known as "Carpenter's Children." In return the boys were to assist at divine service in the choir of Guildhall chapel on festival days. Moreover the parish clerks of London were, by this date, professional lay singing men who had formed a fraternity dedicated to St. Nicholas based in Guildhall chapel. The clerks provided a chaplain to serve the St. Nicholas altar in the chapel and, in this way, the liturgical requirements of the city's rulers were supplied by the fraternity of clerks. On the eve of the Reformation, Guildhall college and chapel were served by a warden, two (or possibly more) priests, one tutor, four boys, and two singing clerks. Moreover in 1525 the city had decided to employ William Lewes, who had recently been a verger at St. Paul's, as the choirmaster and to play the organ. In requiring that the civic mass of the Holy Spirit should be celebrated with solemn music, Woodcock was in keeping with current liturgical practices that were elaborating the music that accompanied the mass.

But in spite of the pious hopes of those who instituted the mass at the mayoral election the citizens of London were not always able in the fifteenth century to elect a new mayor "peacefully and amicably." In the late 1430s there was bitter craft rivalry between the Drapers' and the Tailors' crafts. Each was eager to secure the election of a member of their craft as mayor, and on three successive occasions, although the citizens chose as one of their two candidates the tailor, Ralph Holland, he was always rejected by the aldermen. When this happened for the third time and the mayor, John Paddesley, presented the draper Robert Clopton to the assembled commoners they rioted and shouted out "Nay, Nay, Not this man but Rawlyn Holland."[6] The riot was not quelled until the protesters were rounded up and sent to Newgate. But these events were exceptional and on the whole the Holy Spirit appears to have worked. In the election in 1499 the chronicler Robert Fabyan, who was at the time one of the aldermen, records that "afftir the messe of the holy goost was sungyn and the mayer with his brethyr was comyn into the Inner counseyll chambyr to comon of such matier as theym lykid," and were waiting for the citizens to choose their two candidates, a letter arrived from the king asking the aldermen to select John Percyvall, a tailor, if he should be one of the two candidates. In the event, Percyvall was chosen although Fabyan noted that he had been passed over by the aldermen on previous occasions because it was thought that "he was verray desyrous to have it."[7]

Following his election, Richard Whittington would have had fifteen days to wait before taking up office. On 28 October the mayor went to the Guildhall to take his oath and on the following day he rode in procession to Westminster to be sworn before the king or, in his absence, before the Barons of the Exchequer. It is clear that well before 1406 this "Riding" to Westminster had become a very important event in the civic calendar. When the goldsmith John de Chichester was elected as mayor in 1369 his craft spent considerable sums of money on minstrels, horses, red livery cloth, and painted banners for his "Riding."[8] But it should be noted that the "riding" was always a secular event, enlivened by music and, later, by "disguisings" and "pageants"; the city clergy played no part in the procession. It is for this reason that the "mayor's riding" came to have an enhanced significance after the Reformation that stripped away other, more religious, ceremonies from London life.

John Woodcock's introduction of the mass of the Holy Spirit into the procedure for electing a new mayor for the city of London during the reign of Henry IV (1399–1413) may not have been simply the result of his own personal piety or a desire to bring good order to civic elections. Paul Strohm has recently made a powerful case for the conscious use by Henry IV, who had displaced and murdered his cousin, Richard II, as king of England, of the structures of orthodox religion to bolster his new regime. The threat of "lollardy" was, perhaps, exaggerated in order to legitimize a crack-down on the expressions of dissident opinion of any kind. The Lancastrian kings, he suggests, used religious institutions and structures to legitimize and consolidate their political power.[9] In 1406 John Woodcock, a firm supporter of the new regime, also used religion, in his turn, to enhance the authority of the rulers of the city of London. In this way the mayoral election can be seen as coming to play a part in the evolving "theatre of orthodoxy."[10] Just as the king harnessed the Church to the service of the new regime, so the city of London introduced a mass of the Holy Spirit to bolster the authority of an incoming mayor.

Notes

1. J. S. Roskell, L. Clarke, and C. Rawcliffe, *The House of Commons, 1386–1421*, vol. 2 (Stroud, 1992), 896–99.

2. H. T. Riley, ed. and trans. *Liber Albus: The White Book of the City of London* (London, 1861), 19.

3. Anne Sutton, "Merchants, Music, and Social Harmony: The London Puy and Its French and London Contexts c. 1300," *London Journal* 17 (1992): 1–17.

4. Caroline Barron, *The Medieval Guildhall of London* (London, 1974), 37–38.

5. Caroline Barron, "Church Music in English Towns, 1450–1550: An Interim Report," *Urban History* 29 (2002): 83–91, esp. 87–88; Caroline Barron, "The Political Culture of Medieval London," in *Political Culture in Late Medieval Britain*, ed. Linda Clark and Christine Carpenter (Woodbridge, 2004), 111–33.

6. *The Great Chronicle of London*, ed. A. H. Thomas and I. D. Thornley (Gloucester, 1983), 175; Caroline M. Barron, *London in the Later Middle Ages: Government and People, 1200–1500* (Oxford, 2004), 151.

7. The *Great Chronicle*, 288.

8. L. Jefferson, ed., *Wardens' Accounts and Court Minute Books of the Goldsmiths' Mistery of London, 1334–1446* (Woodbridge, 2003), 125.

9. Paul Strohm, *England's Empty Throne: Usurpation and the Language of Legitimation, 1399–1422* (New Haven, CT, and London, 1998), esp. chap. 2.

10. Michel de Certeau, *The Writing of History* (New York, 1988), 157.

Further Reading

M. Berlin, "Civic Ceremony in Early Modern London," *Urban History Yearbook* (1986): 15–27.

A. Brown, "Civic Ritual: Bruges and the Counts of Flanders in the Later Middle Ages," *English Historical Review* 112 (1997): 277–99.

INDEX